4 Advanced properties **85**

4.1 Introduction to Internal Set Theory 85
4.1.1 Properties of IST 87
4.1.2 External sets 89
4.2 The nature of halflines, neutrices and external numbers . . . 94
4.3 Generalized Dedekind completeness 96
4.4 Flexible sequences and functions 100
4.4.1 Flexible functions 100
4.4.2 Flexible sequences 104
4.5 Idempotent neutrices and ideals 109
4.5.1 Idempotent neutrices 110
4.5.2 Ideals and the product of neutrices 111

5 Sequences. Convergence up to a neutrix **117**

5.1 Notions of convergence for flexible sequences 118
5.1.1 Convergence for infinite sequences 118
5.1.2 Convergence with respect to an initial segment 125
5.2 Operations on flexible sequences 129
5.2.1 Boundedness and monotonicity 129
5.2.2 Operations 131
5.3 Cauchy flexible sequences 135

6 Functions of external numbers **139**

6.1 Limits of flexible functions 140
6.1.1 Relation with convergence for sequences; strong convergence 146
6.2 Flexible continuity 149
6.2.1 Outer continuity 150
6.2.2 Inner continuity 153
6.3 $M \times N$-derivation of flexible functions 156
6.4 Weak extrema and monotonicity 160

7 Integration of functions of external numbers **163**

7.1 Integrals of internal functions on external intervals 163
7.2 Integrals of flexible functions 167
7.3 Elementary properties of integrals 172
7.4 Special integrals and applications 178
7.4.1 Mass and tail of probabilities and integrals 178
7.4.2 On local averaging 180
7.4.3 The concentration lemma and the Laplace method . . 182

8 Flexible systems of linear equations 189

8.1 Flexible systems . . . 189
8.2 Determinants . . . 192
8.3 On Gauss-Jordan elimination . . . 196
8.4 Parameter method . . . 200
8.4.1 Non-singular systems . . . 202
8.4.2 Singular systems with strict rank equal to the number of equations . . . 204
8.4.3 Singular systems with strict rank less than the number of equations . . . 206

9 Applications in asymptotics 211

9.1 Nonstandard Borel-Ritt Theorem . . . 211
9.2 Tools for solution of external equations . . . 213
9.3 Matching principles . . . 225
9.4 An external singular perturbation with canard solutions . . . 230
9.4.1 External differentiable equations and their solutions . 230
9.4.2 The external Riccati-Hermite equation . . . 231
9.4.3 Solving the external Riccati-Hermite equation . . . 232
9.4.4 Description of the canard behaviour . . . 236
9.4.5 Influence of the singular point on the localization of canards . . . 237

10 Applications in other fields 243

10.1 The Sorites paradox in philosophy . . . 244
10.1.1 Forms of the paradox . . . 245
10.1.2 Response proposals . . . 247
10.1.3 External numbers as a model . . . 255
10.2 External recurrence relations and near stability . . . 258
10.3 On the size of fluctuations of the financial market . . . 264
10.4 Further applications of external numbers . . . 269
10.4.1 Near-optimization with uncertainties . . . 269
10.4.2 On statistical estimation of uncertainties . . . 270

11 External numbers as a complete arithmetical solid 275

11.1 The axioms . . . 277
11.1.1 Algebraic axioms . . . 277
11.1.2 Generalized Completeness axiom . . . 280
11.1.3 Arithmetical axioms . . . 281
11.2 A formal construction of the external numbers . . . 282
11.3 The solid $\mathcal{E}$ as a model for the axioms . . . 284
11.4 On the axioms for the external numbers . . . 291

A Background on Nonstandard Analysis 295

A.1 On the foundations of external sets 295
A.2 Set theoretical Nonstandard Analysis 296
A.2.1 ZFC 296
A.2.2 Theories for internal sets: IST and BST 299
A.2.3 Theories for external sets: HST 301
A.2.3.1 HST axioms 303
A.3 Model theoretical nonstandard analysis 305
A.3.1 The superstructure approach 306

B Solutions to selected exercises 309

Bibliography 325

Index 337

Foreword

Nonstandard Analysis and "*the set of x such that ...*"

A superficial look at this book might suggest that it is addressed to "pure mathematicians" and more specifically to specialists in the logical foundations of mathematics. It is not so. This book is for applied mathematicians confronted with the imprecision of models. Let us try to explain what it is all about with the words of the ordinary working mathematician.

In informal mathematical writing we often enunciate seemingly innocent sentences, as for example:

> *The set of complex numbers that are the solution of a polynomial equation with integer coefficients is countable*

which in reality hides all the strength (and weakness) of the (formal) set theory of Zermelo, Fraenkel with Choice (ZFC). Recall that the theory ZFC is a theory where all mathematical objects $1, 2, \ldots, \mathbb{N}, \sqrt{2}, \mathbb{C}, \ldots$ have the status of *set*. A *set* is an object about which one can state properties, for example "to be countable" as above, and sets are patiently constructed from the empty set using a number of axioms. The axiom of *comprehension* (also called of *separation*) states that if P is a proposition written correctly in the language of set theory then:

$$\forall\, X\, \exists\, Y\, \forall\, z\, \big(z \in Y \Leftrightarrow (z \in X \wedge P(z)\big),$$

more commonly written:

$$Y = \big\{z \in X : P(z)\big\},$$

for which we read Y *is the set of elements* z *that satisfy* P. This is the axiom of *comprehension* because it "puts together" all z of X which satisfy P or of *separation* because it "separates" X into two complementary subsets, those whose elements have the property P and those whose elements do not. The informal statement that we proposed at the beginning assumes that "to be a solution of a polynomial equation with integer coefficients" is a proposition that is well written in the language of set theory, so that the complex numbers that have this property do indeed constitute a set, whose cardinal is that of $\mathbb{N}$.

But this property of separation is not a property satisfied by all propositions. Take for example the proposition:

There are few virtuous men.

We will agree that this proposal makes sense but, on the other hand, we will find it more difficult to say that

Ten percent of men are virtuous,

for it supposes that virtuous men constitute a well-defined whole whose cardinality could be calculated and compared to that of the totality of men. Being virtuous is a "vague" property because men are not either *virtuous or not* in an absolute way but are *more or less* virtuous and, thus, it is impossible to draw a precise boundary between those who would be and those who would not be. So we see that there are many ways to talk about "vagueness", some like the first one, that are meaningful, others, like the second, which obviously are not. "Vagueness", when not treated properly, is dangerous because it can lead to contradictions. This is the famous paradox of the heap of sand. A "heap" has the property that when you take a grain out of it there is still a "heap". On the other hand, 1 grain, 2 grains, 3 grains, ... are not a "heap"; let n_0 be the first integer such that n_0 grains constitute a "heap", then $n_0 - 1$ grains do not constitute a "heap", which contradicts the property of the "heap". We see that "heap" is a concept with intrinsic "vagueness" to which the reasoning by recurrence does not apply. This paradox of the "heap" (the educated philosophers say "Sorites paradox") does not date from today since it is attributed to Eubulides of Megara who was born around 400 BCE.

To avoid this problem, set theory has made a distinction. Only collections of objects defined by permitted formulas of set theory (thus susceptible to quantification) are sets. Set theory does not allow one to talk about what is vague, that is its strength (no paradox to be feared), but, and that is its weakness, because "vagueness" is necessary to speak simply of certain indispensable things as the *orders of magnitudes.* All of physics is based on orders of magnitude and the art of neglecting what can be neglected. This is why mathematicians of the past, who were still somewhat physicists, had difficulty separating themselves from infinitesimals in favor of the rigor of $\forall \varepsilon > 0 \, \exists \delta > 0 \cdots$ during the nineteenth century. Physicists, who have fewer problems with formal rigor, have never ceased to use them.

In the middle of the twentieth century the theory of sets was recognized by almost all mathematicians as the foundation of mathematics. It was a point of no return. Thanks to the progresses of formal logic Robinson succeeded in constructing, within set theory, a theory containing expressions like ω is an infinitely large integer, ε is an infinitely small strictly positive number; while keeping their common sense, these have a precise mathematical meaning where paradoxes can no longer occur - unless set theory is not consistent, but that's another story! However, the formalism attached to this new analysis remained rather heavy until Nelson [92] and Hrbáček [64] proposed another approach more radical and easier to understand by the common man. The idea is that the vague concepts, which set theory does not want to take into account, will be supported by a specific language, added to the usual language. This is particularly simple in the case of the integers. Let us see. We consider the set $\mathbb{N}$ provided to us by any usual mathematical treatise (e.g. Bourbaki [21]) and then we decide to axiomatize the concept of "heap". An integer n can be $\mathrm{NH}(n)$ (reads n is a "Non-Heap") whose negation $\neg\mathrm{NT}(n)$ reads n is a "Heap" with the following axioms:

1. $\mathrm{NH}(0)$ (0 is a Non-Heap).
2. $\mathrm{NH}(n) \Rightarrow \mathrm{NH}(n+1)$ (adding 1 to a Non-Heap makes a Non-Heap).
3. $\exists\omega : \neg\mathrm{NH}(\omega)$ (there exists a Heap).

Since we have decided that the elements n are the integers that Mr. Bourbaki provided us, we will certainly not be able to talk about the set of the elements n belonging to $\mathbb{N}$ such that $\mathrm{NH}(n)$ since Non-Heap does not belong to the set language and we thus escape the "heap" paradox. The fact that "heap" is a vague concept whose boundary is not defined is taken into account in the formalism by the fact that not being a set one cannot say what is its smallest element. The miracle is now that:

- As a formal system, ZFC + NH() is relatively consistent with ZFC.
- This system, as simple as it is, is powerful enough to allow us to rewrite in a simple way difficult chapters of classical analysis, provided that one pays attention to *illegal set formation* (a property with the expression "heap" does not define a set).

This is wonderfully explained in the first two chapters of this book which explains in 50 pages the ENA system (for Elementary Nonstandard Analysis), and which does not assume any previous background in logic. After reading these 50 pages, we advise the reader to jump to page 246 where he will find a deep discussion on the "Sorites paradox". In *The Virtue of Simplicity* [137, pages 27–32] Nelson says:

> Much of mathematics is intrinsically complex, and there is delight to be found in mastering complexity. But there can also be an extrinsic complexity arising from unnecessarily complicated ways of expressing intuitive mathematical ideas. Heretofore nonstandard analysis has been used primarily to simplify proofs of theorems. But it can also be used to simplify theories. There are several reasons for doing this. First and foremost is the aesthetic impulse, to create beauty. Second and very important is our obligation to the larger scientific community, to make our theories more accessible to those who need to use them. To simplify theories we need to have the courage to leave results in simple, external form - fully to embrace nonstandard analysis as a new paradigm for mathematics. ...

In practice, orders of magnitude have an intrinsic vagueness. They tend to have the "Sorites" property of being invariant by some shifts, a property shared by their complements. Neutrices respect the Sorites property, for example the limited numbers are stable by adding 1, and the infinitesimals are stable by adding a fixed non-zero infinitesimal ε.

An *external number* is a sum of a real number and a neutrix. So they mostly are external sets and share with the neutrices the Sorites property. If an external number is not reduced to a neutrix, it is called *zeroless*. Then the neutrix part of an external number is infinitely small with respect to the absolute value of the external number, making it a model for errors of measurement or rounding off.

The algebraic laws for external numbers are much stronger than those for the $o(\cdot)$'s and $O(\cdot)$'s and for Van der Corput's neutrices. Some striking properties are total order respecting the operations, a distributive law whose validity, though not total, is completely characterized, and a Dedekind completeness property expressing that external convex subsets (the mathematical models of orders of magnitude), just as intervals, have a left extremity and a right extremity, now in the form of an external number.

It is the stability under sufficiently small modifications given by the Sorites property, in combination with the robustness under operations, which justifies to call the set of external numbers a flexible number system. All in all, we defend that the external numbers form an appropriate and effective model for a calculus of orders of magnitude and for error analysis.

This book contains, firstly, a thorough introduction to the theory of neutrices and external numbers, secondly develops an analysis of them, and thirdly presents applications in various domains. The book is mostly self-contained; for some elaborated proofs in the theoretic part and the applications we refer to the literature. An introduction to elementary nonstandard analysis is contained in the first chapter, an introduction to more advanced nonstandard analysis in Chapter 4, and foundational aspects of nonstandard analysis in Appendix A. We aimed at a didactic and expository presentation, in which the material is developed gradually, with increasing difficulty, and accompanied by many exercises. It focusses more on concepts than computational complexity and is addressed to anyone with interest in asymptotics and its applications. Some maturity in set theory is needed in the parts on Internal Set Theory and in the final chapter on the axiomatic approach.

The applications concern phenomena in which orders of magnitude, uncertainties and imprecisions are important, notably approximations, asymptotics, singular perturbations, stability up to small fluctuations, localization of probabilistic and statistic events, error propagation in linear algebra, near-optimization and the relation between microscopic and macroscopic observations in economy. The applications of Chapter 2 are elementary, and aim to illustrate the potential of the theory. Chapter 9 presents more elaborate applications coming from asymptotics, and the applications of Chapter 10 have their origin in other fields. Some highlights are a treatment of Stirling's formula, and more generally the Laplace method, without epsilontics, a general matching principle which is in particular valid for singular perturbations, the description in the presence of uncertainty of a *canard* solution, i.e. a bounded solution of a singular perturbation losing stability, and a consistent model for the Sorites paradox, which is compared to other approaches. Common asymptotics uses the symbol $=$ to indicate the asymptotic order of a given quantity. In De Bruijn's formalism this corresponds to inclusion $\subseteq$. Here we will consistently adopt this notion of inclusion as to not interfere with algebraic equality which is of fundamental importance in this book.

The algebraic, arithmetic and analytic properties of neutrices and external numbers are presented at several stages. A first elementary calculus with external numbers is developed in Chapter 1, growing out of the calculus for infinitesimals and limited numbers known as the "Leibniz Rules". The algebraic laws for external numbers are derived in Chapter 3, except for a notion of idempotency, necessary to deal with the product of neutrices. This is done in Chapter 4, for it needs some advanced properties, including an extension of Dedekind completeness to the external numbers. The description of the external numbers is completed in the final Chapter 11. This contains an axiomatic system which is, though more elaborate, not unlike the axiomatic definition of the real numbers. Consistency is shown by the construction of a model called a *complete arithmetical solid.* The model characterizes the external numbers almost completely, given a set of neutrices. Both the axiomatic system and the model are formulated outside nonstandard analysis, but one shows that the elementary nonstandard analysis of Chapter 1 is essentially valid.

In the central chapters 5-8 we take advantage of the strong algebraic properties and order properties of the external numbers, and develop "flexible" analysis and linear algebra. As for analysis, we follow a usual textbook exposition, adapting the classical notions of limits, continuity, differentiation and integration. Sequences of external numbers are often called *flexible sequences*, functions of external numbers *flexible functions* and systems of linear equations *flexible systems.*

Chapter 5 deals with flexible sequences. In addition to the classical notion of limit in terms of approaching, we consider *strong convergence*, which means that the sequence enters the limit set (an external number). This phenomenon is fairly universal, and is the basis of a general property of matching different behaviour on adjacent sets. Chapter 6 concerns limits, continuity and differentiation of flexible functions. Chapter 7 considers various methods of integration of flexible functions, in view of applications to asymptotic integrals. In Chapter 8 we study flexible systems, which are linear systems with an arbitrary standard number of variables and an arbitrary standard number of equations, each coefficient coming with its own individual neutrix, modelling different kinds of imprecisions and uncertainties. The usual solution methods by Cramer's rule and Gauss-Jordan elimination are considered, each with its possibilities and limitations. In addition we introduce a *parameter method*, which, by extending the notion of neutrix to more dimensions, in many cases enables us to determine the shape of the domain of uncertainty of the solution.

Chapter 5, and notably Chapters 6 and 8 include many results by Nam Van Tran in his thesis [146]. The presentation of Cramer's rule and Gauss-Jordan elimination also profited from work by Júlia Justino [67, 68]. The algebraic part and Chapter 7 on integration contains material from the thesis of Fouad Koudjeti [75] and [76]. The results on statistics in Chapter 10 come from Jacques Bosgiraud [18, 19]. We are deeply grateful to all of these authors for agreeing to include their results. Part of the remaining material is published earlier in the form of joint articles of the authors [41, 42, 43, 44, 45] and the thesis of the first author [40], but presented here in a different format. We thank the respective publishers for kindly authorizing us to include the material in question.

The first author acknowledges the support of the Centro de Matemática, Aplicações Fundamentais e Investigação Operacional/ Fundação da Faculdade de Ciências da Universidade de Lisboa via the grant UID/MAT/04561/2013 and a post-doctoral from Erasmus Mundus Mobility with Asia–East 14. The first author would also like to thank Inês Isidoro for all her love and support, and help with the figures.

We thank Ulf Clotz (Wuppertal) and Fernando Ferreira (University of Lisbon) for enlightening discussions on foundational matters. We thank Claude Lobry for accepting the request to write the foreword for this book.

As Van der Corput said, there reigns in asymptotics a harmony, which is only partly revealed. We hope to have made a step forward.

Chapter 1

Introduction to Elementary Nonstandard Analysis

1.1 The axiomatic system ZFL and the Leibniz Rules 2
1.2 Internal and external sets, permanence 9
1.3 External Induction and the axiomatic system ENA 16
1.4 Orders of Magnitude 19
1.5 Nonstandard regularity properties of real internal functions 27
1.5.1 S-continuity 27
1.5.2 S-differentiability 29
1.5.3 S-integrability 32

Nonstandard Analysis aims to deal consistently with infinitesimal and infinitely large numbers. It was invented in the 1960's by Abraham Robinson [111, 112] in the setting of Model Theory. Here we use an axiomatic approach to Nonstandard Analysis, based upon the Internal Set Theory, abbreviated IST, of Edward Nelson [92]. Classical mathematics is usually formalized by the axioms ZFC of Zermelo-Fraenkel Set Theory with the Axiom of Choice, in a language which only contains one undefined non-logical symbol, $\in$, for set membership. Internal Set Theory adds to ZFC a new undefined symbol, the predicate st, "standard", and three new axioms. These new axioms express how to deal with the standard mathematical objects, and also with nonstandard mathematical objects, for one of the axioms implies that every infinite standard set has nonstandard elements.

The new axioms are of a logical nature. In fact they are axiom schemes, which originate from laws which describe how Robinsonian nonstandard models of analysis relate to standard, classical analysis. The axioms of IST deal with internal sets. These are sets defined in the language of ZFC, i.e. using only the symbol $\in$. They may be standard or nonstandard; in the latter case they usually depend on a parameter which may take nonstandard values. The axiomatic system IST does not consider the so-called external sets, such as almost all the neutrices and external numbers, which are the subject of the present book. External sets use somewhere in their definition the predicate st in an essential way. Kanovei and Reeken showed that IST can be adapted to handle such sets, which resulted in the axiomatic system HST [71]. Though it may be argued that by avoiding Model Theory a simplification is attained, still at a first contact IST and HST may seem complicated to work with. In Chapter 4 and Appendix A we give a thorough description of these systems.

In the present chapter we introduce a simpler system called ENA, for Elementary Nonstandard Analysis. It is based upon the axioms introduced by Nelson in his book Radically Elementary Probability Theory [94], which is a subsystem of IST. In this system the predicate st can only be attributed to some of the natural numbers. In fact, the existence of nonstandard natural numbers is postulated. Furthermore, one axiom scheme called External Induction, is added. External Induction is similar to the usual Induction principle but with quantification restricted to standard numbers. In the last chapter of [94] Nelson argues that the mathematical content of his simplified axiomatic system is the same as the mathematical content of IST. Since external sets which are subsets of internal sets are of such primordial importance to the present work, we complete ENA with an axiom scheme, not unlike the Axiom of Extensionality of ZFC, which permits defining these external subsets.

We start in Section 1.1 with the axiomatic system ZFL (Zermelo-Fraenkel-Leibniz), introduced by Lutz [84], which is a consequence of ENA and still simpler. It includes calculatory axioms which are commonly called the *Leibniz Rules*, and are sufficient for elementary analysis dealing with orders of magnitude and nonstandard forms of convergence and continuity. In spite of its simplicity ZFL still allows for quite a few applications.

In Section 1.2 we identify internal and external sets, the latter by extending the Axiom of Extensionality of ZFC to external formulas. We consider properties of *permanence*, which are a consequence of the existence of sets of a different nature: a property which is of a different nature as the set on which it is shown to hold *a priori* must still hold outside this set.

In Section 1.3 we introduce Nelson's axiom scheme of External Induction, and show that the Leibniz Rules are a consequence.

In Section 1.4 we define orders of magnitudes of numbers and show that the infinitesimals and the infinitely large numbers permit defining a large variety of orders of magnitude.

In Section 1.5 we define various nonstandard notions of convergence and regularity of functions, with quite simple operational properties.

1.1 The axiomatic system ZFL and the Leibniz Rules

Classical mathematics is usually formalized in Zermelo-Fraenkel Set Theory with Choice, ZFC (a more detailed description of ZFC is given in Appendix A). The language of ZFC consists of a single symbol $\in$, "is element of", for set membership. We expand this system by adding to the language the unary predicate "standard", abbreviated "st". A set X can either be standard ($\mathrm{st}(X)$) or nonstandard ($\neg\mathrm{st}(X)$).

The axiomatic system ZFL (Zermelo-Fraenkel-Leibniz) attributes the predicate "standard" only to some natural numbers.

Definition 1.1.1. *The axiomatic system* ZFL *consists of* ZFC *together with the following axioms.*

Axiom 1.1. $\mathrm{st}(0)$.

Axiom 1.2. $\forall n \in \mathbb{N}\,(\mathrm{st}(n) \Rightarrow \mathrm{st}(n+1))$.

Axiom 1.3. $\forall n, m \in \mathbb{N}\,(\mathrm{st}(n) \wedge \mathrm{st}(m) \Rightarrow \mathrm{st}(n+m))$.

Axiom 1.4. $\forall n, m \in \mathbb{N}\,(\mathrm{st}(n) \wedge \mathrm{st}(m) \Rightarrow \mathrm{st}(n.m))$.

Axiom 1.5. $\forall n, m \in \mathbb{N}\,(\mathrm{st}(n) \wedge \mathrm{st}(m) \Rightarrow \mathrm{st}(n^m))$.

Axiom 1.6. $\exists \omega \in \mathbb{N}\,(\neg \mathrm{st}(\omega))$.

Comments:

1. Leibniz and Newton independently invented Calculus at the end of the 17th century. Both approaches used infinitesimals in some form. In Leibniz's approach we may recognize calculation rules which are close to Axioms 1.3 and 1.4. These axioms are sometimes called the *Leibniz Rules* for the predicate st, below they will be extended to the so-called "limited" real numbers.

2. Experience shows that Nelson's IST [92] (see Chapter 4), and already ZFL, may lead to quite a number of questions and new visions on the nature of mathematics. To begin with, there is apparently a problem of consistency. Axioms 1.1 and 1.2 imply that

$$\mathrm{st}(0), \mathrm{st}(1), \mathrm{st}(2), \mathrm{st}(3), \mathrm{st}(4), \ldots$$

while Axiom 1.6 states that the set $\mathbb{N}$ contains a nonstandard element ω. If one identifies mentally the set $\mathbb{N}$ with

$$0, 1, 2, 3, 4, \ldots \tag{1.1}$$

there is a contradiction. However, such identification is not justified. The set $\mathbb{N}$ is part of the formal set theory ZFC, and has a concrete (formal) definition. Usually it is stated as the smallest inductive set or as the smallest infinite ordinal number [66]. Anyhow, the definition uses a finite number of words, and not the infinite counting of (1.1). Reeb [110] addressed the paradox by calling the counting numbers of (1.1) *naive*, and then stated

$$\text{Les entiers naïfs ne remplissent pas } \mathbb{N}.^1 \tag{1.2}$$

[1]The naive numbers do not fill up $\mathbb{N}$.

3. Observe that ZFL is an extension, but not a modification, of classical mathematics. For instance, we do not modify the set $\mathbb{N}$, but the way of talking about $\mathbb{N}$, which instead of a language with one undefined symbol $\in$ uses now a richer language with two undefined symbols $\in$ and st. These kinds of extensions of theories are called *conservative* (see for example [62]).

4. The problem of the consistency of ZFL is in fact the same as for ZFC. Due to Gödel's incompleteness results [54] it is impossible to prove the consistency of ZFC within itself. In fact, the result is true for every theory containing Peano Arithmetic, hence also for ZFL. However, Nelson [92, 95] showed that IST is relatively consistent with ZFC, meaning that the consistency of ZFC implies the consistency of IST. Then ZFL, being weaker than (i.e. being a subsystem of) IST, is also relatively consistent with ZFC. Otherwise said, if ZFL leads to a contradiction, ZFC would already lead to a contradiction.

5. By (1.2) all natural numbers which are counted, constructive, intuitive, etc. are standard. One may argue that "standard" is a mathematical model, in the sense of set theory, of these informal notions. Conversely, the number ω of Axiom 1.6 does not have a constructive interpretation; its existence is purely formal. Of course, this is also true for $\mathbb{N}$, in fact for all infinite sets of ZFC.

6. Because ZFL is relatively consistent with ZFC, the existence of nonstandard natural numbers within $\mathbb{N}$ is consistent with ZFC. Hence the existence of non-naive natural numbers within $\mathbb{N}$ is consistent with ZFC. Reeb observed in [35] that this gives a second argument in favour of (1.2).

7. Although it is not considered to express a Leibniz Rule, it is convenient to include the exponential Axiom 1.5 into ZFL. This may generate some discussion on whether "st" is an appropriate formalization of "constructive", because e.g. it is possible to count up to 1000, but probably not up to 1000^{1000}. See also [93].

8. The transition from the standard natural numbers to the nonstandard natural numbers reflects the Sorites paradox related to orders of magnitude. Indeed, if n is standard, by Axiom 1.2 its successor $n+1$ is also standard, and as a consequence of the same axiom, if ω is nonstandard, its predecessor $\omega - 1$ is also nonstandard. So in the setting of ZFL, the Sorites paradox can be modelled, respecting its essential feature of invariance under shifts. We will come back to this matter in Chapter 10. For now we would just like to observe that the collections $\{n \in \mathbb{N} : \mathrm{st}(n)\}$ and $\{n \in \mathbb{N} : \neg\mathrm{st}(n)\}$ of $\mathbb{N}$ cannot be sets of ZFC: the second does not have a minimal element, and the first is bounded but does not have a maximum.

Remark 1.1.1. We use the following abbreviations: we write $\exists^{\mathrm{st}} x(\dots)$ instead of $(\exists x)(\mathrm{st}(x) \wedge \dots)$ and $\forall^{\mathrm{st}} x(\dots)$ instead of $(\forall x)(\mathrm{st}(x) \Rightarrow \dots)$. Observe that implicitly all such quantifications are bounded by $\mathbb{N}$, because only natural numbers can be standard.

With the predicate "standard" it is possible to define some orders of magnitude within the set of real numbers $\mathbb{R}$.

Definition 1.1.2. *Let x be a real number.*

1. *x is* limited, *if there exists standard $n \in \mathbb{N}$ (i.e. $n \in \mathbb{N} \wedge \mathrm{st}(n)$) such that $|x| \leq n$.*

2. *x is* unlimited, *or* infinitely large *if x is not limited.*

3. *x is* infinitesimal, *or* infinitely small *if for every standard $n \in \mathbb{N} \setminus \{0\}$ we have $|x| \leq \frac{1}{n}$, in which case we write $x \simeq 0$.*

4. *x is* appreciable *if x is limited but not infinitesimal.*

Clearly, if ω is unlimited, then $\frac{1}{\omega}$ is infinitesimal and if ε is infinitesimal then $\frac{1}{\varepsilon}$ is infinitely large. Also $x < y$, if x is limited and y is unlimited and positive.

Below we show how these orders of magnitude behave under operations.

Theorem 1.1.2. *Let x and y be limited reals. Then the following hold in* ZFL*:*

1. *0 is limited.*

2. *$x + 1$ is limited.*

3. *$x + y$ is limited.*

4. *xy is limited.*

5. *If $y > 0$, then x^y is limited, if well-defined.*

Proof. 1. This is a consequence of Axiom 1.1.

2. Assume x is limited. Then there exists $\mathrm{st}(m) \in \mathbb{N}$ such that $|x| \leq m$. Then $\mathrm{st}(m + 1)$ by Axiom 1.2. Clearly $|x + 1| \leq m + 1$, hence $x + 1$ is limited.

3. Assume x and y are limited. Then there exist $\mathrm{st}(m), \mathrm{st}(n) \in \mathbb{N}$ such that $|x| \leq m$ and $|y| \leq n$. Then $\mathrm{st}(m + n)$ by Axiom 1.3. Moreover $|x + y| \leq |x| + |y| \leq m + n$. Hence $x + y$ is limited.

4. Assume x and y are limited. Then there exist $\mathrm{st}(m), \mathrm{st}(n) \in \mathbb{N}$ such that $|x| \leq m$ and $|y| \leq n$. Then $\mathrm{st}(mn)$ by Axiom 1.4. Also $|xy| \leq mn$. Hence xy is limited.

5. Let x and y be limited, such that $y > 0$. Assume that x^y is well-defined. By definition there exist standard $m, n \in \mathbb{N}$ such that $|x| \leq m$ and $|y| \leq n$. Then $|x^y| \leq m^n$. Now $\mathrm{st}(m^n)$ by Axiom 1.6; hence x^y is limited. □

With some abuse of language, the properties of invariance of being limited under the operations of the previous theorem are called the *Leibniz Rules* for limited reals. We will now give other calculation rules.

Theorem 1.1.3. *The inverse of an appreciable number is also appreciable.*

Proof. By definition, a real number a is appreciable if a is limited but not infinitesimal. Then $\exists^{\text{st}} m, n \in \mathbb{N} \setminus \{0\} \left(\frac{1}{m} \leq |a| \leq n\right)$. Then $\frac{1}{n} \leq \left|\frac{1}{a}\right| \leq m$; hence $\frac{1}{a}$ is appreciable. □

Let $\omega > 0$ be unlimited. Clearly there are many unlimited numbers larger than ω, for example $\omega + 1$, ω^2, e^ω, $\omega!$, ω^ω. The next proposition shows that there are also many unlimited numbers smaller than ω.

Proposition 1.1.4. *Let $\omega > 0$ be unlimited. Then $\omega - m$, $\frac{\omega}{m}$ and $\omega^{\frac{1}{m}}$ are unlimited for every standard natural number $m > 0$. Also $\log(\omega)$ is unlimited.*

Proof. If $\omega - m$ was limited, then $\omega = (\omega - m) + m$ would also be limited, a contradiction. Similarly, if $\frac{\omega}{m}$ was limited, then $\omega = \frac{\omega}{m} m$ would also be limited, a contradiction. If $\omega^{\frac{1}{m}}$ was limited, then $\omega = \left(\omega^{\frac{1}{m}}\right)^m$ would also be limited, again a contradiction. If $\log(\omega)$ was limited, we have also a contradiction, for then $\omega = e^{\log(\omega)}$ would also be limited. □

The next theorem gives calculation rules for infinitesimals.

Theorem 1.1.5. *Let $\delta, \varepsilon > 0$ be infinitesimals, and x be limited. Then*

1. *$\varepsilon + \delta$ is infinitesimal.*
2. *εx is infinitesimal.*
3. *$\varepsilon\delta$ is infinitesimal.*

Proof. 1. Let $k \in \mathbb{N} \setminus \{0\}$ be standard. Then $|\delta|, |\varepsilon| \leq \frac{1}{2k}$, so $|\varepsilon + \delta| \leq |\varepsilon| + |\delta| \leq \frac{1}{2k} + \frac{1}{2k} = \frac{1}{k}$. Hence $\varepsilon + \delta$ is infinitesimal.

2. If $|x| \leq 1$, then εx is infinitesimal, for $|\varepsilon x| \leq \varepsilon$. Assume that $|x| \geq 1$. Suppose that εx is not infinitesimal. Then $\exists^{\text{st}} n \in \mathbb{N} \setminus \{0\} \left(|x\varepsilon| > \frac{1}{n}\right)$. Then $|x| n$ is limited, while $|\varepsilon| > \frac{1}{|x| n}$, a contradiction. Hence $x\varepsilon$ is infinitesimal.

3. By Part 2. □

Proposition 1.1.6. *Let $\varepsilon > 0$ be infinitesimal and $m > 0$ be standard. Then $\varepsilon^{\frac{1}{m}}$ is infinitesimal. Furthermore $e^{-\frac{1}{\varepsilon}}$ and $\frac{1}{\log(\varepsilon)}$ are infinitesimals.*

Proof. See Exercise 1.1.1. □

With the help of infinitesimals we define two notions of closeness.

Definition 1.1.3. *We say that two real numbers x, y are* infinitely close *and write $x \simeq y$ if $x - y$ is infinitesimal. If $y \neq 0$, we say that the numbers x and y are* asymptotic *and write $x \sim y$ if $\frac{x}{y}$ is infinitely close to 1.*

Notation 1.1.7. *We write $x \gtrsim y$ (respectively $x \lesssim y$) as an abbreviation of $x > y \vee x \simeq y$ (respectively $x < y \vee x \simeq y$). Similarly, $x \gtrsim\!\!\!\!\!\not\;\; y$ (respectively $x \lesssim\!\!\!\!\!\not\;\; y$) abbreviates $x > y \wedge x \not\simeq y$ (respectively $x < y \wedge x \not\simeq y$).*

If x is infinitesimal we have $x \simeq 0$. Also, if $x \simeq y$ then there exists $\varepsilon \simeq 0$ such that $x = y + \varepsilon$. With abuse of notation we may write $x \simeq +\infty$, if x is positive and unlimited.

Let $\varepsilon \simeq 0, \varepsilon > 0$ and $\omega \simeq +\infty$. Then $1-\varepsilon \simeq 1+\sqrt{\varepsilon}$ and also $1-\varepsilon \sim 1+\sqrt{\varepsilon}$. Also $\omega + 1 \sim \omega$, but not $\omega + 1 \simeq \omega$.

Proposition 1.1.8. *Let x and y be appreciable. Then $x \simeq y$ if and only if $x \sim y$.*

Proof. See Exercise 1.1.2. □

Infinite closeness is the basis of a nonstandard notion of continuity, as shown in Section 1.5. Here we show that, in this sense, the arithmetic operations are continuous.

Theorem 1.1.9. *Let x, y, ζ, η be real numbers such that $\zeta \simeq x$ and $\eta \simeq y$. Then*

1. *$\zeta + \eta \simeq x + y$.*
2. *If x,y are limited, then $\zeta\eta \simeq xy$.*
3. *If x is appreciable, then $\frac{1}{\zeta} \simeq \frac{1}{x}$.*

Proof. Let $\zeta = x + \delta$ and $\eta = y + \varepsilon$, where δ and ε are infinitesimals.

1. One has $\zeta + \eta = x + \delta + y + \varepsilon \simeq x + y$, because $\delta + \varepsilon \simeq 0$.

2. We have $\zeta\eta = (x+\delta)(y+\varepsilon) = xy + x\varepsilon + \delta y + \delta\varepsilon \simeq xy$, because $x\varepsilon \simeq 0$, $\delta y \simeq 0$, and $\delta\varepsilon \simeq 0$ by Theorem 1.1.5.

3. We have

$$\frac{1}{\zeta} = \frac{1}{x+\delta} = \frac{1}{x} - \frac{\frac{\delta}{x^2}}{1+\frac{\delta}{x}}.$$

Moreover $\frac{1}{x}$ is limited, so $\frac{1}{x^2}$ is limited. Then $\frac{\delta}{x^2}$ is infinitesimal by Theorem 1.1.5. Also $1 + \frac{\delta}{x} \simeq 1$. Hence $\frac{\frac{\delta}{x^2}}{1+\frac{\delta}{x}} \simeq 0$. We conclude that $\frac{1}{\zeta} \simeq \frac{1}{x}$. □

We prove next nonstandard forms of additive and multiplicative closeness for the exponential and the logarithm.

Theorem 1.1.10. *Let $x \simeq 0$. Then $e^x \simeq 1$ and $\log(1+x) \simeq 0$. Also $e^x - 1 \sim x$ and $\log(1+x) \sim x$.*

Proof. Note that e^x is limited because $x \simeq 0$. Using Taylor expansions one has

$$|e^x - 1| \leq |x| \max_{y \in [-x,x]} e^y \leq |x|\, e^{|x|} \simeq 0.$$

Hence $e^x \simeq 1$. Also

$$\left|\frac{e^x - 1 - x}{x}\right| \leq \frac{|x|}{2} \max_{y \in [-x,x]} e^y \leq |x|\, e^{|x|} \simeq 0.$$

Then $\frac{e^x-1}{x} - 1 \simeq 0$, hence $\frac{e^x-1}{x} \simeq 1$, i.e. $e^x - 1 \sim x$.

As for the logarithm,

$$|\log(1+x)| \leq |x| \left|\max_{y\in[-x,x]} \frac{1}{1+y}\right| \leq |x| \frac{1}{1-|x|} \simeq 0.$$

Hence $\log(1+x) \simeq 0$. Finally,

$$\left|\frac{\log(1+x) - x}{x}\right| \leq \frac{|x|}{2} \left|\max_{y\in[-x,x]} -\frac{1}{1+y^2}\right| \leq |x| \simeq 0.$$

So $\frac{\log(1+x)-x}{x} \simeq 0$ and $\frac{\log(1+x)}{x} \simeq 1$. Hence $\log(1+x) \sim x$. □

Exercises

Exercise 1.1.1. *Prove Proposition 1.1.6.*

Exercise 1.1.2. *Prove Proposition 1.1.8.*

Exercise 1.1.3. *Let $\omega > 0$ be unlimited. Order the following numbers: 1000000, $\frac{1}{\omega}$, ω^3, $\log(\omega)$, $e^{-\omega}$, ω^ω, $\sqrt{\omega}$, $\omega!$, e^ω. What can you say about $\sin(\omega)$?*

Exercise 1.1.4. *Give examples of unlimited numbers such that their quotient is unlimited, appreciable, or infinitesimal.*

Exercise 1.1.5. *Give an example where the sum of a family of infinitesimal numbers is appreciable, and an example where such a sum is unlimited. Can the sum of an unlimited family of infinitesimal numbers still be infinitesimal?*

Exercise 1.1.6. *Let $x, y > 0$, $x \simeq y$. Show that $\sqrt{y} \simeq \sqrt{x}$ (Hint: consider the cases $x, y \simeq 0$ and $x, y \gtrsim\!\!\!\!/\; 0$).*

Exercise 1.1.7. *Let x, y be limited and such that $x \simeq y$. Prove that $e^x \simeq e^y$. Give examples of numbers x, y such that $x \simeq y$ and $e^x \not\simeq e^y$.*

Exercise 1.1.8. *Let $x \simeq 0$. Prove that $\sin(x) \sim x$.*

Exercise 1.1.9. *Show a nonstandard version of Euler's formula: if $\omega \simeq +\infty$, then $\left(1+\frac{1}{\omega}\right)^\omega \simeq e$; moreover, if x is limited, then $\left(1+\frac{x}{\omega}\right)^\omega \simeq e^x$.*

Exercise 1.1.10. *Let $\omega \in \mathbb{N}$ be such that $\omega \simeq +\infty$. Let $n \in \mathbb{N}$ be such that $0 \leq n \leq \omega$.*

1. *Let $\frac{n}{\omega} \leq x \leq \frac{n+1}{\omega}$. Show that $e^{\frac{n}{\omega}}\frac{1}{\omega} \leq \int_{\frac{n}{\omega}}^{\frac{n+1}{\omega}} e^x dx \leq e^{\frac{n+1}{\omega}}\frac{1}{\omega}$.*

2. *Show that* $\sum_{0\leq n<\omega}\left(e^{\frac{n+1}{\omega}}-e^{\frac{n}{\omega}}\right)\frac{1}{\omega}\simeq 0$.

3. *Deduce that* $\sum_{0\leq n<\omega}e^{\frac{n}{\omega}}\frac{1}{\omega}\simeq\int_0^1 e^x dx$.

4. *Give an approximation of* $\sum_{0\leq n<\omega}\left(1+\frac{n}{\omega^2}\right)^{\omega}$.

1.2 Internal and external sets, permanence

Definition 1.2.1. *A formula of the language* $\{\in\}$*, i.e. of the language of set theory* ZFC*, is called* internal. *A formula of the language* $\{\in, \mathrm{st}\}$ *is called* external.

The formulas $x = 0$, $\forall n \in \mathbb{N}(n/2 \in \mathbb{N})$, $[0,1) \subset \mathbb{R}$, and $\varepsilon \neq 0$ are all internal because the symbols $0, \mathbb{N}, 2, 1, [.,.), \subset, \mathbb{R}, \neq$ are definable in the language $\{\in\}$. Examples of external formulas are the formulas $\mathrm{st}(0)$, $\varepsilon \simeq 0$, $\forall a \in \mathbb{R}(a$ appreciable $\Rightarrow a$ limited). Observe that to define near equality, and the predicates appreciable and limited, one needs the symbol st.

Sometimes an external formula is equivalent to an internal formula. For example, if

$$\forall x(x \simeq 0 \wedge \mathrm{st}(x) \Rightarrow \forall y > 0(e^y > x)), \tag{1.3}$$

then necessarily $x = 0$. Hence (1.3) is equivalent to

$$\forall x(x = 0 \Rightarrow \forall y > 0(e^y > x)).$$

A set is called *internal* if it is defined by an internal formula. Rigorously speaking, internal sets are sets in the sense of ZFC.

Some misunderstanding may arise from the use of parameters with nonstandard values. For example, what to think of:

"Let $\varepsilon > 0, \varepsilon \simeq 0$. Consider the interval $[0,\varepsilon]$. Clearly $[0,\varepsilon] \subset [0,1]$." (1.4)

One might be inclined to think that the interval $[0,\varepsilon]$ is not an internal set; after all ε is infinitesimal. However, all intervals of the form $[0,\varepsilon]$, where $\varepsilon \in \mathbb{R}$, are internal; indeed, in the definition $[0,\varepsilon] := \{x \in \mathbb{R} : 0 \leq x \leq \varepsilon\}$ the symbol ε is simply a free variable. In (1.4) the property of being external arises only from the declaration of ε as being infinitesimal. Formalizing, we obtain the external formula $\forall\varepsilon > 0(\varepsilon \simeq 0 \Rightarrow [0,\varepsilon] \subset [0,1])$.

Some entities which are not internal sets arise naturally, when dealing with infinitesimals, limited or unlimited numbers. To some extent they correspond to abbreviations of ways of speaking. Since such entities are also essential for a formal definition of the notion of order of magnitude of numbers, we introduce also an axiomatic system for them, where we use part of the system EEST

(Elementary External Set Theory) of Kanovei and Reeken [71]. Consider the following expressions.

$$\begin{aligned} \oslash &= \{x \in \mathbb{R} : x \text{ infinitesimal}\}, \\ \pounds &= \{x \in \mathbb{R} : x \text{ limited}\}, \\ @ &= \{x \in \mathbb{R} : x > 0 \text{ appreciable}\}, \\ \not\!\infty &= \{x \in \mathbb{R} : x > 0 \text{ unlimited}\}. \end{aligned} \tag{1.5}$$

Formally, no set is being defined, because all expressions rely on the use of the predicate st. Also an expression using one of the above entities is often easily substituted by a formula without them. For instance $\oslash \subset [-1, 1]$ is an abbreviation of $\forall x \in \mathbb{R}(x \simeq 0 \Rightarrow -1 < x < 1)$.

The expressions in (1.5) are examples of collections of internal elements defined by external formulas Φ. We wish to recognize the collection $Y = \{x \in a : \Phi(x)\}$ as a "pre-external" set. A pre-external set which is not internal is called an *external* set. To this end we must distinguish between quantifications over internal sets and quantifications over such pre-external sets, so we denote the former by lower-case letters and the latter by upper-case letters. The next axiom is sufficient for elementary use of such collections.

Axiom 1.7. *Let z be an internal set and Φ be a formula using the symbols $\in$ and* st, *allowing for quantifications only with respect to internal sets. Then*

$$\exists Y(x \in Y \Leftrightarrow (x \in z \wedge \Phi(x))). \tag{1.6}$$

Definition 1.2.2. *The collection $Y = \{x \in z : \Phi(x)\}$ of* (1.6) *is called a* pre-external *set. A pre-external set which is not internal is called an* external *set.*

Observe that if the only non-logical symbol occurring in (1.6) is $\in$, the axiom corresponds to the usual Axiom of Extensionality of ZFC, and then the set $Y = \{x \in z : \Phi(x)\}$ is internal indeed.

We will not be pedantic with the use of upper-case letters for external sets.

The fact that the entities defined by (1.5) are external sets in the sense of Axiom 1.7 is seen as follows. Firstly, they are definable by a formula using the symbol st:

$$\begin{aligned} \oslash &= \left\{x \in \mathbb{R} : \forall^{\text{st}} n \left(|x| < \frac{1}{n}\right)\right\}, \\ \pounds &= \left\{x \in \mathbb{R} : \exists^{\text{st}} n (|x| < n)\right\}, \\ @ &= \left\{x \in \mathbb{R} : \exists^{\text{st}} n \left(\frac{1}{n} < |x| < n\right)\right\}, \\ \not\!\infty &= \left\{x \in \mathbb{R} : \forall^{\text{st}} n (|x| > n)\right\}. \end{aligned} \tag{1.7}$$

Secondly, all sets are bounded from below, without greatest lower bound. So they cannot be internal. To recognize whether a pre-external set is indeed external, as above it often suffices to verify if a common property is not satisfied. For instance $\{n \in \mathbb{N} : n \simeq +\infty\}$ is external, because it is a subset of $\mathbb{N}$, with no minimal element.

Remark 1.2.1. Unless otherwise said, throughout this chapter all functions and sequences are supposed to be internal; i.e. their definition does not include the symbol st, nor any property that reduces to st.

Even though ZFL together with Axiom 1.7 extends, and does not modify classical mathematics, we observe a major practical difference. In classical mathematics all sets are of the same nature, while there are now several types: internal sets, standard sets (some natural numbers) and external sets. A somewhat similar situation occurs in topology, but now the distinction between several types of sets is extended to the whole of mathematics. This is put into good use, by the so-called properties of *overspill* or *permanence*. The most obvious distinction is sometimes called *Cauchy's Principle* [124].

Theorem 1.2.2 (Cauchy's Principle). *No external set is internal.*

Corollary 1.2.3. *Let I be an internal set and E be an external set.*

1. *If $I \subseteq E$, then $I \subset E$.*
2. *If $E \subseteq I$, then $E \subset I$.*

An internal property $i(x)$, which is verified for all elements of an external set E, still holds for some elements outside of E; in fact $I := \{x : i(x)\}$ is internal and such that $E \subseteq I$, which implies that $E \subset I$. Similarly, an external property $e(x)$, which is verified for all elements of an internal set I gives rise to an external set $E := \{x : e(x)\}$ such that $I \subseteq E$, which implies that $I \subset E$; hence the property still holds for some elements outside of I.

Example 1.2.1.

1. Let $\varepsilon \simeq 0$, $\varepsilon > 0$. Consider the sequence $(n\varepsilon)_{n\in\mathbb{N}}$. Clearly $n\varepsilon \simeq 0$ as long as n is limited. Then $n\varepsilon < 1$. By Cauchy's Principle there exists an unlimited n such that still $n\varepsilon < 1$ (we may specify such a number, for instance $n = \left[\frac{1}{2\varepsilon}\right]$, but that is not needed).
2. Let $m \in \mathbb{N}$ be limited. Then $n\varepsilon < \frac{1}{m}$ for all standard n. By Cauchy's Principle there exists unlimited n such that still $n\varepsilon < \frac{1}{m}$.

Definition 1.2.3. [38]

1. *Let $(A_n)_{n\in\mathbb{N}}$ be an internal sequence of (internal) sets. Let $G := \bigcup_{\mathrm{st}(n)\in\mathbb{N}} A_n$. Then G is called a* pregalaxy. *If G is external it is called a* galaxy.
2. *Let $(B_n)_{n\in\mathbb{N}}$ be an internal sequence of (internal) sets. Let $H := \bigcap_{\mathrm{st}(n)\in\mathbb{N}} B_n$. Then H is called a* prehalo. *If H is external it is called a* halo.

Pregalaxies and prehalos are pre-external sets indeed, for

$$y \in G \Leftrightarrow \exists^{\mathrm{st}} n (x \in A_n),$$

and

$$y \in H \Leftrightarrow \forall^{\mathrm{st}} n (x \in B_n).$$

Conversely, if $\phi(n, x)$ is an internal formula, where $n \in \mathbb{N}$ and $x \in X$, for some internal set X, we may define $A_n = B_n = \{x \in X : \phi(n, x)\}$. The set $\{x \in X : \exists^{\mathrm{st}} n\, \phi(n, x)\}$ is a pregalaxy, for

$$\{x \in X : \exists^{\mathrm{st}} n\, \phi(n, x)\} = \bigcup_{\mathrm{st}(n) \in \mathbb{N}} A_n,$$

and the set $\{x \in X : \forall^{\mathrm{st}} n\, \phi(n, x)\}$ is a prehalo, for

$$\{x \in X : \forall^{\mathrm{st}} n\, \phi(n, x)\} = \bigcap_{\mathrm{st}(n) \in \mathbb{N}} B_n.$$

Without restriction of generality, we may suppose that the sequence $(A_n)_{n \in \mathbb{N}}$ is increasing for inclusion, and the sequence $(B_n)_{n \in \mathbb{N}}$ is decreasing for inclusion, for we may always put $A'_n = \bigcup_{m \leq n} A_m$ and $B'_n = \bigcap_{m \leq n} B_m$; then $\bigcup_{\mathrm{st}(n)} A'_n = \bigcup_{\mathrm{st}(n)} A_n$ and $\bigcap_{\mathrm{st}(n)} B'_n = \bigcap_{\mathrm{st}(n)} B_n$.

Observe that a pregalaxy is a galaxy, i.e. external, if $(A_n)_{n \in \mathbb{N}}$ is strictly increasing and a prehalo is a halo if $(B_n)_{n \in \mathbb{N}}$ is strictly decreasing.

Example 1.2.2. The following external sets are galaxies: ${}^{\sigma}\mathbb{N} := \{n \in \mathbb{N} : \mathrm{st}(n)\}$, $\pounds$ (the *principal* galaxy), @, $\{x \in \mathbb{R} : 0 \lnsim x \lnsim 1\}$. For $\varepsilon \simeq 0$, $\varepsilon > 0$ the *ε-galaxy* $\varepsilon\pounds$ is defined by

$$\varepsilon\pounds = \left\{x \in \mathbb{R} : \exists^{\mathrm{st}} n \left(|x| \leq n\varepsilon\right)\right\}.$$

Elements of the *ε-galaxy* are also called *real numbers of order* ε.

Example 1.2.3. The following external sets are halos: $\{n \in \mathbb{N} : n \simeq +\infty\}$, $\oslash$, $\not\oslash$, $\{x \in \mathbb{R} : 0 \lesssim x < 1\}$. For $\varepsilon \simeq 0, \varepsilon > 0$ the *ε-halo* $\varepsilon\oslash$ is defined by

$$\varepsilon\oslash := \left\{x \in \mathbb{R} : \forall^{\mathrm{st}} n \left(|x| \leq \frac{\varepsilon}{n}\right)\right\}.$$

The external set $E := \{x \in \mathbb{R} : 0 \lnsim x \lesssim 1\}$ is neither a galaxy, nor a halo, as a consequence of Theorem 1.2.5 below. This set may be written in the form

$$E = \bigcup_{\mathrm{st}(m) \in \mathbb{N}} \bigcap_{\mathrm{st}(n) \in \mathbb{N}} \left[\frac{1}{m}, 1 + \frac{1}{n}\right].$$

Repeating the operations $\bigcup_{\mathrm{st}(n) \in \mathbb{N}}$ and $\bigcap_{\mathrm{st}(n) \in \mathbb{N}}$ a naive number of times, yields "external Borel sets".

In practice, galaxies are in one way or the other defined by the word "limited", i.e. related to the set of standard numbers ${}^{\sigma}\mathbb{N}$, and halos tend to be defined by the word "infinitesimal". This way to recognize a halo or a galaxy is formalized in the next theorem.

Theorem 1.2.4 (Representation theorem [38]). *Let X be an internal set.*

1. *A subset $G \subseteq X$ is a pregalaxy if and only if $G = f^{-1}({}^{\sigma}\mathbb{N})$, where $f : X \to \mathbb{N}$ is an internal function. It is a galaxy if f is onto.*
2. *A subset $H \subseteq X$ is a prehalo if and only if $H = f^{-1}(\oslash)$, where $f : X \to \mathbb{R}$ is an internal function. It is a halo if f is onto.*

Proof. 1. Let $(A_n)_{n\in\mathbb{N}}$ be an internal increasing sequence such that $G = \bigcup_{\mathrm{st}(n)\in\mathbb{N}} A_n$. Let $\nu \in \mathbb{N}, \nu \simeq +\infty$ be arbitrary. Define $f : X \to \mathbb{N}$ by

$$f(x) = \begin{cases} \min\{n : x \in A_n\}, & x \in \bigcup_{n\in\mathbb{N}} A_n \\ \nu, & \text{otherwise.} \end{cases}$$

Then $G = f^{-1}({}^{\sigma}\mathbb{N})$.

As for the converse, put $A_n = f^{-1}\{0, ..., n\}$. Then $G = \bigcup_{\mathrm{st}(n)\in\mathbb{N}} A_n$ is a pregalaxy. Observe that for G to be internal, it must be equal to A_n for some $\mathrm{st}(n)$. This is impossible if f is onto. Hence, in this case G is a galaxy.

2. Let $(B_n)_{n\in\mathbb{N}}$ be an internal increasing sequence such that $H = \bigcap_{\mathrm{st}(n)\in\mathbb{N}} B_n$. We may assume that $B_0 = X$ and $\bigcap_{n\in\mathbb{N}} B_n = \emptyset$. Define $f : X \to \mathbb{R}$ by

$$f(x) = \min\left\{\frac{1}{n} : x \notin B_n\right\} = \frac{1}{\max\{n : x \in B_n\}}.$$

Then

$$\begin{aligned} x \in H &\Leftrightarrow \forall^{\mathrm{st}} n (x \in B_n) \\ &\Leftrightarrow \max\{n : x \in B_n\} \simeq +\infty \\ &\Leftrightarrow \min\left\{\frac{1}{n} : x \notin B_n\right\} \simeq 0 \\ &\Leftrightarrow f(x) \simeq 0. \end{aligned}$$

Hence $H = f^{-1}(\oslash)$. As for the converse, put $B_n = f^{-1}\left[-\frac{1}{n}, \frac{1}{n}\right]$. Then $H := \bigcap_{\mathrm{st}(n)\in\mathbb{N}} B_n$ is a prehalo. If H was internal, it should be equal to B_ν for some unlimited $\nu \in \mathbb{N}$. This is impossible if f is onto. Hence in this case H must be a halo. □

Of course, the complement of a galaxy within an internal set is a halo, and vice versa.

Theorem 1.2.5 (The Fehrele Principle). *No halo is a galaxy.*

Proof. Let G be a galaxy, and $(A_n)_{n\in\mathbb{N}}$ be an increasing internal sequence such that $G = \bigcup_{\mathrm{st}(n)\in\mathbb{N}} A_n$. Let H be a halo and $(B_n)_{n\in\mathbb{N}}$ be a decreasing internal sequence such that $H = \bigcap_{\mathrm{st}(n)\in\mathbb{N}} B_n$.

Assume that $G \subseteq H$. We show that $G \subset H$. Because the sequences $(A_n)_{n\in\mathbb{N}}$ and $(B_n)_{n\in\mathbb{N}}$ are internal, the set $I := \{n \in \mathbb{N} : A_n \subseteq B_n\}$ is internal. Clearly $A_n \subseteq B_n$ for all standard n. Then ${}^{\sigma}\mathbb{N}$ is contained in I. By Cauchy's Principle ${}^{\sigma}\mathbb{N} \subset I$. Then there exists $\nu \in \mathbb{N}$ unlimited such that $A_\nu \subseteq B_\nu$. Hence $G \subset A_\nu \subseteq B_\nu \subset H$. We conclude that $G \neq H$. □

Corollary 1.2.6 (Robinson's Lemma). *Let $(a_n)_{n\in\mathbb{N}}$ be an internal sequence such that $a_n \simeq 0$ for all standard n. Then there exists ν unlimited and such that $a_n \simeq 0$ for all $n \leq \nu$.*

Proof. The set ${}^{\sigma}\mathbb{N}$ is a galaxy, while $H := \{\nu \in \mathbb{N} : \forall n \leq \nu\, (a_n \simeq 0)\}$ is a prehalo. We conclude by Cauchy's Principle or the Fehrele Principle. □

Robinson's Lemma [112] was stated before the Fehrele Principle [38], but in a sense it amounts to a special case[2].

By Part 2 of Theorem 1.1.5 it holds that $n\varepsilon \simeq 0$ for each st(n). By Robinson's Lemma there exists some unlimited natural number ν such that $n\varepsilon \simeq 0$, for all $n \leq \nu$ (it is not difficult to exhibit such a ν).

As an example, let $\omega \simeq +\infty$ and x be limited. Then

$$\left(1 + \frac{x}{\sqrt{\omega}}\right)^{\omega} = e^{\omega \log\left(1+\frac{x}{\sqrt{\omega}}\right)} = e^{\sqrt{\omega}x - \frac{x^2}{2}(1+\varepsilon(x))},$$

where $\varepsilon(x) \simeq 0$. Hence for limited x

$$e^{-\sqrt{\omega}x}\left(1 + \frac{x}{\sqrt{\omega}}\right)^{\omega} \simeq e^{-\frac{x^2}{2}}. \tag{1.8}$$

By the Fehrele Principle there exists some unlimited ξ such that (1.8) is still true for all $|x| \leq \xi$. To prove this near-equality directly, and to exhibit such a ξ, already some non-trivial work is needed.

Surprising as it may seem, galaxies and halos, theoretical instruments of nonstandard set theory, intervene naturally in problems in applied theories such as asymptotics or singular perturbations. The classical study of these problems consists of sometimes difficult *ad hoc* methods, as appear in for instance [46, 97, 98], and many others. One may defend that the nonstandard method is simplifying and structuring, due to the existence of actual small and large numbers, the possibility to distinguish domains of asymptotic behaviour in terms of halos and galaxies, and the existence of general methods of matching local asymptotic behaviour through principles of permanence [36]. We will see some examples in Chapter 2 and develop some theory in Chapter 9.

[2] Reeb (in a seminar) called it the Fehrele Principle, after Strasbourg's mythical perpetual student Fehrele who was proud to have learnt that "A Esel isch ke Kolibri". This principle expresses quite well another Alsatian adagium "Fîhr nix, ebb's; diss gibbt's" [85].

Exercises

Exercise 1.2.1. *Let $\omega \in \mathbb{N}$ be nonstandard. Is $\{0, ..., \omega\}$ finite? Does there exist an internal $1-1$-mapping onto $\{1, ..., \omega\}$? Does there exist an external $1-1$-mapping onto $\{1, ..., \omega\}$? Is ${}^{\sigma}\mathbb{N}$ finite?*

Exercise 1.2.2. *Let $\varepsilon > 0$, $\varepsilon \simeq 0$. Sketch the graphs of the following real internal functions f defined by: $f(x) = \varepsilon x$, $f(x) = \frac{x}{\varepsilon}$, $f(x) = e^{\frac{x}{\varepsilon}}$, $f(x) = \sin\left(\frac{x}{\varepsilon}\right)$.*

Exercise 1.2.3. *Is the function $f : \mathbb{R} \to \mathbb{R}$ defined by $f(x) = 0$, if x is limited, and $f(x) = 1$, if x is unlimited, continuous? Is it external?*

Exercise 1.2.4. *For every galaxy given in Example 1.2.2, exhibit an internal sequence, so that the galaxy is written as a union with standard indices. Do the same for the halos of Example 1.2.3 with intersection instead of union.*

Exercise 1.2.5. *Let $\varepsilon \simeq 0$, $\varepsilon > 0$, $\omega \simeq +\infty$. Indicate if the following sets are internal or external. Indicate for each set an element which is inside and an element which is outside. Which sets are galaxies and which are halos? Which of the sets are groups for addition?*

1. $\varepsilon \pounds$.
2. $\omega \oslash$.
3. $\{x : \exists^{\mathrm{st}} n \left(|x| \leq \omega^n\right)\}$.
4. $\{x : \forall^{\mathrm{st}} n \left(|x| \leq \varepsilon^n\right)\}$.
5. $\left\{x : \exists^{\mathrm{st}} n \left(|x| \leq e^{-\frac{\omega}{n}}\right)\right\}$.
6. $\left\{x : \forall n \simeq +\infty \left(|x| \leq \varepsilon^{\frac{1}{n}}\right)\right\}$.
7. $\{x : \exists^{\mathrm{st}} n \left(f(x) \leq n\right)\}$, *where $f : \mathbb{R} \to \mathbb{R}$ is defined by $f(x) = 0$, for all x.*
8. $\left\{y : \exists x \geq 1 \left(x \text{ is limited and } y \leq 1 - \frac{1}{x}\right)\right\}$.
9. $\{y : \exists x (x \text{ is unlimited and } y \simeq \sin(x))\}$.
10. $\left\{(x, y) \in \mathbb{R}^2 : x \simeq 0,\ y \text{ arbitrary}\right\}$.

Exercise 1.2.6. *Consider in $\mathbb{R}^2$ the set $\oslash \times \pounds$ and the lines l_a defined by*

$$l_a = \{(x, y) : y = ax\} \qquad (a \in \mathbb{R}).$$

Why is $l_a \cap (\oslash \times \pounds)$ always external? For which values of a is the intersection a galaxy? For which values of a is the intersection a halo? Deduce that $\oslash \times \pounds$ is neither a galaxy nor a halo.

Exercise 1.2.7. *Let $\varepsilon \simeq 0$. Show, by using Cauchy's Principle, and explicitly, that there exists an unlimited $\nu \in \mathbb{N}$ such that $\nu\varepsilon \leq \sqrt{\varepsilon}$.*

Exercise 1.2.8. *Let $\omega \in \mathbb{N}$ be such that $\omega \simeq +\infty$.*

1. *Let $n \in \mathbb{N}$ be such that $\frac{n}{\omega}$ is limited and $\frac{n}{\omega} \leq x \leq \frac{n+1}{\omega}$. Show that $e^{-x} \sim e^{-\frac{n}{\omega}}$ and $\int_{\frac{n}{\omega}}^{\frac{n+1}{\omega}} e^{-x}dx \sim e^{-\frac{n}{\omega}}\frac{1}{\omega}$.*

2. *Derive for limited $a \geq 0$ the approximation*

$$\sum_{0 \leq n \leq a\omega} e^{-\frac{n}{\omega}}\frac{1}{\omega} \sim \int_0^a e^{-x}dx,$$

implying that

$$\sum_{0 \leq n \leq a\omega} e^{-\frac{n}{\omega}}\frac{1}{\omega} \simeq \int_0^a e^{-x}dx. \tag{1.9}$$

3. *Show that the approximation (1.9) still holds for some unlimited $a > 0$.*

4. *Show that $\sum_{n \geq 0} e^{-\frac{n}{\omega}}\frac{1}{\omega} \simeq \int_0^{\infty} e^{-x}dx = 1$.*

Exercise 1.2.9. *Let $f : [0,1] \times \mathbb{R} \to \mathbb{R}$ be a function of class C^1 such that $f(x,0) = 0$, $\frac{\partial f}{\partial y}(x,y) > 0$ is appreciable for all $x \in [0,1]$, y limited. Show that*

1. *for all $x \in [0,1]$, the set $\{(x,y) : f(x,y) \simeq 0\}$ is a halo;*

2. *$\{(x,y) : 0 \leq x \leq 1 \wedge f(x,y) \simeq 0\}$ is a halo;*

3. *$\{(x,y) : 0 \leq x \leq 1 \wedge f(x,y)$ is appreciable$\}$ is a galaxy.*

1.3 External Induction and the axiomatic system ENA

The axiomatic system Elementary Nonstandard Analysis ENA consists of the axioms presented by Nelson in [94], to which we add Axiom 1.7 on external sets. We give here the full list of axioms, in the language $\{\in, \mathrm{st}\}$.

Definition 1.3.1. *The axioms of* ENA *are the axioms of* ZFC *together with the following axioms.*

Axiom 1.8. $\mathrm{st}(0)$.

Axiom 1.9. $\forall n \in \mathbb{N}\,(\mathrm{st}(n) \Rightarrow \mathrm{st}(n+1))$.

Axiom 1.10. $\exists \omega \in \mathbb{N}\,(\neg\mathrm{st}(\omega))$.

Axiom 1.11 (External Induction)**.** *Let Φ be a formula. Then*

$$(\Phi(0) \wedge \forall^{\mathrm{st}} n(\Phi(n) \Rightarrow \Phi(n+1))) \Rightarrow \forall^{\mathrm{st}} n(\Phi(n)).$$

Axiom 1.12. *Let z be an internal set and Φ be a formula. Then there exists a set Y such that*

$$x \in Y \Leftrightarrow (x \in z \wedge \Phi(x)).$$

We recall that $Y = \{x \in X : \Phi(x)\}$ is possibly not a set in the sense of ZFC, in which case it is called an *external set*. If we do not consider Axiom 1.12 we denote the resulting system by ENA$^-$.

In Peano Arithmetic [99] Induction takes the form: Let ϕ be a formula of the language $\{0, 1, +, \cdot\}$. Then

$$(\phi(0) \wedge \forall n\, (\phi(n) \Rightarrow \phi(n+1))) \Rightarrow \forall n(\phi(n). \tag{1.10}$$

In ZFC the assertion (1.10) comes as a theorem, and is a consequence of among others the Axiom of Infinity; the property ϕ is now formulated in the language $\{\in\}$. As a consequence, in ENA induction holds on the whole of $\mathbb{N}$ for internal formulas, and on the external set of all standard natural numbers ${}^{\sigma}\mathbb{N}$ for external formulas.

In ENA we recognize Axioms 1.1, 1.2 and 1.6 of ZFL. In fact ENA permits proving the remaining axioms of ZFL, hence is stronger than ZFL.

Theorem 1.3.1. *The axiomatic system* ENA *implies the axioms of* ZFL.

Proof. We only need to prove that Axioms 1.3, 1.4 and 1.5 hold in ENA.

(i) Let n be standard. Clearly $n + 0$ is standard. Suppose that $\text{st}(n + m)$. Then $\text{st}(n + (m + 1))$ by Axiom 1.9. By External Induction $\text{st}(n + m)$ for all standard m. This proves Axiom 1.3.

(ii) Let n be standard. By Axiom 1.8, the element $0 = n \cdot 0$ is standard. Suppose that $\text{st}(nm)$. Then $n(m+1) = nm+n$ is standard by (i). By External Induction $\text{st}(nm)$ for all standard m. This proves Axiom 1.4.

(iii) Let n be standard. By Axioms 1.8 and 1.9 the element 1 is standard. Then $n^0 = 1$ is standard. Suppose that $\text{st}(n^m)$. Then $n^{m+1} = n^m n$ is standard by (ii). By External Induction $\text{st}(n^m)$ for all standard m. This proves Axiom 1.5. □

With the Axiom scheme of External Induction one proves that all standard natural numbers come before the nonstandard natural numbers.

Theorem 1.3.2. *Let n be standard and $m \leq n$. Then m is standard.*

Proof. See Exercise 1.3.1. □

To apply permanence, it is needed to recognize whether sets are internal or external. Below are some criteria, proved with the aid of External Induction.

Theorem 1.3.3. *1. A set G is a galaxy if and only if there exists an internal sequence $(C_n)_{n \in \mathbb{N}}$, which is strictly increasing at least for the standard indices, such that $G = \bigcup_{\text{st}(n) \in \mathbb{N}} C_n$.*

2. *A set H is a halo if and only if there exists an internal sequence $(D_n)_{n\in\mathbb{N}}$, which is strictly decreasing at least for the standard indices, such that $H=\bigcap_{\mathrm{st}(n)\in\mathbb{N}} D_n$.*

Proof. 1. Assume first that G is external. Let $(A_k)_{k\in\mathbb{N}}$ be an increasing internal sequence such that $G=\bigcup_{\mathrm{st}(k)\in\mathbb{N}} A_k$. Put $C_0=A_0$, and suppose that n is such that $(C_m)_{m\leq n}$ is strictly increasing, with $C_n=A_{\overline{k}}$ for some $\overline{k}\in\mathbb{N}$. If there exists $p\in\mathbb{N}$ such that $A_{\overline{k}}\subset A_p$, let $\overline{p}$ be minimal with this property. Then put $C_{n+1}=A_{\overline{p}}$. If such a $\overline{p}$ does not exist, we define $C_{n+1}=C_n$. We prove that $\mathrm{st}(\overline{k})$ whenever $\mathrm{st}(n)$ by External Induction. The property is clearly true for $n=0$. Suppose that it is valid for n. If $\overline{p}$ is unlimited, then $(A_k)_{\overline{k}\leq k\leq \overline{p}-1}$ must be stationary, implying that $G=A_{\overline{k}}$ is internal, a contradiction. Hence $\overline{p}$ is standard. We conclude by External Induction. Then $G=\bigcup_{\mathrm{st}(k)\in\mathbb{N}} A_k=\bigcup_{\mathrm{st}(n)\in\mathbb{N}} C_n$, where $(C_n)_{n\in\mathbb{N}}$ is internal and strictly increasing at least for the standard indices.

Conversely, let $(C_n)_{n\in\mathbb{N}}$ be an internal sequence (of internal sets), which is strictly increasing at least for the standard indices, and $G=\bigcup_{\mathrm{st}(n)\in\mathbb{N}} C_n$. Then $C_n\subset G$ for all $\mathrm{st}(n)$. If G were internal, by Cauchy's Principle still $C_\nu\subset G$ for some unlimited ν, a contradiction. Hence G must be external and therefore a galaxy.

2. See Exercise 1.3.2. □

A representation theorem for galaxies and halos will be stated in the form of a corollary to Theorem 1.2.4.

Corollary 1.3.4. *Let X be an internal set.*

1. *A subset $G\subseteq X$ is a galaxy if and only if there exists an internal function $f:X\to\mathbb{N}$ which is onto ${}^\sigma\mathbb{N}$ and such that $G=f^{-1}({}^\sigma\mathbb{N})$.*

2. *A subset $H\subseteq X$ is a halo if and only if there exists an internal function if $f:X\to\mathbb{R}$ which is onto $\oslash$ and such that $H=f^{-1}(\oslash)$.*

In practice, to recognize a set as a halo, it may be convenient to prove that its complement is a galaxy.

Exercises

Exercise 1.3.1. *Prove Theorem 1.3.2.*

Exercise 1.3.2. *Prove Part 2 of Theorem 1.3.3.*

Exercise 1.3.3. *Let X be an internal set. Let $f:\mathbb{R}\to\mathbb{R}$ be an internal function. Prove that $f(\pounds)$ is a galaxy, and $f(\oslash)$ is a halo, if f is $1-1$.*

Exercise 1.3.4. *Let $f:\mathbb{R}\to\mathbb{R}$ be defined by $f(x)=\arctan(x)$. Let $\varepsilon\simeq 0$, $\varepsilon>0$. Investigate whether the following sets are internal, galaxies or halos.*

1. $f^{-1}(\oslash)$.
2. $f^{-1}(1+\oslash)$.
3. $f^{-1}\left(\{x \in \mathbb{R} : -\frac{\pi}{2} \lessapprox x \lnapprox \frac{\pi}{2}\}\right)$.
4. $f(\mathbb{R}^+)$.
5. $f(\varepsilon \pounds)$.

Exercise 1.3.5. *Let $\omega > 0$ be unlimited. Euler's formula $\left(1+\frac{x}{\omega}\right)^\omega \simeq e^x$ is true for limited x (see Exercise 1.1.9) and also for some unlimited x by permanence, but it is not true, e.g. for $x = 2\omega$. We define the error function R_ω by*

$$R_\omega(x) := e^x - \left(1+\frac{x}{\omega}\right)^\omega.$$

1. *Show that $\{x : R_\omega(x) \simeq 0\}$ is a halo.*
2. *Show that $\{x : R_\omega(x)$ is limited$\}$ is a galaxy.*
3. *Conclude that $\{x : R_\omega(x)$ is appreciable$\}$ is a galaxy.*
4. *Verify that $\{x \leq 0 : R_\omega(x)$ is appreciable$\} = -2\omega + \pounds$.*

1.4 Orders of Magnitude

Definition 1.4.1. *An* order of magnitude *is a convex part of* $\mathbb{R}$.

The "trivial" orders of magnitude are the internal intervals, but the "true" orders of magnitude are the external convex sets, which possess intrinsic properties of vagueness. Observe that the external sets $\oslash, \pounds, @, \not\oslash, \varepsilon\pounds, \varepsilon\oslash$ (where $\varepsilon \neq 0$ is an arbitrary infinitesimal) are all convex, hence orders of magnitude. The vagueness appears because they do not have a greatest lower bound nor a lowest upper bound: one cannot get hold of a concrete point where they start or where they end. The vagueness may also be expressed by the observation that they are stable under at least some shifts, additions and multiplications: for each of these external sets E there exist $d > 0$ and $p \neq 1$ such that $x+d, xp \in E$ for all $x \in E$. Another example of an external set with imprecise boundaries is $\{x : 0 \lessapprox x \lnapprox 1\}$, which is stable under infinitesimal shifts. Also, let f be an increasing real continuous function, ranging from $-\infty$ to $+\infty$. Then for all $y \in \mathbb{R}$ the set $\{x : f(x) \simeq y\}$ is an order of magnitude, and it follows from the continuity of f that the set is external, and invariant under at least some non-zero shifts.

It is convenient to use the interval notation for orders of magnitudes. For example,

$$\begin{aligned} \{x : 0 \lesssim x \not\lesssim 1\} &:= [\oslash, 1+\oslash)\,, \\ \not\infty &:= (\pounds, \infty)\,, \\ @ &:= (\oslash, \pounds]\,. \end{aligned}$$

Definition 1.4.2. *Arithmetic operations on orders of magnitude are the* Minkowski operations, *given by*

$$\begin{aligned} A + B &:= \{x + y : x \in A \wedge y \in B\}\,, \\ AB &:= \{xy : x \in A \wedge y \in B\}\,, \\ A^B &:= \{x^y : x \in A \wedge y \in B \wedge x^y \textit{ is well-defined}\}\,. \end{aligned}$$

We may identify a real number a with its singleton $\{a\}$. Observe that some care is needed for the exponentiation of negative numbers. We have $\pounds^{1/3} = \pounds$, but $\pounds^{1/2} = (\pounds^{+})^{1/2} = \pounds^{+}$.

The results of the arithmetic operations applied to the basic orders of magnitude $\oslash, \pounds, @, \not\infty$ are resumed in the following tables.

$+$	$\oslash$	$@$	$\pounds$	$\not\infty$
$\oslash$	$\oslash$	$@$	$\pounds$	$\not\infty$
$@$	$@$	$@$	$\pounds$	$\not\infty$
$\pounds$	$\pounds$	$\pounds$	$\pounds$	$\not\infty$
$\not\infty$	$\not\infty$	$\not\infty$	$\not\infty$	$\not\infty$

(1.11)

$\cdot$	$\oslash$	$@$	$\pounds$	$\not\infty$
$\oslash$	$\oslash$	$\oslash$	$\oslash$	$\mathbb{R}$
$@$	$\oslash$	$@$	$\pounds$	$\not\infty$
$\pounds$	$\oslash$	$\pounds$	$\pounds$	$\mathbb{R}$
$\not\infty$	$\mathbb{R}$	$\not\infty$	$\mathbb{R}$	$\not\infty$

(1.12)

exp	$\oslash^{+}$	$@$	$\pounds^{+}$	$\not\infty$
$\oslash^{+}$	$[0,1]$	$\oslash^{+}$	$[0,1]$	$\oslash^{+}$
$@$	$1+\oslash$	$@$	$@$	$(0,+\infty)$
$\pounds^{+}$	$[0,1+\oslash]$	$\pounds^{+}$	$\pounds^{+}$	$\mathbb{R}^{+}$
$\not\infty$	$[1,+\infty)$	$\not\infty$	$[1,+\infty)$	$\not\infty$

(1.13)

The tables (1.11) and (1.12) follow from the calculation rules of Theorem 1.1.2. The verifications of some entries in table (1.13) are not always immediate (see Exercise 1.4.2). Note that

$$e^{\pounds} = @, \qquad \log @ = \pounds. \tag{1.14}$$

Definition 1.4.3. *A* neutrix *is an order of magnitude which is a group for addition. An* external number α *is the sum of a real number* a *and a neutrix* N *in the sense of Definition 1.4.2, i.e.*

$$\alpha := a + N = \{a + x : x \in N\}.$$

An external number which is not reduced to a neutrix is called zeroless.

This book is largely devoted to the calculus of neutrices and external numbers; we will thoroughly study their algebraic properties in Chapter 3. Observe that every neutrix contains zero, and that zeroless numbers do not contain zero indeed.

There are several types of neutrices. Only two neutrices, $\{0\}$ and $\mathbb{R}$ are internal, but among the external sets there are many neutrices. The set of limited numbers $\pounds$, the set of infinitesimals $\oslash$ and all multiples $\varepsilon\pounds$ and $\varepsilon\oslash$, with $\varepsilon \neq 0$ are external neutrices. By the Fehrele Principle $\pounds$ and $\oslash$ cannot be isomorphic, for there is no bijection between them. There are more non-isomorphic neutrices, and some have practical relevance, like the microhalo and microgalaxy defined below.

Definition 1.4.4. *Let* $\varepsilon \simeq 0$, $\varepsilon > 0$.

1. *The* microhalo M_ε *is defined by* $M_\varepsilon := \bigcap_{\mathrm{st}(n)} [-\varepsilon^n, \varepsilon^n]$.

2. *The* microgalaxy m_ε *is defined by* $m_\varepsilon := \bigcup_{\mathrm{st}(n)} \left[-e^{\frac{-1}{n\varepsilon}}, e^{\frac{-1}{n\varepsilon}}\right]$.

Clearly there cannot exist an internal bijection between the halo M_ε and the galaxy m_ε, let alone an (internal) isomorphism. Observe that $M_\varepsilon = \varepsilon M_\varepsilon$, for

$$M_\varepsilon = \bigcap_{\mathrm{st}(n)} [-\varepsilon^n, \varepsilon^n] = \bigcap_{\mathrm{st}(n)} [-\varepsilon^{n+1}, \varepsilon^{n+1}] = \bigcap_{\mathrm{st}(n)} \varepsilon[-\varepsilon^n, \varepsilon^n] = \varepsilon M_\varepsilon;$$

hence also $M_\varepsilon = \frac{1}{\varepsilon} M_\varepsilon$. Moreover $Q(n) := \frac{e^{\frac{-1}{\varepsilon(n+1)}}}{e^{\frac{-1}{\varepsilon n}}} \simeq +\infty$, for all $\mathrm{st}(n)$, because $e^{\frac{-1}{\varepsilon(n+1)}} = e^{\frac{-1}{\varepsilon n}} e^{\frac{1}{\varepsilon(n+1)n}}$. By the Fehrele Principle there exists $\omega \simeq +\infty$ such that $Q(n) > \omega$ for all $\mathrm{st}(n)$. Hence

$$\begin{aligned} m_\varepsilon &= \bigcup_{\mathrm{st}(n)} \left[-e^{\frac{-1}{n\varepsilon}}, e^{\frac{-1}{n\varepsilon}}\right] \\ &= \bigcup_{\mathrm{st}(n)} \left[-e^{\frac{-1}{\varepsilon(n+1)}}, e^{\frac{-1}{\varepsilon(n+1)}}\right] \\ &\supseteq \omega \bigcup_{\mathrm{st}(n)} \left[-e^{\frac{-1}{n\varepsilon}}, e^{\frac{-1}{n\varepsilon}}\right] \\ &= \omega m_\varepsilon \supseteq m_\varepsilon. \end{aligned}$$

We see that both M_ε and m_ε are stable by multiplication by an unlimited number, which is false for both $\pounds$ and $\oslash$. This already shows that M_ε and m_ε are not isomorphic to $\pounds$ nor $\oslash$ as a ring.

The microhalo is related to the set of *flat* functions, i.e. functions with Taylor expansion identically zero [35, 130]. The microgalaxy intervenes in the theory of singular perturbations, or slow-fast systems. Sometimes the set of times where the behaviour of a solution is fast is a microgalaxy [35, 13]. Also the set of parameters where such a system has the so-called *canard* solutions, i.e. solutions which lose their stability, but remain limited, often is a microgalaxy [8] (see also Chapter 2 and Chapter 9).

Definition 1.4.5. *Let M, N be neutrices, with $M \subseteq N$. Then N is the maximum with respect to inclusion and we write* $\max(M,N) = N$.

Theorem 1.4.1. *Let M, N be neutrices. Then $-M$, $M + N$ and MN are neutrices. Moreover*

1. $-M = M$.
2. $M + N = \max(M, N)$.
3. $2M = M$.
4. $\pounds M = M$.
5. $\frac{1}{1+M} = 1 + M$, *if* $M \subseteq \oslash$.

Proof. If M contains some element, being a group, it also contains its symmetrical. This proves Part 1. As for Part 2, without restriction of generality we may assume that $M \subseteq N$. By the group property $N \subseteq M + N \subseteq N + N = N$. Hence $M + N = \max(M, N)$. As a consequence $2N = N + N = N$ which proves Part 3. To prove Part 4, observe that $M \subseteq \pounds M$. Conversely, $0M = 0 \subseteq M$. Assume that $\mathrm{st}(n) \in \mathbb{N}$ is such that $nM = M$. Let $x \in (n+1)M$. Then $x = (n+1)y = ny + y$ with $y \in M$. By hypothesis $ny \in M$. Then $x \in M$ by the group property. Hence $ny \in M$ for all $\mathrm{st}(n) \in \mathbb{N}$ by External Induction. Then $\pounds M \subseteq M$ by convexity. Finally, let $M \subseteq \oslash$ be a neutrix and $\varepsilon \in N$. Then $\varepsilon \simeq 0$. By applying the geometric series, we find

$$1 - 2|\varepsilon| \leq \frac{1}{1+\varepsilon} \leq 1 + 2|\varepsilon|.$$

Hence $1 + N = 1 - 2N \leq \frac{1}{1+N} \leq 1 + 2N = 1 + N$, which proves Part 5. □

In practice, calculations with external numbers tend to be quite evident, with some obvious simplifications, where we may apply Theorem 1.4.1 and the tables (1.11)-(1.13). Here are some examples, with $\omega \simeq +\infty$.

$$\omega^2 + \pounds + (\omega + \oslash) = \omega^2 + \omega + \pounds.$$

We apply Part 2 of Theorem 1.4.1 which can be summarized by "the largest neutrix absorbs the smallest", in this case, $\pounds + \oslash = \pounds$. Below we apply Part 1 of Theorem 1.4.1 corresponding to "minus is plus, for neutrices":

$$\omega^2 + \oslash - (\omega + \pounds) = \omega^2 - \omega - \pounds = \omega^2 - \omega + \pounds.$$

The product of two zeroless external numbers leads initially to three neutrices, which by Part 2 of Theorem 1.4.1 is equal to the biggest, e.g.

$$(\omega^2 + \pounds)(\omega + \oslash) = \omega^3 + \omega^2 \oslash + \omega\pounds + \pounds\oslash = \omega^3 + \omega^2 \oslash.$$

The factor that comes from multiplying the two neutrices, in this case $\pounds\oslash = \oslash$, is always absorbed. We apply the property that "division amounts to multiplication" of Part 5 of Theorem 1.4.1 to simplify the quotient

$$\begin{aligned} \frac{\omega^2 + \pounds}{\omega + \oslash} &= \frac{\omega^2 + \pounds}{\omega\left(1 + \frac{\oslash}{\omega}\right)} \\ &= \left(\omega + \frac{\pounds}{\omega}\right)\left(1 + \frac{\oslash}{\omega}\right) \\ &= \omega + \oslash. \end{aligned}$$

The next theorem resumes some elementary properties of the calculus of external numbers.

Theorem 1.4.2. *Let α, β be external numbers. Then $-\alpha$, $\alpha + \beta$, $\alpha \cdot \beta$ are external numbers. Moreover, $+$ and $\cdot$ are associative and commutative operations over the external numbers. If $\alpha = a + A$ with a a real number and A a neutrix, then $-\alpha = -a + A$. If α is zeroless, then $\frac{1}{\alpha}$ is also an external number and then $a \notin A$. Moreover*

1. $\frac{A}{\alpha} = \frac{A}{a} \subseteq \oslash$.
2. $\frac{1}{\alpha} = \frac{1}{a} + \frac{A}{a^2} = \frac{\alpha}{a^2}$.
3. $\frac{\beta}{\alpha} = \frac{\alpha\beta}{a^2}$.

Proof. We only prove that $a \notin A$ if α is zeroless and the last three formulas, the remaining properties are straightforward. If α is zeroless and $a \in A$ then by the group property also $-a \in A$ and then $0 = a - a \in \alpha + A = \alpha$, a contradiction. To prove Part 1, suppose that there exist $b \in A$ and $\mathrm{st}(n)$ such that $\left|\frac{b}{a}\right| > \frac{1}{n}$. Then $|a| < n\,|b| \in A$ by Part 4 of Theorem 1.4.1, a contradiction. Hence $\frac{A}{a} \subseteq \oslash$. Also

$$\frac{\alpha}{a} = 1 + \frac{A}{a} \subseteq 1 + \oslash.$$

Then $\frac{a}{\alpha} \subseteq 1 + \oslash$ by Part 5 of Theorem 1.4.1, and using Part 4 of Theorem 1.4.1,

$$\frac{A}{\alpha} = \frac{A}{a}\frac{a}{\alpha} \subseteq \pounds\frac{A}{a} = \frac{A}{a}.$$

Clearly $\frac{A}{a} \subseteq \frac{A}{\alpha}$; hence $\frac{A}{a} = \frac{A}{\alpha}$. Finally by Part 5 of Theorem 1.4.1,

$$\frac{1}{\alpha} = \frac{1}{a\left(1+\frac{A}{a}\right)} = \frac{1}{a}\left(1+\frac{A}{a}\right) = \frac{1}{a} + \frac{A}{a^2} = \frac{\alpha}{a^2}.$$

This proves Part 2. Part 3 follows immediately. □

As for the properties of distributivity and order, more care is needed. We define

$$\alpha \leq \beta \Leftrightarrow \forall x \in \alpha \exists y \in \beta (x \leq y),$$

where α, β are external numbers. In this sense β is *positive* if $0 \leq \beta$. i.e. if $\exists y \in \beta(0 \leq y)$. Observe that with this order relation external neutrices are positive even though they always contain negative elements.

Distributivity does not hold in full generality, for $\oslash = \pounds((1+\oslash)-1)$, while $\pounds(1+\oslash) - \pounds 1 = \pounds - \pounds = \pounds + \pounds = \pounds$. So we have inclusion instead of equality. The previous example is a typical example where distributivity does not hold: multiplication of almost opposite elements, by a rather imprecise element. Classically it is well-known that asymptotic approximations do not behave well for subtraction, in particular for the derivative. We address the problem of distributivity in Chapter 3, and arrive at a full characterization (Theorem 3.3.4). For now we prove the following special case which often occurs in applications.

Theorem 1.4.3. *Let α, β and γ be external numbers, where β and γ are positive. Then $\alpha(\beta + \gamma) = \alpha\beta + b\gamma$.*

Proof. Let $a, b, c \in \mathbb{R}$ and A, B, C be neutrices such that $\alpha = a + A$, $\beta = b + B$ and $\gamma = c + C$. We may assume that $b, c \geq 0$ and also that $B \subseteq C$. Then $A(b+c) = Ab + Ac$, $aB + aC = aC$ and $AB + AC = AC$. So

$$\begin{aligned}
\alpha(\beta+\gamma) &= (a+A)(b+B+c+C) \\
&= (a+A)((b+c)+C) \\
&= a(b+c) + aC + A(b+c) + AC \\
&= ab + ac + aB + aC + Ab + Ac + AB + AC \\
&= ab + aB + Ab + AB + ac + aC + Ac + AC \\
&= (a+A)(b+B) + (a+A)(c+C) \\
&= \alpha\beta + \alpha\gamma.
\end{aligned}$$

□

Neutrices and external numbers enable us to express closeness in a similar way to the functional $O(\cdot)$ and $o(\cdot)$ notation. Let a, b be real numbers. Instead of $a \simeq b$ we may write $a \in b + \oslash$ and $b \in a + \oslash$, and if $a, b \neq 0$, instead of $a \sim b$ we may write $a \in b(1+\oslash)$ and $b \in (1+\oslash)a$. It is tempting to write $a = b(1+\oslash)$ or $b = (1+\oslash)a$, just as in the common notation $f(x) = O(\cdot)$ or $f(x) = o(\cdot)$ of classical asymptotics; this happened in earlier articles, and in concrete cases it hardly leads to misunderstandings. However the present book is somewhat more formal, and we chose the usual set theoretical notations $\in$ and $\subseteq$. The equality sign is used only when we are sure about the equality of the external numbers in question.

We use the notation of external numbers also in the following.

$$\sum_{i=0}^{\infty}(1+\oslash)\,x_i = \left\{\sum_{i=0}^{\infty}(1+\varepsilon_i)x_i : \varepsilon : \mathbb{N}\to\mathbb{R} \text{ internal } \wedge \forall i\in\mathbb{N}(\varepsilon_i \simeq 0)\right\}.$$

We prove a lemma on asymptotic approximations, which is a form of Cauchy's multiplicative approximation lemma: sums of asymptotic terms of the same sign are asymptotic.

Theorem 1.4.4 (Cauchy's multiplicative approximations lemma). *Let $\sum_{i=0}^{\infty} x_i$ be a convergent series of positive real numbers. Then*

$$\sum_{i=0}^{\infty}(1+\oslash)x_i \subseteq (1+\oslash)\sum_{i=0}^{\infty} x_i. \tag{1.15}$$

Proof. Let $\varepsilon : \mathbb{N}\to\mathbb{R}$ be internal and such that $\varepsilon_i \simeq 0$ for all $i\in\mathbb{N}$. Let $\varepsilon = \sup_{i\in\mathbb{N}}|\varepsilon_i|$. Then $\forall i\in\mathbb{N}(|\varepsilon_i|<a)$ holds for all positive appreciable a. By Cauchy's Principle it holds for some infinitesimal δ. Hence

$$(1-\delta)\sum_{i=0}^{\infty}x_i \le \sum_{i=0}^{\infty}(1+\varepsilon_i)x_i \le (1+\delta)\sum_{i=0}^{\infty}x_i,$$

which implies convergence of the series $\sum_{i=0}^{\infty}(1+\varepsilon_i)x_i$, and also that

$$\sum_{i=0}^{\infty}(1+\varepsilon_i)x_i = (1+\varepsilon)\sum_{i=0}^{\infty}x_i,$$

for some $\varepsilon \simeq 0$. This proves (1.15). □

An analogous approximation lemma holds for integrals. Let f be a positive continuous real function and $a,b\in\mathbb{R}$. We define

$$\int_a^b (1+\oslash)f(x)dx =$$

$$\left\{\int_a^b (1+\varepsilon(x))\,f(x)dx : \varepsilon : [a,b]\to\mathbb{R} \text{ internal } \wedge \forall x\in\mathbb{R}(\varepsilon(x)\simeq 0)\right\}.$$

Theorem 1.4.5 (Multiplicative approximation of functions). *Let f be a positive continuous real function and $a,b\in\mathbb{R}$. Then*

$$\int_a^b (1+\oslash)f(x)dx \subseteq (1+\oslash)\int_a^b f(x)dx. \tag{1.16}$$

Proof. Let $\varepsilon : [a,b] \to \mathbb{R}$ be internal and such that $\varepsilon(x) \simeq 0$ for all $x \in [a,b]$. Let $\varepsilon = \sup_{x\in[a,b]} |\varepsilon(x)|$. Then $\forall x \in \mathbb{R}(|\varepsilon(x)| < a)$ holds for all positive appreciable a. By Cauchy's Principle it holds for some infinitesimal δ. Hence

$$(1-\delta)\int_a^b f(x)dx \le \int_a^b (1+\varepsilon(x))f(x)dx \le (1+\delta)\int_a^b f(x)dx.$$

We conclude that $\int_a^b (1+\varepsilon(x))f(x)dx = (1+\varepsilon)\int_a^b f(x)dx$ for some $\varepsilon \simeq 0$. This proves (1.16). □

Analogous results hold for improper integrals, provided that they are convergent. The approximation of integrals will be further developed in Chapter 7.

Exercises

Exercise 1.4.1. *Determine* $e^{\oslash}$, $e^{-\not\infty}$, $\log(\oslash^+\backslash\{0\})$, $\log(\pounds^+\backslash\{0\})$. *If* $\varepsilon \simeq 0$, $\varepsilon > 0$, *Determine* $\frac{1}{1+\varepsilon\pounds}$, $e^{\pounds\varepsilon}$, $@^{\varepsilon}$, $\sqrt{1+\varepsilon\pounds}$.

Exercise 1.4.2. *Verify formula* (1.14) *and table* (1.13).

Exercise 1.4.3. *Exhibit a neutrix which is not isomorphic to none of the neutrices mentioned so far.*

Exercise 1.4.4. *Show that Cauchy's theorem on multiplicative approximations can be reformulated as follows. Let* $(x_i)_{i\in\mathbb{N}}, (\xi_i)_{i\in\mathbb{N}}$ *be two sequences of positive real numbers, such that* $\sum_{i=0}^{\infty} x_i$ *is convergent and* $\xi_i \sim x_i$ *for all* $i \in \mathbb{N}$. *Then* $\sum_{i=0}^{\infty} \xi_i$ *is convergent and* $\sum_{i=0}^{\infty} x_i \sim \sum_{i=0}^{\infty} \xi_i$.

Exercise 1.4.5. *Let* $\omega \in \mathbb{N}$ *be unlimited and let* u *be limited and such that* $u\sqrt{\omega} \in \mathbb{N}$. *Show that* $(\omega+1)(\omega+2)\cdots(\omega+u\sqrt{\omega}) \sim \omega^{u\sqrt{\omega}}e^{u^2}$.

Exercise 1.4.6. *Let* $\omega \simeq +\infty$. *Prove that*

$$\int_{-\frac{1}{\sqrt{\omega}}}^{\frac{1}{\sqrt{\omega}}} e^{-\omega v}(1+v)^{\omega}dv = (1+\oslash)\sqrt{\frac{2\pi}{\omega}}\cdot(\mathcal{N}(1)-\mathcal{N}(-1)),$$

where

$$\mathcal{N}(x) = \frac{1}{\sqrt{2\pi}}\int_{-\infty}^{x} e^{-\frac{y^2}{2}}dy$$

is the standard normal distribution. (Hint: use the substitution $v = \frac{u}{\sqrt{\omega}}$.*)*

Exercise 1.4.7. *Consider the differential equation* $\frac{dy}{dt} = y$. *For how long will the solution with initial condition* $y(0) = \varepsilon \simeq 0$, $\varepsilon > 0$ *be infinitely close to the solution starting in* 0 *(which is identically* 0*)? Answer the same question for the equation* $\frac{dy}{dt} = y^2$.

1.5 Nonstandard regularity properties of real internal functions

This section deals with S-continuity, S-differentiation and S-integration. In a sense these notions correspond to the respective classical notions up to an arbitrary infinitesimal, one could also say up to the neutrix $\oslash$. Chapters 6 and 7 extend these notions to approximations and imprecisions with respect to general classes of neutrices.

1.5.1 S-continuity

Definition 1.5.1. *A real function f is S-*continuous at $x \in \mathbb{R}$ *if for every $y \simeq x$ it holds that $f(y) \simeq f(x)$. If $D \subseteq \mathbb{R}$, we say that $f : D \to \mathbb{R}$ is* S-continuous *if f is S-continuous at all $x \in D$.*

Let $a \in \mathbb{R}$ be limited. Consider the linear function $f_a : \mathbb{R} \to \mathbb{R}$ defined by $f_a(x) = ax$. Let $x \in \mathbb{R}$ be arbitrary, and $y \simeq x$. By Theorem 1.1.9 one has $f_a(y) = ay \simeq ax = f_a(x)$. So f_a is S-continuous at x. Consider now the quadratic function $g : \mathbb{R} \to \mathbb{R}$ defined by $g(x) = x^2$. Again applying Theorem 1.1.9, for limited $x \in \mathbb{R}$ and $y \simeq x$ we find that $y^2 \simeq x^2$, so at these points x is S-continuous. However g is not S-continuous at unlimited points x, for

$$g\left(x + \frac{1}{x}\right) = x^2 + 2 + \frac{1}{x^2} \not\simeq x^2 = g(x).$$

It follows from Exercise 1.1.7 that the exponential function is S-continuous at limited points. The logarithmic function is S-continuous at all $x \gtrsim 0$, for if $\varepsilon \simeq 0$, then

$$\log(x + \varepsilon) = \log\left(x\left(1 + \frac{\varepsilon}{x}\right)\right) = \log(x) + \log\left(1 + \frac{\varepsilon}{x}\right) \simeq \log(x).$$

Definition 1.5.2. *A real function f is said to be* of class S^0 *on some set $D \subseteq \mathbb{R}$, if it is limited and S-continuous at limited $x \in D$. If $D = \mathbb{R}$, we simply say that f is a* function of class S^0.

Observe that the linear functions f_a (where a is limited), the quadratic function g and the exponential function are of class S^0, and that the logarithmic function is of class S^0 on the appreciables. In the axiomatic system IST one may show that at standard points standard functions are continuous if and only if they are S-continuous [35]; also standard functions are continuous on the whole of $\mathbb{R}$ if and only if they are of class S^0.

An example of a continuous function which is not S-continuous is the function f given by $f(x) = \sin(\omega x)$, where $\omega \simeq +\infty$. Indeed, $\frac{\pi}{2\omega} \simeq 0$ while

$f\left(\frac{\pi}{2\omega}\right) = \sin\left(\frac{\pi}{2}\right) = 1 \not\simeq 0 = \sin(0) = f(0)$. An example of an S-continuous function which is not continuous is the function g given by

$$g(x) = \begin{cases} 0, & x \leq 0 \\ \frac{1}{\omega}, & x > 0. \end{cases}$$

Functions of class S^0 behave particularly well under operations.

Theorem 1.5.1. *1. The composition, the sum and the product of two functions of class S^0 are of class S^0.*

2. The multiplicative inverse of an appreciable function of class S^0 is of class S^0.

3. A function that is infinitely close to a function of class S^0 for limited arguments is of class S^0.

Proof. Let f, g be real functions of class S^0 and x, y be limited, with $y \simeq x$.

1. The values $f(x)$, $f(y)$, $g(x)$ and $g(y)$ are all limited. Then $f(g(y))$ and $f(g(x))$ are limited. Also $g(y) \simeq g(x)$; hence $f(g(y)) \simeq f(g(x))$. Hence $f \circ g$ is of class S^0. Moreover $f(x)+g(x)$ and $f(x)g(x)$ are limited, while $f(y)+g(y) \simeq f(x)+g(x)$ and $f(y)g(y) \simeq f(x)g(x)$ by Theorem 1.1.9. Hence both $f+g$ and fg are of class S^0.

2. If $f(x)$ is appreciable, then $\frac{1}{f(x)}$ is also appreciable, while $\frac{1}{f(y)} \simeq \frac{1}{f(x)}$ by Theorem 1.1.9. Hence $\frac{1}{f}$ is of class S^0.

3. Let h be a real function such that $h(x) \simeq f(x)$ for limited $x \in \mathbb{R}$. Then $h(x)$ is limited for limited $x \in \mathbb{R}$, and for $y \simeq x$ it holds that $h(y) \simeq f(y) \simeq f(x) \simeq h(x)$. Hence h is of class S^0. □

The Mean Value Theorem holds for continuous functions. Moreover, continuous functions take maxima and minima on compact intervals. We show that approximate versions of these results hold for S-continuous functions.

Theorem 1.5.2. *Let $a, b \in \mathbb{R}$ with $a < b$ and $f : [a, b] \to \mathbb{R}$ be S-continuous.*

1. There exists $c \in [a, b]$ such that $f(c) \simeq 0$.

2. There exist $\lambda, \mu \in [a, b]$ such that $f(x) \gtrsim f(\lambda)$ for all $x \in [a, b]$ and such that $f(x) \lesssim f(\mu)$ for all $x \in [a, b]$.

Proof. Let $\omega \in \mathbb{R}^+$ be such that $b - a$ is a multiple of $\frac{1}{\omega}$.

1. Let $k \leq \frac{\omega}{b-a}$ be maximal such that $f\left(a + \frac{k}{\omega}\right) \leq 0$. Then $f\left(a + \frac{k+1}{\omega}\right) > 0$, while $f\left(a + \frac{k+1}{\omega}\right) \simeq f\left(a + \frac{k}{\omega}\right)$. Hence $f\left(a + \frac{k}{\omega}\right) \simeq 0$.

2. Let λ be such that $f(\lambda) = \min\left\{f\left(a + \frac{k}{\omega}\right) : 0 \leq k \leq \omega(b-a)\right\}$. Let $x \in [a, b]$, and $\kappa \in \mathbb{N}$ be maximal such that $a + \frac{\kappa}{\omega} \leq x$. Then $f(x) \simeq f\left(a + \frac{\kappa}{\omega}\right) \geq f(\lambda)$. Hence $f(x) \gtrsim f(\lambda)$. Define μ by

$$f(\mu) = \max\left\{f\left(a + \frac{k}{\omega}\right) : 0 \leq k \leq \omega(b-a)\right\}.$$

The proof that $f(x) \lesssim f(\mu)$ for all $x \in [a, b]$ is analogous. □

Functions of class S^0 may also be defined in a discrete setting, i.e. for internal functions defined on a discrete set $X \subset \mathbb{R}$, such that the distance of two successive points of X is always infinitesimal.

Exercises

Exercise 1.5.1. *Show that every polynomial of standard degree and limited coefficients is of class S^0.*

Exercise 1.5.2. *Where is the function $x \mapsto \frac{1}{1-x^2}$ S-continuous?*

Exercise 1.5.3. *Let the function € be defined by €$(x) = (1+\delta x)^{\frac{x}{\delta x}}$, where x is supposed to be a multiple of the infinitesimal δx, i.e. $x \in \mathbb{X} := \{k\delta x : k \in \mathbb{Z}\}$; such a function expresses composed interest rates (of 1%).*

1. *Show that € is of class S^0.*
2. *Let the function F be defined for $x \in \mathbb{X}^+$ by $F(x) := \prod_{0\le\xi<x}(1+\xi\delta x)$. Show that $F(x) \simeq e^{\frac{x^2}{2}}$ for limited x, and conclude that F is of class S^0.*

1.5.2 *S*-differentiability

Definition 1.5.3. *Let $a, b \in \mathbb{R}$ with $a < b$ and $f : [a, b] \to \mathbb{R}$. We say that f is* S-differentiable *at $x \in [a, b]$ if there exists a limited value $\phi(x)$ such that for all $y \simeq x, y \in [a, b]$, $y \neq x$ one has $\frac{f(y)-f(x)}{y-x} \simeq \phi(x)$. Then $\phi(x)$ is called an* S-derivative *of f at x. The function f is called* S-differentiable *if it is S-differentiable at all $x \in [a, b]$. An internal function $\phi : [a, b] \to \mathbb{R}$ is called an S-derivative of f if $\phi(x)$ is an* S-derivative of *f at x for all $x \in [a, b]$.*

Definition 1.5.4. *Let $a, b \in \mathbb{R}$ with $a < b$ and $f : [a, b] \to \mathbb{R}$ be S-differentiable with S-derivative ϕ. If both f and ϕ are of class S^0 on $[a, b]$, the function f is said to be* of class S^1 *on $[a, b]$. If f and ϕ are of class S^0, the function f is said to be* of class S^1.

It follows that S-derivatives are not unique since we may always add an internal infinitesimal function. Exercise 1.5.4 gives examples of differentiable functions of class S^1 with derivatives which are also S-derivatives.

Observe that one may choose some fixed $y \simeq x$, and then $\frac{f(y)-f(x)}{y-x}$ is already an S-*derivative* of f at x. If we take $y-x > 0$ to be a fixed infinitesimal, we obtain in this way an S-*derivative* of f at all points $x \in \text{dom}(f)$ such that $y \in \text{dom}(f)$. All internal functions infinitely close to such a difference quotient are S-derivatives. If f is of class S^1 with S-derivative ϕ, for all limited x, and for all y, z such that $y \simeq z \simeq x, y \neq z$ we have $\frac{f(z)-f(y)}{z-y} \simeq \phi(x)$. The next theorem shows that the usual rules of derivation can be obtained for S-derivatives.

Theorem 1.5.3. *Let $a, b \in \mathbb{R}$ with $a < b$, $f : \mathbb{R} \to \mathbb{R}$ be of class S^1 with S-derivative ϕ and $g : \mathbb{R} \to \mathbb{R}$ be of class S^1 with S-derivative ψ. Then*

1. *$f + g$ is of class S^1 with S-derivative $\phi + \psi$.*

2. *fg is of class S^1 with S-derivative $\phi g + f\psi$.*

3. *$\frac{1}{g}$ is of class S^1 on the set where it is appreciable with S-derivative $-\frac{\psi}{g^2}$.*

4. *$\frac{f}{g}$ is of class S^1 on the set where g is appreciable with S-derivative $\frac{\phi g - f\psi}{g^2}$.*

5. *$f \circ g$ is of class S^1 with S-derivative $(\phi \circ g) \cdot \psi$.*

Proof. Let x, y be limited and such that $y \simeq x$ and $y \neq x$. In each of the above cases we verify that the given expressions are S-derivatives indeed. It follows from Theorem 1.5.1 that they are at least of class S^0; hence the original functions are of class S^1.

1. We have

$$\frac{f(y) + g(y) - (f(x) + g(x))}{y - x} = \frac{f(y) - f(x)}{y - x} + \frac{g(y) - g(x)}{y - x} \simeq \phi(x) + \psi(x).$$

2. We have

$$\begin{aligned}\frac{f(y)g(y) - f(x)g(x)}{y - x} &= \frac{f(y)g(y) - f(x)g(y)}{y - x} + \frac{f(x)g(y) - f(x)g(x)}{y - x} \\ &= \frac{f(y) - f(x)}{y - x} g(y) + f(x)\frac{g(y) - g(x)}{y - x} \\ &\simeq \phi(x)g(x) + f(x)\psi(x),\end{aligned}$$

where in the last step we use that $\frac{f(y)-f(x)}{y-x}$ and $g(y)$ are limited, $g(y) \simeq g(x)$ and Theorem 1.1.9.

3. Assume that $g(x)$ is appreciable. Then $g(y) \simeq g(x)$ is also appreciable, and

$$\frac{\frac{1}{g(y)} - \frac{1}{g(x)}}{y - x} = \frac{-(g(y) - g(x))}{y - x} \frac{1}{g(y)g(x)} \simeq -\frac{\psi(x)}{g^2(x)},$$

where we use that $\frac{g(y)-g(x)}{y-x}$ and $\frac{1}{g(y)}$ are limited, $\frac{1}{g(y)} \simeq \frac{1}{g(x)}$ and Theorem 1.1.9.

4. By Parts 2 and 3.

5. If $g(y) \neq g(x)$, then

$$\frac{f(g(y)) - f(g(x))}{y - x} = \frac{f(g(y)) - f(g(x))}{g(y) - g(x)} \frac{g(y) - g(x)}{y - x}$$

Now $g(y) \simeq g(x)$ by S-continuity of g, hence $\frac{f(g(y))-f(g(x))}{g(y)-g(x)} \simeq \phi(g(x))$, and we obtain by Theorem 1.1.9 that $\frac{f(g(y))-f(g(x))}{y-x} \simeq \phi(g(x))\psi(x)$. If $g(y) = g(x)$, then $\psi(x) \simeq \frac{g(y)-g(x)}{y-x} = 0$. Because $\phi(g(x))$ is limited,

$$\frac{f(g(y)) - f(g(x))}{y - x} = 0 \simeq \phi(g(x))\psi(x). \qquad \square$$

Definition 1.5.5. *Let f be a real function and $c \in \mathbb{R}$. We say that f has an* S-local minimum *at c if there exist $a, b \in \mathrm{dom(f)}$, with $a \lessapprox c \lessapprox b$ such that $f(x) \geq f(c)$ for all $x \in [a, b]$. We say that f has an* S-local maximum *at c if there exist $a, b \in \mathrm{dom}(f)$, with $a \lessapprox c \lessapprox b$ such that $f(x) \leq f(c)$ for all $x \in [a, b]$.*

Theorem 1.5.4. *Let f be a function of class S^1, with S-derivative ϕ. Let $c \in \mathbb{R}$, and suppose that f has a local minimum, or a local maximum at c. Then $\phi(c) \simeq 0$.*

Proof. We consider only the case where f has a local minimum at c. Let $\varepsilon \simeq 0, \varepsilon > 0$. Then $\phi(c) \simeq \frac{f(c+\varepsilon)-f(c)}{\varepsilon} \geq 0$ and $\phi(c - \varepsilon) \simeq \frac{f(c)-f(c-\varepsilon)}{\varepsilon} \leq 0$. Hence $\phi(c) \simeq \phi(c - \varepsilon) \simeq 0$. $\square$

If f is of class S^0 and has an S-derivative of class S^1, the function is said to be of class S^2. In this way we may define functions of class S^n for all $\mathrm{st}(n)$. Like functions of class S^0, considering appropriate difference quotients, such functions may also be defined in a discrete setting [135].

Exercises

Exercise 1.5.4. *Prove that the constant a is an S-derivative of the linear function f_a, given by $f_a(x) = ax$, for limited a. Prove that the function $x \mapsto 2x$ is an S-derivative of the linear function $x \mapsto x^2$, and the function $x \mapsto \frac{1}{x}$ is an S-derivative of the logarithmic function.*

Exercise 1.5.5. *Let $\omega \simeq +\infty$. Show that $x \mapsto \frac{x}{\omega}\sin(\omega x)$ is of class S^0, but not of class S^1.*

Exercise 1.5.6. *Let $\delta x \simeq 0$, $\delta x > 0$. Let $\mathbb{X} = \{k\delta x : k \in \mathbb{Z}\}$. If $f : \mathbb{X} \to \mathbb{R}$, put $\delta f(x) = f(x + \delta x) - f(x)$.*

1. *Let the function € be defined by $€(x) = (1 + \delta x)^{\frac{x}{\delta x}}$ (see Exercise 1.5.3). Show that $\frac{\delta €(x)}{\delta x} = €(x)$. Conclude that € is of class S^1. Is it of class S^n, for all $\mathrm{st}(n)$?*

2. *Let the function F be defined for $x \in \mathbb{X}^+$ by $F(x) = \prod_{0 \leq \xi < x}(1 + \xi\delta x)$ (see Exercise 1.5.3). Show that $\frac{\delta F(x)}{\delta x} = xF(x)$. Conclude that F is of class S^1.*

1.5.3 S-integrability

Definition 1.5.6. *Let $a,b \in \mathbb{R}$ be limited with $a \lnsim b$. Let $\delta x \simeq 0$, $\delta x > 0$ be such that $\frac{b-a}{\delta x} \in \mathbb{N}$. Then $[a \cdot\cdot b] := \left\{a + k\delta x : 0 \leq k \leq \frac{b-a}{\delta x}\right\}$ is called an* infinitesimal discretization *of $[a,b]$ (with stepsize δx). We write $[a \cdot\cdot b) = [a \cdot\cdot b] \setminus \{b\}$.*

Definition 1.5.7. *Let $a,b \in \mathbb{R}$ be limited with $a \lnsim b$. Let $f : [a,b] \to \mathbb{R}$ be a real function and $[a \cdot\cdot b]$ be an infinitesimal discretization of $[a,b]$ with stepsize δx. Suppose that for all $x \in [a \cdot\cdot b)$*

$$m_x := \inf\{f(y) : x \leq y \leq x + \delta x\}, \quad M_x := \sup\{f(y) : x \leq y \leq x + \delta x\}$$

are well-defined. Then $\sum_{x\in[a\cdot\cdot b)} m_x \delta x$ is called a lower sum *for f on $[a,b]$, and $\sum_{x\in[a\cdot\cdot b)} M_x \delta x$ is called an* upper sum *for f on $[a,b]$.*

Definition 1.5.8. *Let $a,b \in \mathbb{R}$ be limited and such that $a \lnsim b$. Let $f : [a,b] \to \mathbb{R}$ be a real function. If for all infinitesimal discretizations of $[a,b]$, the upper sum and lower sum are well defined, limited, and infinitely close, the function f is called* S-Riemann-integrable *on $[a,b]$, and a Riemann sum $\sum_{a\leq x<b} f(x)\delta x$ is called an* S-integral *of f.*

Lemma 1.5.5. *Let $a,b \in \mathbb{R}$ be with $a \lnsim b$.*

1. *Let $\varepsilon : [a \cdot\cdot b] \to \mathbb{R}$ be such that $\varepsilon(x) \simeq 0$ for all $x \in [a \cdot\cdot b)$. Then there exists $\eta \simeq 0$ such that*

$$\sum_{a\leq x<b} \varepsilon(x)\delta x = \eta(b-a). \tag{1.17}$$

   If a,b are limited, $\sum_{a\leq x<b} \varepsilon(x)\delta x \simeq 0$.

2. *Let $f : [a,b] \to \mathbb{R}$ be Riemann-integrable and such that $f(x) \simeq 0$ for all $x \in [a,b]$. Then there exists $\zeta \simeq 0$ such that $\int_a^b f(x)dx = \zeta(b-a)$. If a,b are limited, then $\int_a^b f(x)dx \simeq 0$.*

Proof. As for the first part, let $\mu := \max\{|\varepsilon(x)| : x \in [a \cdot\cdot b)\}$. Then

$$\left|\sum_{a\leq x<b} \varepsilon(x)\delta x\right| \leq \mu \sum_{a\leq x<b} \delta x = \mu(b-a).$$

This implies (1.17) for some $\eta \simeq 0$. If a and b are limited, clearly $\eta(b-a) \simeq 0$.

The proof of the second part is analogous, working with $\sigma := \sup\{|f(x)| : x \in [a,b]\}$; observe that σ is infinitesimal by Cauchy's Principle, being less than all positive appreciable real numbers. □

Theorem 1.5.6. *Let $a, b \in \mathbb{R}$ be limited and such that $a \lessapprox b$. Let $f : [a, b] \to \mathbb{R}$ be a function of class S^0. Then f is S-Riemann-integrable on $[a, b]$, and*

$$\sum_{a \le x < b} m_x \delta x \simeq \sum_{a \le x < b} f(x) \delta x \simeq \sum_{a \le x < b} M_x \delta x. \tag{1.18}$$

If f is also Riemann-integrable, then

$$\sum_{a \le x < b} f(x) \delta x \simeq \int_a^b f(x) \delta x. \tag{1.19}$$

Proof. Let $[a \cdot\cdot b]$ be an infinitesimal discretization of $[a, b]$ with stepsize δx. Let $m := \min\{m_x : x \in [a \cdot \cdot b)\}$ and $M := \max\{M_x : x \in [a \cdot \cdot b)\}$. Then m, M are limited and

$$m(b-a) \le \sum_{a \le x < b} m(x) \delta x \le \sum_{a \le x < b} M(x) \delta x \le M(b-a).$$

Hence both the lower and upper sums are limited. Also $m_x \simeq M_x$ for all $x \in [a \cdot \cdot b)$. Applying Part 1 of Lemma 1.5.5,

$$\sum_{a \le x < b} M_x \delta x - \sum_{a \le x < b} m_x \delta x = \sum_{a \le x < b} (M_x - m_x) \delta x \simeq 0.$$

This implies (1.18).

If f is Riemann-integrable, we choose an infinitesimal stepsize δx, and then

$$\sum_{a \le x < b} m_x \delta x \le \int_a^b f(x) dx \le \sum_{a \le x < b} M_x \delta x.$$

This implies (1.19). □

Theorem 1.4.4 shows that common functions of analysis, such as polynomials with standard degree and limited functions, rational and root functions built upon them, and exponential and logarithmic functions are S-integrable, on appropriate intervals. The possibility of making infinitesimal approximations of Riemann sums by Riemann integrals is useful in many asymptotic calculations (see also the exercises at the end of this subsection).

We now present an analogue to the Fundamental Theorem of Calculus.

Definition 1.5.9. *Let $a, b \in \mathbb{R}$ be limited and such that $a \lessapprox b$. Let $f : [a, b] \to \mathbb{R}$ be an S-Riemann-integrable function on $[a, b]$. Let $\delta x \simeq 0, \delta x > 0$ be such that $\frac{b-a}{\delta x} \in \mathbb{N}$. An S-*primitive *of f is a function $\Phi : [a \cdot \cdot b] \to \mathbb{R}$ such that*

$$\frac{\Phi(x+h) - \Phi(x)}{h} \simeq f(x),$$

for all $x \in [a \cdot \cdot b), h \simeq 0, h > 0$ such that $x + h \in [a \cdot \cdot b]$.

Theorem 1.5.7. *Let $a, b \in \mathbb{R}$ be limited and such that $a \lnapprox b$. Let $f : [a \cdot\cdot b] \to \mathbb{R}$ be a function of class S^0. Define $F : [a \cdot \cdot b] \to \mathbb{R}$ by*

$$F(x) = \sum_{a \leq y < x} f(y)\delta x.$$

Then F is an S-differentiable S-primitive of f. Also, for every S-primitive Φ of f

$$\sum_{a \leq x < b} f(x)\delta x \simeq \Phi(b) - \Phi(a). \tag{1.20}$$

Moreover, if f is also continuous on $[a, b]$, every restriction of a primitive Ψ of f to $[a \cdot \cdot b]$ is an S-primitive of f.

Proof. Let $x \in [a \cdot \cdot b)$ and $h \simeq 0, h > 0$ be such that $x + h \in [a \cdot \cdot b)$. By Part 1 of Lemma 1.5.5, for some $\eta \simeq 0$, we have $\sum_{x \leq y < x+h}(f(y) - f(x))\delta x = \eta h$. Hence

$$\begin{aligned}\frac{F(x+h) - F(x)}{h} &= \frac{1}{h} \sum_{x \leq y < x+h} f(y)\delta x \\ &= \frac{1}{h} \sum_{x \leq y < x+h} (f(y) - f(x))\delta x + \frac{1}{h} \sum_{x \leq y < x+h} f(x)\delta x \\ &= \frac{1}{h}\eta h + \frac{1}{h} f(x)h \simeq f(x),\end{aligned}$$

so F is an S-primitive of F.

Let Φ be an S-primitive of f. Then $\frac{\Phi(x+\delta x) - \Phi(x)}{\delta x} - f(x) \simeq 0$ for all $x \in [a \cdot \cdot b)$. By Lemma 1.5.5

$$\begin{aligned}&\sum_{a \leq x < b} f(x)\delta x \\ &\quad= \sum_{a \leq x < b} \frac{\Phi(x + \delta x) - \Phi(x)}{\delta x}\delta x - \sum_{a \leq x < b} \left(\frac{\Phi(x + \delta x) - \Phi(x)}{\delta x} - f(x) \right) \delta x \\ &\quad\simeq \sum_{a \leq x < b} \Phi(x + \delta x) - \Phi(x) = \Phi(b) - \Phi(a).\end{aligned}$$

This implies (1.20).

Finally, let Ψ be a primitive of f. Then by Part 2 of Lemma 1.5.5

$$\begin{aligned}\frac{\Psi(x+h) - \Psi(x)}{h} &= \frac{1}{h} \int_x^{x+h} f(y)dy \\ &= \frac{1}{h} \int_x^{x+h} (f(y) - f(x))dy + \frac{1}{h} \int_x^{x+h} f(x)dy \simeq f(x).\end{aligned}$$ □

For asymptotic approximations of integrals many times we dispose of an approximation of the integrand which is valid only locally. This suffices in the case of a sufficiently strong global bound. The same is true for series. The next lemma gives appropriate conditions. They are related to Lebesgue's Dominated Convergence theorem [6], which is also known for improper Riemann-integrals.

Definition 1.5.10. *A convergent series* $\sum_{n=0}^{\infty} u_n$ *is* S-convergent *if* $\sum_{n=\omega}^{\infty} u_n \simeq 0$ *for all* $\omega \simeq +\infty$. *A convergent improper Riemann-integral* $\int_a^{\infty} f(x)dx$ *with a limited is* S-convergent *if* $\int_{\omega}^{\infty} f(x)dx \simeq 0$ *for all* $\omega \simeq +\infty$.

Typical S-convergent series are $\sum_{n=0}^{\infty} a^n$, with $|a| \lessapprox 1$, for $\sum_{n=\omega}^{\infty} a^n = \frac{a^{\omega}}{1-a} \simeq 0$ for all $\omega \simeq +\infty$. Using primitives, we see that the integral $\int_0^{\infty} e^{-x}dx$ is S-convergent, such as are the integrals $\int_0^{\infty} \frac{1}{x^p}dx$ for $p \gtrapprox 1$. The latter implies that $\sum_{n=0}^{\infty} \frac{1}{n^p}$ is S-convergent for $p \gtrapprox 1$.

Lemma 1.5.8 (Dominated Approximation).

1. *Let* $\sum_{n=0}^{\infty} u_n$ *and* $\sum_{n=0}^{\infty} v_n$ *be convergent series of real numbers. If* $u_n \simeq v_n$ *for standard* $n \in \mathbb{N}$ *and if there exists an S-convergent series of positive real numbers* $\sum_{n=0}^{\infty} w_n$ *such that* $|u_n|, |v_n| \leq w_n$ *for all* $n \in \mathbb{N}$, *then* $\sum_{n=0}^{\infty} u_n \simeq \sum_{n=0}^{\infty} v_n$.

2. *Let a be limited and* $\int_a^{\infty} f(x)dx$ *and* $\int_a^{\infty} g(x)dx$ *be convergent improper Riemann-integrals of real functions f and g. If* $f(x) \simeq g(x)$ *for standard* $x \in \mathbb{R}$ *and if there exists an S-convergent improper Riemann-integral* $\int_a^{\infty} h(x)dx$ *of a positive real function h such that* $|f(x)|, |g(x)| \leq h(x)$ *for all* $x \geq a$, *then*

$$\int_a^{\infty} f(x)dx \simeq \int_a^{\infty} g(x)dx. \tag{1.21}$$

Proof. We prove only Part 2; the proof of Part 1 is similar.

Let $b > a$ be limited. Then by Part 2 of Lemma 1.5.5 it holds that $\int_a^b f(x)dx \simeq \int_a^b g(x)dx$. By the Fehrele Principle there exists $\omega \simeq +\infty$ such that $\int_a^{\omega} f(x)dx \simeq \int_a^{\omega} g(x)dx$. Now

$$\left|\int_{\omega}^{\infty} f(x)dx\right|, \left|\int_{\omega}^{\infty} g(x)dx\right| \leq \int_{\omega}^{\infty} h(x)dx \simeq 0.$$

Combining, we obtain (1.21). □

Exercises

Exercise 1.5.7. *Let* $\delta x \simeq 0$, $\delta x > 0$, $\mathbb{X} = \{n\delta x : n \in \mathbb{N}\}$, $\varepsilon \simeq 0$. *With always* $x \in \mathbb{X}$, *give an infinitesimal approximation of the following sums:*

1. $\sum\limits_{0 \leq x < 1} x^3 \delta x$.

2. $\sum\limits_{1\leq x<2} \frac{1}{x+1}\delta x$.

3. $\sum\limits_{0\leq x<1} x^2(1+\varepsilon e^{\sin(x)})\delta x$.

4. $\sum\limits_{0\leq x<1} (1+x\delta x)^{\frac{1}{\delta x}}\delta x$.

Exercise 1.5.8. *Let $\delta x \simeq 0$, $\delta x > 0$, $\mathbb{X} = \{n\delta x : n \in \mathbb{N}\}$. We suppose that always $x \in \mathbb{X}$. Using permanence principles, give an approximation of each of the following sums.*

1. $\sum\limits_{1\leq x<\infty} \frac{1}{x^2}\delta x$.

2. $\sum\limits_{-\infty<x<+\infty} e^{-\frac{x^2}{2}}\delta x$.

3. $\sum\limits_{0\leq x<\frac{1}{\delta x}} (1-x\delta x)^{\frac{1}{\delta x}}\delta x$.

Exercise 1.5.9. *Let $\omega \simeq +\infty, \omega \in \mathbb{N}$ be a square. Put $\delta x = \frac{1}{\omega}$, and consider the near-interval $[0\cdot\cdot 1]$ with increment δx. We define $f : [0\cdot\cdot 1] \to \mathbb{R}$ by*

$$f(x) = \begin{cases} \frac{1}{\sqrt{\delta x}}, & \frac{x}{\sqrt{\delta x}} \in \mathbb{N}\setminus\{0\} \\ 0, & \text{otherwise.} \end{cases}$$

Let $F(x) = \sum_{0\leq y<x} f(y)\delta x$ be a discrete primitive of f. The function F is a sort of "singular measure", as follows from below:

1. *Show that F is S-continuous, but not its "discrete density" f.*

2. *Show that the support of f has infinitesimal measure.*

3. *What is the value of $F(1)$?*

Exercise 1.5.10. *Let $\omega \simeq +\infty$. Consider the Fourier sum*

$$S_\omega(x) = \frac{4}{\pi}\sum_{n=0}^{\frac{\omega-1}{2}} \frac{\sin(2n+1)x}{2n+1}.$$

Put $x = \frac{u}{\omega}$ with u limited. Write $S_\omega(x)$ as a Riemann sum. What is the increment of integration? Conclude that

$$S_\omega\left(\frac{u}{\omega}\right) \simeq \frac{2}{\pi}\int_0^u \frac{\sin(t)}{t}dt.$$

N.B. The exercise is related to the Gibbs–Wilbraham phenomenon *(see for example [149]). The sum is the Fourier transform of a locally constant function, with alternate values 1 and -1. At the discontinuity 0, the sum S_ω does not jump from -1 to 1, but has a slightly bigger jump from $-\frac{2}{\pi}\int_0^\pi \frac{\sin(t)}{t}dt$ to $\frac{2}{\pi}\int_0^\pi \frac{\sin(t)}{t}dt$.*

Exercise 1.5.11. *If we restrict ourselves to discrete functions and differences $\delta x \simeq 0$, we have a simpler analogue to the Fundamental Theorem of Calculus. Let $a, b \in \mathbb{R}$ be limited with $a \lnsim b$. Let $f : [a \cdot\cdot b] \to R$ be a function. Define $F : [a \cdot\cdot b] \to \mathbb{R}$ by $F(x) = \sum_{a \leq y < x} f(y)\delta x$. Prove that for all $x \in [a \cdot\cdot b)$*

1. $\frac{F(x+\delta x)-F(x)}{\delta x} = f(x)$.

2. $\sum\limits_{a \leq x < b} f(x)\delta x = F(b) - F(a)$.

Chapter 2

Some models and calculations involving imprecisions

2.1	Validity of asymptotic approximation by a Taylor polynomial ..	40
2.2	Mass and tail of a random variable	47
	2.2.1 The Mass Concentration Lemma	47
	2.2.2 Application: Stirling's formula	49
2.3	Jumps in singular perturbations	51
2.4	On linear equations	54

Asymptotic behaviour is behaviour under the influence of small or large parameters. Classical asymptotics formally involves limits, when going to infinity or approaching some point [31, 97]. Still there is some need felt in localizing asymptotic behaviour, notably with respect to boundary layers and free boundaries of singular perturbations, and domain decompositions in partial differential equations. The localization of asymptotic behaviour has an intrinsic vagueness, and the setting of Nonstandard Analysis permits us to take this vagueness into account by the use of external sets.

This chapter provides some examples of the use of external sets in the modelling of orders of magnitude and the localization of asymptotic behaviour. A typical tool is an *external equation*: given a function f, and representing an order of magnitude by an external number α, we try to find the set of solutions of the equation $f(x) \in \alpha$. Such equations will be studied more systematically in Chapter 9. Here we study some special cases, which deal with sets of close approximation, a mathematical model of the notion of the statistical confidence interval, which happens to be related to the localization of the principal contribution of asymptotic integrals (see also Chapter 7) and the localization of jumps in singular perturbations. We end with a systematic treatment of linear equations with only one variable. This is perhaps the simplest class of external equations; still there are already three independent uncertainties to deal with.

2.1 Validity of asymptotic approximation by a Taylor polynomial

Let $y, a \in \mathbb{R}$. We model "a is a good approximation of y" by $a \simeq y$. Then for a given real function f the domain of good approximation by a real function g is modelled by $\{x \in \mathbb{R} : g(x) \simeq f(x)\}$. In this section we let f be an analytic function and g be a Taylor polynomial. In this context we also search for an explanation of the regularities in the configuration of the remainders of Taylor polynomials observed in Figures 2.1 and 2.2.

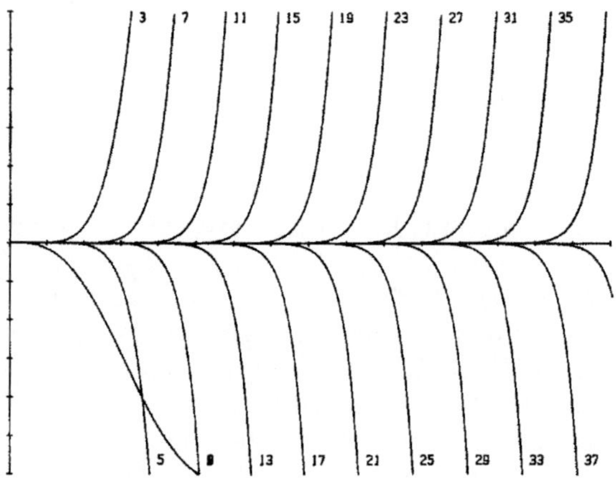

FIGURE 2.1: Remainders $R_n(x)$ of the Taylor expansion of $\sin(x)$ for $n = 1$ to $n = 41$ [130].

Figure 2.1 shows the remainders of the Taylor polynomials $\sum_{k=0}^{n}(-1)^k \frac{x^{2k+1}}{(2k+1)!}$ of the sine function from degree 1 to 41. One observes that, say, from degree 3 and up, a Taylor polynomial follows closely the sine function, and then takes off. Also from, say, degree 5 and up, the polynomials leave the sine function in a remarkably similar way, showing a linear progression. Before leaving the sine function, the resolution of the picture does not permit to distinguish the sine function from the approximating Taylor polynomial. So on the level of this graphics it is a good approximation. We point out the correspondence towards the mathematical model of "good approximation" stated above: a human observation cannot distinguish an infinitesimal from zero.

A configuration with similar regularity is shown in Figure 2.2, which shows the remainders of the approximation of the exponential function by its Taylor polynomials for $n = 5$ to $n = 22$. Again we observe that the graphs of the

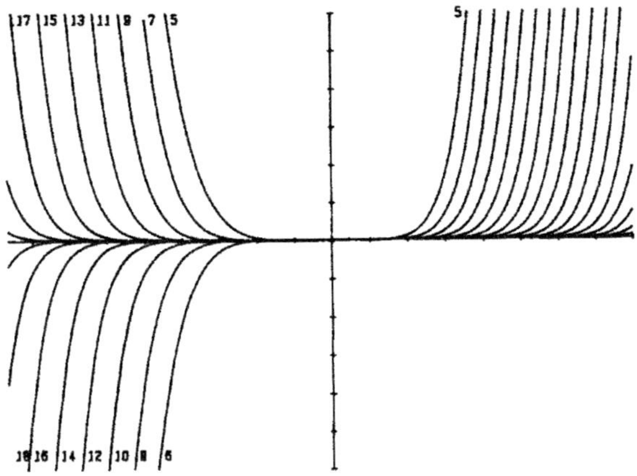

FIGURE 2.2: Remainders $e^x - \sum_{k=0}^{n} \frac{x^k}{k!}$ of the Taylor expansion of e^x. The numbers indicate the degrees [130].

remainders starting at $(0,0)$ remain glued to the horizontal axis, and then leave it in almost parallel equidistant curves. The values of x range from -8 to $+8$, the values of the remainders from -6 to 6. Remainders of Taylor polynomials of the exponential function have a very similar pattern of constant shifts. Theorem 2.1.1 aims at a more detailed analysis of Figure 2.2. Let ω be an infinitely large integer. On the sets $\frac{\omega}{e} + \frac{\log(\omega)}{2e} + £$, where they are appreciable, they behave as exponential functions themselves, of type e^{eu} [130].

Definition 2.1.1. *Let f be a real analytic function and $(P_n)_{n \in \mathbb{N}}$ be the sequence of its Taylor polynomials. For $x \in \operatorname{dom}(f)$ and $n \in \mathbb{N}$ we let $T_n(x) := \frac{f^{(n)}(0)}{n!} x^n$ be the* general term *of the Taylor series of f, $R_n(x) := f(x) - P_n(x)$ be the* error *after truncation of the Taylor series at n, and $A_n(x) := \frac{R_n(x)}{T_{n+1}(x)}$ be the* approximation factor. *The* domain of infinitesimal approximation *of f by P_n will be denoted by $I_n := \{x \in \mathbb{R} : R_n(x) \simeq 0\}$.*

The next theorem considers the approximation of the exponential function e^x by $P_m(x) = \sum_{n=0}^{m} \frac{x^n}{n!}$, where $m \in \mathbb{N}$. The exponential function being entire, for all $x \in \mathbb{R}$,

$$R_m(x) = \sum_{n=m+1}^{\infty} \frac{x^n}{n!}. \tag{2.1}$$

Theorem 2.1.1. *Let R_m be given by* (2.1).

1. *If m is standard, then $I_m = \emptyset$.*

2. *If* $m = \omega - 1$, *with* ω *an infinitely large integer, and* $\omega' = \frac{\omega}{e} + \frac{\log(\omega)}{2e}$, *we have*

$$I_{\omega-1} = (-\omega' + \pounds, \omega' + \pounds). \tag{2.2}$$

3. *For limited* u *we have*

$$\begin{aligned} R_{\omega-1}(\omega' + u) &\simeq \frac{e}{e-1}e^{eu} \\ R_{\omega-1}(-\omega' + u) &\simeq \frac{e}{1+e}e^{-eu}. \end{aligned} \tag{2.3}$$

4. *For limited* u *we have*

$$\begin{aligned} R_{\omega}((\omega+1)' + u) &\simeq R_{\omega-1}\left(\omega' + \frac{1}{e} + u\right) \\ R_{\omega}(-(\omega+1)' + u) &\simeq -R_{\omega-1}\left(-\omega' - \frac{1}{e} + u\right). \end{aligned} \tag{2.4}$$

Before giving the proof, we use the formulas of Theorem 2.1.1 to make a mathematical model of the configuration observed in Figure 2.2.

To begin with, we consider x to be positive. We cannot model the observed progression of the domains where the graph of the polynomial coincides with the graph of the exponential function with standard indices, for the domain where a polynomial of standard degree is nearly equal to the exponential function is always equal to the infinitesimals. Also the parallelism and the almost constant progression are not verified by standard remainder functions: differences are standard polynomials, which do neither display parallelism nor constant progression. This suggests that the mathematical model describing the pictures must be asymptotical, involving infinitely large degrees.

We propose to model "optically coinciding" by "infinitesimal distance" and "vertically distinct from the horizontal axis but still in the figure" by "appreciable". Formula (2.2) states that the set where the remainders are infinitesimal is not constant, but depends increasingly on the degrees. Formulas (2.3) and (2.4) state that as long as the remainders are appreciable, there is indeed near-parallelism, with nearly-equal horizontal progression. This progression is nearly equal to $\frac{1}{e}$, and with this shift the remainders are almost equal to the same exponential function $u \mapsto \frac{e}{e-1}e^{eu}$. It is to be expected that if we extend the picture to higher values of x, the phenomenon continues to be observed, justifying the asymptotic approach. Interestingly enough, even for relatively low values of n and x, the nonstandard asymptotic description seems to fit better than a description using standard values. On the other hand, for the vertical coordinate the choice of limited values seems justified to explain the parallelism; note that the curves representing the remainders eventually will cross: polynomials of higher degree grow faster.

We now turn to the proof of Theorem 2.1.1. It involves the solution of some external equations, which correspond to equalities up to some order of magnitude, like being infinitesimal or appreciable. We solve them in a mixed heuristic-systematic strategy. In Chapter 9 we return to the problem in question and present a more systematic strategy based on the possibility to replace given equations by simpler equations with the same set of solutions.

Proof of Theorem 2.1.1. 1. If $m = 0$, the remainder is given by $R_0 = e^x - 1$, which is infinitesimal if and only if $x \simeq 0$.

Let $m \geq 1$. Let $x \geq 0$. It follows from the well-known expression for the remainder of Taylor polynomials in terms of higher-order derivatives that for some θ with $0 < \theta < 1$

$$R_{m-1}(x) = \frac{x^m}{m!} e^{\theta x}. \tag{2.5}$$

As long as x is limited $e^{\theta x}$ is appreciable, and because $m!$ is appreciable we have $R_{m-1}(x) \in \oslash \Leftrightarrow x^m \in \oslash \Leftrightarrow x \in \oslash$; this is proved similarly for $x < 0$. For unlimited positive x the expression (2.5) is clearly unlimited, so $R_{m-1}(x)$ can never be infinitesimal. For unlimited negative x we use the formula for the remainder in integral form for e^{-x}, with the mth derivative of the exponential equal to $(-1)^m e^{-x}$,

$$R_{m-1}(x) = \frac{(-1)^m}{(m-1)!} \int_0^x (x-t)^m e^{-t} dt. \tag{2.6}$$

It follows from the Lemma of Dominated Approximation (i.e. Lemma 1.5.8, see also Exercise 2.1.1) that

$$\int_0^x \left(1 - \frac{t}{x}\right)^m e^{-t} dt \simeq \int_0^x e^{-t} dt \simeq 1. \tag{2.7}$$

So

$$\begin{aligned} R_{m-1}(x) &= \frac{(-1)^m x^m}{(m-1)!} \int_0^x \left(1 - \frac{t}{x}\right)^m e^{-t} dt \\ &\in (1+\oslash)\frac{(-1)^m x^m}{(m-1)!} \int_0^x e^{-t} dt \\ &= (1+\oslash)\frac{(-1)^m x^m}{(m-1)!}. \end{aligned}$$

Again the remainder can never be infinitesimal for unlimited x. Combining, we derive that $I_m = \oslash$.

2. Because the Taylor series is convergent, we have

$$R_{\omega-1}(x) = \sum_{n=\omega}^{\infty} \frac{x^n}{n!} = \frac{x^\omega}{\omega!} \sum_{k=0}^{\infty} \frac{x^k}{(\omega+1) \cdot \ldots \cdot (\omega+k)}.$$

So $T_\omega(x) = \frac{x^\omega}{\omega!}$ and $A_\omega(x) = \sum_{k=0}^{\infty} \frac{x^k}{(\omega+1)\cdot\ldots\cdot(\omega+k)}$.

We carry out the proof in the following way. (i) We show that all elements $x \in I_{\omega-1}$ satisfy $\left|\frac{x}{\omega}\right| \lnapprox 1$. (ii) For all such x we derive a convenient approximation of the approximation factor. (iii) With this approximation it is seen that the equation $R_{\omega-1}(x) \in \oslash$ may be substituted by the simpler equation $T_\omega(x) \in \oslash$, the solution of which is found by using the again simpler auxiliary equation $|T_\omega(x)| \in @$. (iv) The latter equation has a direct solution using external numbers.

(i) Using Stirling's formula: $\omega! \in (1+\oslash)\omega^\omega e^{-\omega}\sqrt{2\pi\omega}$ (see also Section 2.2.2) we have that

$$T_\omega(x) \in (1+\oslash)\sqrt{2\pi\omega}\left(\frac{xe}{\omega}\right)^\omega .$$

This means that $T_\omega(x)$ is exponentially small and of the form a^ω with $|a| \lnapprox 1$ if $\left|\frac{x}{\omega}\right| \lnapprox \frac{1}{e}$ and exponentially large and of the form b^ω with $|b| \gnapprox 1$ if $\left|\frac{x}{\omega}\right| \gnapprox \frac{1}{e}$. For $\left|\frac{x}{\omega}\right| \gtrsim 1$ we use (2.6) with $m = \omega$ and find

$$|R_{\omega-1}(x)| = \frac{|x|^\omega}{(\omega-1)!}\int_0^x \left(1-\frac{t}{x}\right)^\omega e^{-t}dt.$$

Now

$$\frac{((1+\oslash)\omega)^\omega}{(\omega-1)!} \subseteq ((1+\oslash)e)^\omega,$$

which is infinitely large, while for $x \geq 0$, for some $\varepsilon \simeq 0$

$$\int_0^x \left(1-\frac{t}{x}\right)^\omega e^{-t}dt \geq \int_0^x \left(1-\frac{t}{\omega(1+\varepsilon)}\right)^\omega e^{-t}dt \simeq \int_0^x e^{-2t}dt \simeq \frac{1}{2}, \quad (2.8)$$

(see Exercise 2.1.1) and for $x \leq 0$

$$\left|\int_0^x \left(1-\frac{t}{x}\right)^\omega e^{-t}dt\right| \geq \int_0^x e^{-t}dt \simeq 1.$$

Because $|x|^\omega$ is monotone, as a consequence $R_{\omega-1}(x)$ is certainly infinitely large for $\left|\frac{x}{\omega}\right| \gtrsim 1$. So it is sufficient to study the case $\left|\frac{x}{\omega}\right| \lnapprox 1$.

(ii) The approximation factor $A_\omega(x)$ has the obvious bound

$$|A_\omega(x)| \leq \sum_{k=0}^{\infty}\left|\frac{x^k}{(\omega+1)\cdots(\omega+k)}\right| \leq \sum_{k=0}^{\infty}\left|\frac{x}{\omega}\right|^k .$$

The right-hand member is an S-converging series $\left|\frac{x}{\omega}\right| \lnapprox 1$. For these values, by the Lemma of S-Dominated Approximation we find

$$A_\omega(x) \simeq \begin{cases} \frac{1}{1-\frac{x}{\omega}}, & x \geq 0 \\ \frac{1}{1+\frac{x}{\omega}}, & x < 0. \end{cases}$$

Indeed, for limited k, $\frac{x^k}{(\omega+1)\cdots(\omega+k)} \simeq \left(\frac{x}{\omega}\right)^k$, and for all k

$$\left|\frac{x^k}{(\omega+1)\cdots(\omega+k)}\right| < \left(\frac{1+|x|}{\omega}\right)^k ,$$

and $\sum_{k=0}^{\infty}\left(\frac{1+|x|}{\omega}\right)^k$ is S-convergent. So by the Lemma of Dominated Approximation

$$A_\omega(x) \simeq \sum_{k=0}^{\infty}\left(\frac{x}{\omega}\right)^k \simeq \frac{1}{1-\frac{x}{\omega}}. \tag{2.9}$$

This implies (2.1).

(iii) As a consequence of (2.9), for $\left|\frac{x}{\omega}\right| \lessapprox 1$ the approximation factor $A_\omega(x)$ is appreciable. So for these values, the equation $R_{\omega-1}(x) \in \oslash$ corresponds to the equation $T_\omega(x) = \frac{x^\omega}{\omega!} \in \oslash$. It happens that it is easier to solve $T_\omega(x) \in \pm@$, and use the monotonicity of x^ω.

(iv) We solve $T_\omega(x) \in \pm@$ for $x > 0$; the case $x < 0$ is similar. Then $T_\omega(x) \in @$. Observe that $T_\omega(x)$ is increasing in x and that by the observations above $T_\omega(x)$ can only be appreciable for values x such that $\frac{x}{\omega} \simeq \frac{1}{e}$. It follows from the Taylor expansion with the remainder in differential form (2.5) to the order 2 that for $\varepsilon \simeq 0$ we have $e^\varepsilon \in 1+\varepsilon+£e^2$. Using $\omega! \in \left(\frac{\omega}{e}\right)^\omega \omega^{1/2}@$ and $\log(@) = £$ we have

$$\begin{aligned}
\frac{x^\omega}{\omega!} \in @ &\Leftrightarrow x \in \omega!^{\frac{1}{\omega}} @^{\frac{1}{\omega}} \\
&\Leftrightarrow x \in \frac{\omega}{e}\omega^{\frac{1}{2\omega}} @^{\frac{1}{\omega}} \\
&\Leftrightarrow x \in \frac{\omega}{e} e^{\frac{\log(\omega)}{2\omega}} e^{\frac{\log(@)}{\omega}} \\
&\Leftrightarrow x \in \frac{\omega}{e} e^{\frac{\log(\omega)}{2\omega}+\frac{£}{\omega}} \\
&\Leftrightarrow x \in \frac{\omega}{e}\left(1+\frac{\log(\omega)}{2\omega}+\frac{£}{\omega}+£\left(\frac{\log(\omega)}{2\omega}\right)^2\right) \\
&\Leftrightarrow x \in \frac{\omega}{e}+\frac{\log(\omega)}{2e}+£.
\end{aligned}$$

Hence for $x \geq 0$

$$R_{\omega-1}(x) \in \oslash \Leftrightarrow T_\omega(x) \in \oslash \Leftrightarrow 0 \leq x < \frac{\omega}{e}+\frac{\log(\omega)}{2e}+£.$$

For $x < 0$ we find analogously, as long as $\frac{x}{\omega}$ is limited

$$R_{\omega-1}(x) \in \oslash \Leftrightarrow T_\omega(x) \in \oslash \Leftrightarrow -\frac{\omega}{e}-\frac{\log(\omega)}{2e}+£ < x < 0.$$

Combining, we derive (2.2).

3. For $x = \omega' + u$ with limited u, the approximation factor $A_\omega(x)$ is appreciable; indeed, (2.9) yields with $\frac{\omega'+u}{\omega} \simeq \frac{1}{e}$ that

$$A_\omega(\omega'+u) \simeq \frac{e}{e-1}.$$

So $R_{\omega-1}(x) \in @ \Leftrightarrow T_\omega(x) \in @ \Leftrightarrow x \in \omega' + \pounds$. We determine first an approximation for $R_{\omega-1}(\omega')$. Using Stirling's formula and Euler's formula

$$\begin{aligned} T_\omega(\omega') &= \frac{\left(\frac{\omega}{e} + \frac{\log(\omega)}{2e}\right)^\omega}{\omega!} \\ &= \frac{\left(\frac{\omega}{e}\right)^\omega \left(1 + \frac{\log(\omega)}{2\omega}\right)^\omega}{\omega!} \\ &\in \frac{\left(\frac{\omega}{e}\right)^\omega \sqrt{\omega}(1+\oslash)}{\omega!} \\ &\subseteq \frac{(1+\oslash)}{\sqrt{2\pi}} = \frac{1}{\sqrt{2\pi}} + \oslash. \end{aligned}$$

Hence $R_{\omega-1}(\omega') = T_\omega(\omega')A_\omega(\omega') \simeq \frac{e}{\sqrt{2\pi}(e-1)}$. Also

$$\frac{R_{\omega-1}(\omega'+u)}{R_{\omega-1}(\omega')} \simeq \frac{T_\omega(\omega'+u)}{T_\omega(\omega')} = \left(1 + \frac{u\frac{\omega}{\omega'}}{\omega}\right)^\omega \simeq e^{eu}.$$

Hence $R_{\omega-1}(\omega'+u) \simeq \frac{e}{e-1}e^{eu}$. In an analogous way we derive that $R_{\omega-1}(-\omega'+u) \simeq (-1)^{\omega-1}\frac{e}{1+e}e^{-eu}$. Combining, we derive (2.3).

4. Notice that $(\omega+1)' - \omega' \simeq \frac{1}{e}$. Then (2.4) follows from the S-continuity of $u \mapsto e^{eu}$ at limited arguments. □

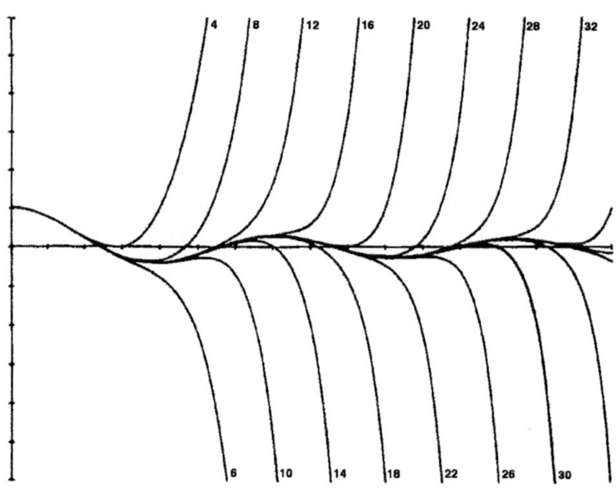

FIGURE 2.3: The Bessel function J_0 and the remainders $R_n(x)$ of its Taylor expansion; the numbers indicate the degrees [130].

Exercises

Exercise 2.1.1. *Prove formulas* (2.7) *and* (2.8).

Exercise 2.1.2. *Prove the analogue for Theorem 2.1.1 for the Taylor-expansion* $\sum_0^\infty (-1)^n \frac{x^{2n}}{(n!)^2}$ *of the Bessel function* J_0 *(see Figure 2.3).*

2.2 Mass and tail of a random variable

It is unlikely that a random variable takes a value which is very far from its mean, while it is probable that it takes values near to its mean. A mathematical way of expressing this is by the *statistical confidence interval.* For example, we generally forecast that 95% of the results of some test lie in the interval $[\mu - 2\sigma, \mu + 2\sigma]$, where μ is the mean and σ is the standard deviation of the sample. It is an *ad hoc* property: intervals corresponding to 90% and 99% are also used in practice.

We intend to formulate an intrinsic mathematical model for the confidence interval, by the notion of the *mass* of a random variable. The complement of the mass is called the *tail*, and is an intrinsic model for the queues used in probability estimates, representing outcomes far from the mean.

With the notion of mass we formulate the *Mass Concentration Lemma*, which relates the mass with both the mean and the standard deviation, and is simultaneously a Chebyshev inequality and a rudimentary Central Limit Theorem. It is universal for finite probabilities; for infinite probabilities the mean and the standard deviation need to exist.

We give some applications to asymptotic approximations of integrals; for instance it yields a proof of Stirling's formula avoiding epsilontics.

2.2.1 The Mass Concentration Lemma

Definition 2.2.1. *Let* $(\Omega, \Pr)$ *be a probability space and* $x : \Omega \to \mathbb{R}$ *be a random variable. A real number* λ *is an element of the* tail $\mathcal{Q}_x$ *of* x *if either* $\Pr\{x \geq \lambda\} \simeq 0$ *or* $\Pr\{x \leq \lambda\} \simeq 0$. *The* mass $\mathcal{M}_x$ *of a random variable* x *is the complement in* $\mathbb{R}$ *of* $\mathcal{Q}_x$, *i.e.*

$$\mathcal{M}_x := \bigcup_{\substack{\Pr\{x \geq \lambda_+\} \gtrsim 0 \\ \Pr\{x \leq \lambda_-\} \gtrsim 0}} [\lambda_-, \lambda_+].$$

Example 2.2.1. Consider a random variable with standard normal distribution function

$$\mathcal{N}(y) = \frac{1}{\sqrt{2\pi}} \int_{-\infty}^{y} e^{-\frac{x^2}{2}} dx.$$

Then $\mathcal{N}(y) \simeq 0$ for all $y \simeq -\infty$, $0 \lessapprox \mathcal{N}(y) \lessapprox 1$ for all limited y, and $\mathcal{N}(y) \simeq 1$ for all $y \simeq +\infty$. Hence $\mathcal{M} = \pounds$.

The density of a normally distributed random variable z with mean μ and standard deviation σ is $\frac{1}{\sqrt{2\pi}\sigma}e^{-\frac{(x-\mu)^2}{2\sigma^2}}$. The substitution $\frac{x-\mu}{\sigma} \mapsto x$ turns the distribution into standard normal. As a consequence the mass of z equals $\mathcal{M}_z = \mu + \pounds\sigma$. The Mass Concentration Lemma below establishes that $\mathcal{M}_z \subseteq \mu + \pounds\sigma$ for every random variable z, provided the mean μ and standard deviation σ are well-defined. Its proof uses Chebyshev's inequality.

For $a > 0$, $b \in \mathbb{R}$ we say that b is of the *order* of a if $\frac{b}{a}$ is limited.

Theorem 2.2.1. *Let y be a positive random variable in a probability space $(\Omega, \Pr)$ with finite mean μ. Then each element of the mass of y is of order μ.*

Proof. Assume first that $\mu = 0$. Let $\lambda > 0$. By Chebyshev's inequality

$$\Pr\{y \geq \lambda\} \leq \frac{\mathrm{E}(y)}{\lambda} = 0.$$

Hence $\lambda \in \mathcal{Q}_y$, and also $\Pr\{y = 0\} = 1$. Hence $\mathcal{M}_x = \{0\}$. Clearly 0 is of the order of 0.

Secondly, assume that $\mu > 0$. Let λ be unlimited. Again by the Chebyshev inequality

$$\Pr\{y \geq \lambda \mathrm{E}(y)\} \leq \frac{\mathrm{E}(y)}{\lambda \mathrm{E}(y)} \leq \frac{1}{\lambda} \simeq 0.$$

So $\lambda \mathrm{E}\, y \in \mathcal{Q}$, implying that $\mathcal{M}_y \subseteq \pounds \mathrm{E}(y)$. □

Theorem 2.2.2 (Mass Concentration Lemma). [132] *Let x be a random variable in a probability space $(\Omega, \Pr)$ with finite mean μ and standard deviation σ. Then the distance between a point of its mass and μ is of order σ.*

Proof. Let a be a point in the mass of $x - \mu$. Then $\Pr\{|x - \mu| \geq a\} \gtrapprox 0$, so $\Pr\{(x-\mu)^2 \geq a^2\} \gtrapprox 0$. By Theorem 2.2.1 it follows that a^2 is of order $\mathrm{E}(x-\mu)^2 = \mathrm{Var}(x) = \sigma^2$. Hence a is of order σ. □

Note that the mean and the standard deviation of random variables defined on finite probability spaces are always well-defined. In this context, the Mass Concentration Lemma becomes a universal result. We recall that the book Radically Elementary Probability Theory of Nelson [94] presents advanced stochastics in a finite nonstandard setting.

The Mass Concentration Lemma has some alternative formulations. Let $(\Omega, \Pr)$, x, μ and σ be as above. Then

1. $\mathcal{M} \subseteq \mu + \pounds\sigma$.

2. Almost surely the difference of the outcome of a random variable with respect to its mean is of the order of the standard deviation. Here almost surely is defined as in [94]: An event $A \subseteq \Omega$ is said to be almost sure if

for all appreciable $\varepsilon > 0$ there is a (measurable) internal event I with $I \subseteq A$ such that $\Pr(I) > 1 - \varepsilon$. Note that in a finite probability space all internal events are measurable.

3. $\mathrm{PR}\{x \in \mu + \pounds\sigma\} \subseteq 1 + \oslash$. The "external probability function" PR has been introduced in [34, Chapter 7] to measure external sets by external numbers. Generally speaking, almost sure events $A \subseteq \Omega$ always satisfy $\mathrm{PR}(A) \subseteq 1 + \oslash$.

Chapter 7 contains a version for densities (Lemma 7.4.1), which is useful for approximating asymptotic integrals.

2.2.2 Application: Stirling's formula

For a proof of Stirling's formula without epsilontics we use the Mass Concentration Lemma, and the possibility to integrate internal functions, and also external functions built from internal functions, but which depend on an external number, over an external interval. This is the subject of Chapter 7.

Theorem 2.2.3 (Stirling's formula). *Let $\omega \in \mathbb{N}$ be unlimited. Then*

$$\omega! \sim \omega^\omega e^{-\omega}\sqrt{2\pi\omega}.$$

Proof. We use the Laplace transform $\omega! = \int_0^\infty e^{-t}t^\omega dt$. Then $\frac{e^{-t}t^\omega}{\omega!}$ is a density, with mean $\omega + 1$ and standard deviation $\sqrt{\omega + 1}$. By the Mass Concentration Lemma its mass $\mathcal{M}$ satisfies

$$\mathcal{M} \subseteq \omega + 1 + \pounds\sqrt{\omega + 1} = \omega + \pounds\sqrt{\omega}.$$

As a consequence $1 \in 1+\oslash = \int_{\omega+\pounds\sqrt{\omega}} \frac{e^{-t}t^{\omega}}{\omega!}dt$. Put $t := \omega + u\sqrt{\omega}$. Then we obtain a new density with mass equal to $\pounds$. Hence

$$\begin{aligned}
1 &= \int_0^{\infty} \frac{e^{-t}t^{\omega}}{\omega!}dt \\
&\in \int_{\omega+\pounds\sqrt{\omega}} \frac{e^{-t}t^{\omega}}{\omega!}dt \\
&= \frac{\omega^{\omega}e^{-\omega}\sqrt{\omega}}{\omega!}\int_{\pounds} e^{-\sqrt{\omega}u}\left(1+\frac{u}{\sqrt{\omega}}\right)^{\omega} du \\
&= \frac{\omega^{\omega}e^{-\omega}\sqrt{\omega}}{\omega!}\int_{\pounds} e^{-\sqrt{\omega}u}e^{\omega\log\left(1+\frac{u}{\sqrt{\omega}}\right)} du \\
&\subseteq \frac{\omega^{\omega}e^{-\omega}\sqrt{\omega}}{\omega!}\int_{\pounds} e^{-\sqrt{\omega}u+\sqrt{\omega}u-\frac{u^2}{2}(1+\oslash)} du \\
&\subseteq \frac{\omega^{\omega}e^{-\omega}\sqrt{\omega}}{\omega!}\int_{\pounds} \left(e^{-\frac{u^2}{2}}+\oslash\right) du \\
&= \frac{\omega^{\omega}e^{-\omega}\sqrt{\omega}}{\omega!}\left(\int_{\pounds} e^{-\frac{u^2}{2}} du + \int_{\pounds} \oslash du\right) \\
&= \frac{\omega^{\omega}e^{-\omega}\sqrt{\omega}}{\omega!}\left(\sqrt{2\pi}+\oslash+\oslash\right) \\
&= \frac{\omega^{\omega}e^{-\omega}\sqrt{\omega}}{\omega!}\sqrt{2\pi}(1+\oslash).
\end{aligned}$$

Hence $\omega! \in (1+\oslash)\omega^{\omega}e^{-\omega}\sqrt{2\pi\omega}$, implying Stirling's formula. □

Exercises

Exercise 2.2.1. *Let $\omega \in \mathbb{N}$ be unlimited. Determine the mass of the following probability distributions.*

1. *The Cauchy distribution on $\mathbb{R}$ with density $\frac{1}{\pi(x^2+1)}$.*
2. *The exponential distribution on $\mathbb{R}^+$ with density e^{-x}.*
3. *The homogeneous distribution on $[0,1]$ with density 1.*
4. *The binomial distribution on $\{0,\ldots,\omega\}$.*
5. *The Poisson distribution on $\mathbb{N}$ given by $\frac{e^{-\omega}\omega^k}{\omega!}$.*

Exercise 2.2.2. *Extend Theorem 2.2.3 to the Gamma function.*

2.3 Jumps in singular perturbations

Informally, a real function exhibits a jump if it goes from one value to a second value which is not close to it within a very small time-interval. In this sense S-continuous functions do not jump, but the function f defined by $f(x) = \arctan\left(\frac{x}{\varepsilon}\right)$, $\varepsilon \simeq 0$, $\varepsilon \neq 0$ does jump in spite of being continuous. Singular perturbations, which in their simplest form are given by the equation

$$\varepsilon \frac{dx}{dt} = f(t, x), \tag{2.10}$$

with $\varepsilon \simeq 0$, $\varepsilon > 0$, typically have solutions with jumps. For jumps it is interesting to study the interval of time when the solutions are really jumping, but due to some natural vagueness associated with the notion, the duration of the jump may not be characterized by a real number. Following Francine Diener in [37] we give a definition of jump and duration in terms of external sets. Then for the most common singular perturbations the duration of the jump is equal to $\varepsilon\pounds$. However there exist interesting examples [37, 12] for which one finds other galaxies.

We start with a formal definition of jump.

Definition 2.3.1. *Let f be a real function and $s \in \mathbb{R}$. We say that f has a* positive jump *in s if there exists an interval $[a, b]$ such that*

1. *$a < s < c$,*
2. *$f(s)$ is limited,*
3. *$a \simeq b$,*
4. *$f(a) \underset{\not\simeq}{<} f(s) \underset{\not\simeq}{<} f(b)$,*
5. *f is increasing on $[a, b]$.*

The set $E_s := \{t - s : f$ has a jump at each point $u \in [s, t]\}$ is called the thickness of the jump *at s. The set $f(E_s)$ is called the* jump *of f at s. If there exist c and d such that $f(E_s) = \{y : c \underset{\not\simeq}{<} y \underset{\not\simeq}{<} d\}$ we say that f* jumps from $c + \oslash$ to $d + \oslash$. *If $d \simeq +\infty$, we say that f* jumps from $c + \oslash$ to $\not\pounds$. *Similarly if $c \simeq -\infty$, we say that f* jumps from $-\not\pounds$ to $d + \oslash$. *With some abuse of language, we may also say that f jumps from $c + \oslash$ to s, or from s to $d + \oslash$. The* negative jump *is defined in an analogous manner.*

In Definition 2.3.1 the elements c and d are clearly not unique.

Example 2.3.1. Let $\varepsilon \simeq 0$, $\varepsilon > 0$ and let $f : \mathbb{R} \to \mathbb{R}$ defined by $f(x) = e^{-\frac{x}{\varepsilon}}$. This function jumps at 0 from $\not\pounds$ to 0, and the thickness of the jump is $\varepsilon\pounds$. In fact, the thickness of a jump is determined by solving the external equation

$$e^{-\frac{x}{\varepsilon}} \in @ \Leftrightarrow -\frac{x}{\varepsilon} \in \pounds \Leftrightarrow x \in \varepsilon\pounds.$$

We now study the jumps of the solutions of the equation (2.10) in more detail. If we write

$$\frac{dx}{dt} = \frac{f(t,x)}{\varepsilon}, \tag{2.11}$$

we see that the solutions are almost vertical ("fast") if $f(t,x) \not\simeq 0$, but can have derivatives which are limited (in this case solutions are "slow") if $f(t,x) \simeq 0$. There exists an intermediate area in which $f(t,x) \simeq 0$ and $\frac{f(t,x)}{\varepsilon} \simeq \infty$. This makes it difficult to perform a rigorous analysis of this type of equation, in which the well-known Tikhonov Theorem [126] is a valuable tool. We study here the simple case where the "slow curve" is horizontal, in fact a part of the horizontal axis. As it turns out, many cases can be reduced to this case by a change of variable.

Let a, b be limited and such that $a \lessapprox b$. Let $c \gtrapprox 0$. Let f be an internal function of class C^1, at least defined for $a \leq t \leq b$, $|x| \leq c$, where it is S-continuous in t and S-differentiable in x, with S-continuous partial derivative ("of class S^1").

We suppose that $f(t,0) \simeq 0$, and $f_2'(t,0) \in -@$ for $a \leq t \leq b$. This means that the slow curve $x = 0$ is *attractive*. We justify this by the following argument.

To start with, $f_2'(t,0) \in -@$ for $a \leq t \leq b, x \simeq 0$ by S-continuity. By the Fehrele Principle there exists $d \gtrapprox 0$ such that $f_2'(t,0) \in -@$ for $a \leq t \leq b$, $-d \leq x \leq d$; without restriction of generality we may assume that $d = c$. An example of such a function is the linear function $f(t,x) = ax$, where $a < 0$ is appreciable. In general we have the representation

$$\begin{aligned} f(t,x) &= f(t,0) + \int_0^x f_2'(t,z)dz \\ &\in \oslash + x f_2'(t,\overline{z}) \quad (\overline{z} \text{ between } 0 \text{ and } x) \\ &\subseteq \oslash - @x. \end{aligned} \tag{2.12}$$

So we see that

$$f(t,x) = \begin{cases} \lessapprox 0, & x \gtrapprox 0 \\ \simeq 0, & x \simeq 0 \\ \gtrapprox 0, & x \lessapprox 0. \end{cases} \tag{2.13}$$

With the help of (2.11), we see that solutions starting above the slow curve have an negative, infinitely large derivative, so will approach the slow curve very fast. Also solutions starting beneath the slow curve approach the slow curve very fast, for they have a positive, infinitely large derivative. The next proposition states that such solutions have jumps indeed, and determines their thickness.

Proposition 2.3.1. *Let a, b be limited and such that $a \lessapprox b$. Let $c \gtrapprox 0$. Let f be an internal function of class C^1, at least defined for for $a \leq t \leq b$, $|x| \leq c$, where it is S-continuous in t and S-differentiable in x, with S-continuous* *partial derivative. Let* λ *be a solution of* (2.10) *such that* $\lambda(a) = c$. *Then* λ *jumps from* c *to* $\oslash$ *with thickness* $\varepsilon \pounds^+$.

Proof. By (2.12) the quasi-linear function $f(t, y)$ of equation (2.11) has a lower bound of the form $g(t, x) = -Kx - \gamma$ with $K > 0$ limited and $\gamma \simeq 0$ and an upper bound of the form $g(t, x) = -Lx + \delta$ with $L > 0$ limited and $\delta \simeq 0$ as long as x is positive; the equations $\frac{dx}{dt} = \frac{g(t,x)}{\varepsilon}$ and $\frac{dx}{dt} = \frac{h(t,x)}{\varepsilon}$ yield explicit lower and upper solutions, both elements of $e^{\frac{-@t}{\varepsilon}} + \oslash$. Now

$$e^{\frac{-@t}{\varepsilon}} + \oslash \in @ \Leftrightarrow e^{\frac{-@t}{\varepsilon}} \in @ \Leftrightarrow \frac{-@t}{\varepsilon} \in \pounds^- \Leftrightarrow t \in \varepsilon \pounds^+.$$

Hence the thicknesses of the (half)jumps of the lower and upper solutions both are equal to $\varepsilon \pounds^+$. As a consequence, also the thickness of the jumps of the true solution λ is of the form $\varepsilon \pounds^+$, and on this set we have $\lambda(t) \in e^{\frac{-@t}{\varepsilon}} + \oslash$. □

If in the above $f_2'(t, 0) \in @$ for $a \leq t \leq b$, by an argument based on reversing time one proves that solutions jump away from the slow curve, which is called *repulsive.*

In Chapter 9 we return to the problem of approaching a slow curve, now in terms of convergence and strong convergence to external sets.

Example 2.3.2 (J.-L Callot [23]). Let $\varepsilon \simeq 0$, $\varepsilon > 0$. Consider the equation

$$\varepsilon \frac{dx}{dt} = tx \tag{2.14}$$

The slow curve is $x = 0$, which is attractive for $t \lessapprox 0$ limited and repulsive for $t \gtrapprox 0$ limited. Let $t_0 \lessapprox 0$ and λ be a solution such that $\lambda(t_0)$ is positive appreciable. By Proposition 2.3.1 the solution λ jumps to $\oslash$ with thickness $\varepsilon \pounds^+$. We may verify this directly by solving the equation. We find

$$x(t) = x(t_0) e^{\frac{t^2 - t_0^2}{2\varepsilon}}. \tag{2.15}$$

So solutions $x(t)$ are symmetric with respect to $t = 0$. Hence $x(t) \simeq 0$ for $x < -t_0 + \varepsilon \pounds^+$, when it jumps again, in fact from $\oslash$ to $\not\oslash$ with thickness $\varepsilon \pounds$ (see also Exercise 2.3.2).

A solution that follows the attractive part of a slow curve and then the repulsive part of the slow curve at least during an appreciable period of time is called a *canard.* The first canards were detected by E. Benoît, F. and M. Diener and J.-L. Callot, using nonstandard methods [8]. These were solutions of the Van der Pol equation, with a certain visual aspect of a duck. The name "canard" is also justified because the existence of solutions with a bounded unstable component during appreciable time seemed very unlikely. Later, Callot found that the simple equation (2.14) also exhibits the phenomenon.

Exercises

Exercise 2.3.1. *Let $\varepsilon \simeq 0$, $\varepsilon > 0$. Determine the jumps of* $\arctan\left(\frac{x}{\varepsilon}\right)$ *and* $e^{\frac{x^3}{\varepsilon}}$ *and their thickness.*

Exercise 2.3.2. *Solve the differential equation $\varepsilon\frac{dx}{dt} = tx$, where $\varepsilon \simeq 0$, $\varepsilon > 0$. Show that the solution λ such that $\lambda(-1) = 1$ has a jump at -1. This solution has one more jump. Where? Determine the thickness of the jump using the explicit formula (2.15) and with the methods of the proof of Proposition 2.3.1.*

Exercise 2.3.3. *Let $\varepsilon \simeq 0$, $\varepsilon > 0$. Let $\lambda(t)$be the solution of $\varepsilon\ddot{x} = (\dot{x})^2$ such that $\lambda(0) = 1$. Show that $\lambda(t) = -\varepsilon \log\left(t + e^{\frac{-1}{\varepsilon}}\right)$ satisfies the initial condition. Show that λ jumps from $\not\infty$ to $\oslash$, and that the thickness of the jump is equal to $\left]-e^{\frac{-1}{\varepsilon}} + e^{\frac{-\not\infty}{\varepsilon}}, e^{\frac{-@}{\varepsilon}}\right]$.*

Exercise 2.3.4. *Let a be appreciable and $\varepsilon \simeq 0$, $\varepsilon > 0$.*

1. *Consider the equation*

$$\frac{dx}{dt} \in \oslash - @x.$$

   Show that the solution λ with $\lambda(0) = a$ satisfies $\lambda(t) \in ae^{-@t} + \oslash$ for limited t.

2. *Consider the equation*

$$\frac{dx}{dt} \in \frac{\oslash - @x}{\varepsilon}.$$

   Show that the solution λ with $\lambda(0) = a$ satisfies $\lambda(t) \in ae^{\frac{-@t}{\varepsilon}} + \oslash$ for $t \in \varepsilon\pounds$.

2.4 On linear equations

In this section we study the most elementary equations with external numbers, which are the linear equations. The study is already somewhat intricate, due to the fact that three uncertainties are involved. The presence of four uncertainties and non-linearities make the study of quadratic equations still less elementary, and we confine ourselves to some exercises.

Not all equations with external numbers can be solved in terms of equalities. For instance, no external number, or even set of external numbers, satisfies the equation $\oslash\xi = \pounds$ since one should have $\xi \subseteq \pounds$ and $\oslash\pounds = \oslash \subset \pounds$. This means that sometimes we find solutions in terms of inclusions instead of equalities.

Let $\alpha = a + A$, $\beta = b + B$ and $\gamma = c + C$ be external numbers. For $x \in \mathbb{R}$, we consider linear equations of the type

$$\alpha + \beta x \subseteq \gamma. \tag{2.16}$$

If not empty, the set $S := \{x \in \mathbb{R} : \alpha + \beta x \subseteq \gamma\}$ is said to be the *maximal solution* of (2.16). It is the *exact solution* if $\alpha + \beta S = \gamma$. If the equation is solvable, we will see that S is always an external number ξ. Then equation (2.16) is equivalent to

$$(\alpha - c) + \beta x \subseteq C.$$

So we may study as well equations of the form

$$\alpha + \beta x \subseteq C. \tag{2.17}$$

Writing the equation in full detail, we obtain necessary conditions for the existence of solutions. Indeed (2.17) can be rewritten as

$$a + A + (b + B)x \subseteq C \Leftrightarrow a + bx + A + Bx \subseteq C. \tag{2.18}$$

So we must have

$$A \subseteq C, \tag{2.19}$$

and then (2.18) reduces to

$$a + bx + Bx \subseteq C. \tag{2.20}$$

Definition 2.4.1. *The equation* (2.17) *is called* homogeneous *if* $a \in C$. *It is called* singular *if* $b \in B$ *and* non-singular *if* β *is zeroless.*

Definition 2.4.2. *Let* $\alpha = a + A$ *be an external number. If* α *is zeroless, we define the* relative uncertainty *of* α *to be the neutrix* $\mathcal{R}(\alpha) = \frac{A}{a}$. *If* α *is a neutrix, we define* $\mathcal{R}(\alpha) = \mathbb{R}$.

The propositions below describe the maximal solution of (2.17) in all possible cases.

Proposition 2.4.1. *If the equation* (2.17) *is homogeneous and non-singular and condition* (2.19) *is satisfied, the maximal solution is given by the external number* $\xi := \frac{C}{\beta}$.

Proof. One has $\beta \frac{C}{\beta} = (1 + \mathcal{R}(\beta))C$. Because $(1 + \mathcal{R}(\beta)) \subseteq (1 + \oslash)$, by Part 1 of Theorem 1.4.2 we have $(1 + \mathcal{R}(\beta))C = C$. □

Proposition 2.4.2. *If the equation* (2.17) *is homogeneous and singular and condition* (2.19) *is satisfied, the maximal solution is given by the neutrix*

$$C : B := \{x \in \mathbb{R} : Bx \subseteq C\}.$$

Proof. We may assume that $a = 0$ so the equation (2.17) reduces to $Bx \subseteq C$. For every limited k we have $kBx \subseteq C$ whenever $Bx \subseteq C$. So $\pounds(C : B) = C : B$. Hence $C : B$ is a neutrix. □

Non-homogeneous singular equations cannot be solved, for by (2.20) they reduce to $Bx \subseteq C - a$, which amounts to a neutrix having to be a subset of a zeroless number.

To solve non-homogeneous non-singular equations we need an extra condition. An obvious solution of $a + bx \in C$ is $x = -\frac{a}{b}$. To satisfy (2.20) we must have

$$B\frac{a}{b} \subseteq C. \tag{2.21}$$

Proposition 2.4.3. *If the equation (2.17) is non-homogeneous non-singular and conditions (2.19) and (2.21) are satisfied, the maximal solution is given by the external number* $\xi := -\frac{\alpha+C}{\beta}$.

Proof. By (2.20) we have

$$a+b\frac{-(\alpha+C)}{\beta}+B\frac{-(\alpha+C)}{\beta} = a+(-a+C)\left(1+\frac{B}{b}\right)+B\frac{a}{b} \subseteq A+C+C = C.$$

The solution is obviously maximal. □

Finally we deal directly with homogeneous equations of the type

$$\beta x \subseteq \gamma. \tag{2.22}$$

Proposition 2.4.4. *Let* $\beta = b + B$ *be zeroless and* $\gamma = c + C$. *If* γ *is zeroless, then (2.22) has a solution if and only if* $\mathcal{R}(\beta) \subseteq \mathcal{R}(\gamma)$. *If well defined, the maximal solution is given by* $\frac{\gamma}{\beta}$. *If* $\gamma = C$, *the equation (2.22) is solvable and the maximal solution is given by* $\frac{C}{\beta}$.

Proof. Assume $\mathcal{R}(\beta) \subseteq \mathcal{R}(\gamma)$. Then $\frac{\gamma}{\beta}$ is a solution, for

$$\beta\frac{\gamma}{\beta} = (1 + \mathcal{R}(\beta))\gamma \subseteq (1 + \mathcal{R}(\gamma))\gamma = \gamma.$$

Conversely, suppose $x \in \mathbb{R}$ satisfies (2.22). Then there exists $b' \in \beta$ and $c' \in \gamma$ such that $x = \frac{c'}{b'}$. Hence

$$\begin{aligned}(b' + B)\frac{c'}{b'} \subseteq c' + C &\Rightarrow (1 + \mathcal{R}(\beta))c' \subseteq c'(1 + \mathcal{R}(\gamma)) \\ &\Rightarrow 1 + \mathcal{R}(\beta) \subseteq 1 + \mathcal{R}(\gamma) \Rightarrow \mathcal{R}(\beta) \subseteq \mathcal{R}(\gamma).\end{aligned}$$

Hence (2.22) has a solution if and only if $\mathcal{R}(\beta) \subseteq \mathcal{R}(\gamma)$.

In the latter case $\beta\frac{\gamma}{\beta} = (1+\mathcal{R}(\beta))(1+\mathcal{R}(\gamma))\gamma = (1+\mathcal{R}(\gamma))\gamma = \gamma$. If $\gamma = C$ clearly $\beta\frac{C}{\beta} = (1 + \mathcal{R}(\beta))C = C$. □

Exercises

Exercise 2.4.1. *Verify that in the case where α, γ are zeroless, if $A \subseteq C$, then $\gamma - \alpha$ is a solution of (2.17). Is this solution unique?*

Exercise 2.4.2. *Solve, if possible, each of the following equations.*

1. $\pounds + x \subseteq \oslash$.
2. $\oslash + x \subseteq \pounds$.
3. $1 + \varepsilon\pounds + x \subseteq \oslash$, *where* $\varepsilon \simeq 0$.
4. $-2 + \varepsilon \oslash x \subseteq 1 + \oslash$.
5. $\pounds x \subseteq \omega + \oslash$, *where* $\omega \simeq +\infty$.
6. $\omega + \oslash x \subseteq \pounds$, *where* $\omega \simeq +\infty$.
7. $(1 + \varepsilon\pounds)x \subseteq \omega + \sqrt{\omega}\oslash$.
8. $5 + \pounds\varepsilon^{\not\propto} + x \subseteq 4 + \pounds e^{-\frac{@}{\delta}}$, *where* ε, δ *are different positive infinitesimals.*

Exercise 2.4.3. *Solve, if possible, the following polynomial equations. If not, explain why there is no solution:*

1. $x^2 - \oslash x + 1 + \frac{\pounds}{\omega} \subseteq \oslash$.
2. $x^2 - \oslash x + 1 + \frac{\pounds}{\omega} \subseteq \pounds$.
3. $x^2 - \oslash x + \frac{\pounds}{\omega} \subseteq \frac{\pounds}{\omega}$.
4. $x^2 - \oslash x - 1 \subseteq \frac{\pounds}{\omega}$.

Exercise 2.4.4. *A neutrix I is called* idempotent *if $I \cdot I = I$. Let $\omega \simeq \infty$, $\omega > 0$.*

1. *Show that $\oslash$ and $\pounds$ are idempotent.*
2. *Show that $\omega\oslash$, $\frac{\oslash}{\omega}$, $\omega\pounds$ and $\frac{\omega}{\pounds}$ are not idempotent.*
3. *Let $I \neq 0$ be an idempotent neutrix. Let $d \in \mathbb{R}$ be such that $d > 0$. Show that the set of solutions of the equation $x(x - d) \in I$ is given by $\xi = I$ for $d \in I$ and $\xi_1 = \frac{I}{d}$, $\xi_2 = d + \frac{I}{d}$ for $d > I$.*
4. *Solve $x(x - 1) \in \oslash$, $x(x - \omega) \in \pounds$ and $x(x - \omega) \in \pounds\omega$.*

Chapter 3

Neutrices and external numbers

3.1	External numbers and operations	59
3.2	Algebraic properties for addition and multiplication	68
	3.2.1 External numbers and regular semigroups	68
	3.2.2 Properties of neutral and inverse elements	69
3.3	Distributivity	72
	3.3.1 Distributivity with neutrices	76
	3.3.2 Distributivity with zeroless external numbers	77
	3.3.3 Application: Binomial formulas	82

In this chapter we study the basic algebraic properties of the external numbers. The algebraic properties of the external numbers lead to the introduction of several algebraic structures in [41, 43, 44]. On the whole these properties are to be read in analogy with similar properties of the real numbers. In fact, one can see an external number as a real number plus an imprecision.

In Section 3.1 we give the basic definitions and properties of neutrices and external numbers, illustrated with some examples. We also define an order relation and prove compatibility with the operations. In Section 3.2 we show how algebraic properties of the external numbers are related with usual algebraic structures such as regular semigroups. In particular we show that neutral and inverse elements are well-behaved. Distributivity happens to be more delicate and involved and we dedicate Section 3.3 to characterize its validity.

3.1 External numbers and operations

Definition 3.1.1. *A* (scalar) neutrix *is an additive convex subgroup of* $\mathbb{R}$.

The only internal sets which are scalar neutrices are $\{0\}$ and $\mathbb{R}$. This means that a theory of scalar neutrices is trivial in classical mathematics. In nonstandard analysis there is a proliferation of scalar neutrices, due to the possibility to have external sets. In Chapter 8 we consider additive convex subgroups of $\mathbb{R}^d$ with $d > 1$. They will also be called neutrices. If there is no ambiguity we will drop the modifier scalar. Then scalar neutrices are neutrices in the sense of Definition 1.4.3.

The following criterion is useful in verifying that a given set is a neutrix.

Proposition 3.1.1. *Let $N \subseteq \mathbb{R}$. Then N is a neutrix if and only if N is convex and $2N = -N = N$.*

Proof. If N is a neutrix, it must be convex by definition. By Theorem 1.4.1 we have $2N = N$ and $-N = N$. Conversely, assume that N is convex and $2N = -N = N$. We only need to verify that N is closed under addition. Let $x, y \in N$. Then $|x + y| \leq 2\max(|x|, |y|) \in N$. Hence $x + y \in N$. □

Example 3.1.1. 1. Let $\pounds$ be the set of limited numbers, and $\oslash$ be the set of infinitesimal numbers as given in (1.5). Clearly $2\pounds = -\pounds = \pounds$ and $2\oslash = -\oslash = \oslash$; hence $\pounds$ and $\oslash$ are neutrices.

2. For $\varepsilon \simeq 0$, $\varepsilon \neq 0$, multiples $\varepsilon\pounds$ of $\pounds$ and $\varepsilon\oslash$ of $\oslash$ are external neutrices. As groups, the *ε-galaxies* $\varepsilon\pounds$ are isomorphic to $\pounds$ and the *ε-halos* $\varepsilon\oslash$ are isomorphic to $\oslash$.

Remark 3.1.2. For $\varepsilon \simeq 0$, $\varepsilon > 0$ the *ε-microhalo* and the *ε-microgalaxy* of Definition 1.4.4 are neutrices. Observe that

$$M_\varepsilon = \{x : \forall^{\mathrm{st}} n \in \mathbb{N}\, (|x| < \varepsilon^n)\}.$$

and

$$m_\varepsilon = \left\{x : \exists^{\mathrm{st}} n \in \mathbb{N} \left(|x| < e^{-\frac{1}{n\varepsilon}}\right)\right\}.$$

The elements of M_ε are therefore infinitely large powers of ε and we may write $M_\varepsilon = \pounds\varepsilon^{\not\infty}$. Also we may write $m_\varepsilon = \pounds e^{-\frac{@}{\varepsilon}}$, with @ the set of positive appreciable numbers, as given in (1.5). See Exercise 3.1.1.

The microhalo and the microgalaxy are related with asymptotic developments (Poincaré and Gevrey developments [100, 101, 52, 53]). Standard functions (see Chapter 4) which have an asymptotic power series with all coefficients equal to zero take values in the microhalo M_ε, for all $\varepsilon \simeq 0$, $\varepsilon > 0$. For some indices these values are even in the microgalaxy m_ε if (and only if) the expansion satisfies the so-called *Gevrey-bound of order* 1 on the error (see Exercise 4.1.4). The Gevrey-bound of order 1 on the expansion $\sum a_n x^n$ indicates that one always has that $a_n \leq Kn!$ for some $K > 0$.

Let $\varepsilon \simeq 0$, $\varepsilon > 0$. As mentioned in Part 2 of Example 3.1.1 the groups $\pounds$ and $\varepsilon\pounds$ are isomorphic. A possible isomorphism is simply the map φ defined by $\varphi(x) = \varepsilon x$. The same map also shows that the groups $\oslash$ and $\varepsilon\oslash$ are isomorphic. However, by the Fehrele Principle, there is no (internal) isomorphism ϕ such that $\phi(\pounds) = \oslash$. This implies that no ε-galaxy is isomorphic to an ε-halo. Also there is no (internal) isomorphism ψ such that $\psi(m_\varepsilon) = M_\varepsilon$. The halo M_ε is also not isomorphic to any η-halo, for $\eta \simeq 0$, $\eta \neq 0$. Similarly, the galaxy m_ε is not isomorphic to any δ-galaxy, for $\delta \simeq 0$, $\delta \neq 0$. This is a consequence of Theorem 3.1.3 and Exercise 3.1.1.

Theorem 3.1.3. *Let N be a neutrix. If there exists $\omega \simeq +\infty$ such that $\omega N = N$, then there is no internal isomorphism φ such that $\varphi(\pounds) = N$, and no internal isomorphism ψ such that $\psi(\oslash) = N$.*

Proof. Let φ be an injective internal homomorphism such that $\varphi(\pounds) \subseteq N$. Let $y = \varphi(1)$. One shows by External Induction that $\varphi(n) = ny$, for all standard $n \in \mathbb{N}$. By Cauchy's Principle there exists $\nu \simeq +\infty$ such that $\varphi(\nu)$ is well-defined and still $\varphi(\nu) = \nu y$. We may assume that $\nu \leq \omega$. Then $\varphi(\nu) \in N$; hence φ is not surjective over N. We conclude that $\pounds$ and N are not isomorphic.

Let ψ be an injective internal homomorphism such that $\psi(\oslash) = N$. By Cauchy's Principle there exists $a > 0$ appreciable such that $\psi(a) \notin N$ is well-defined. Observe that $\psi(a) = \psi\left(\frac{a}{2} + \frac{a}{2}\right) = \psi\left(\frac{a}{2}\right) = 2\psi\left(\frac{a}{2}\right)$, and with External Induction we may prove that $\frac{1}{n}\psi(a) = n\psi(\frac{a}{n})$, for all standard n. By Cauchy's Principle there exists $\nu \leq \omega$ such that $\psi\left(\frac{a}{\nu}\right) = \frac{1}{\nu}\psi(a) \notin N$, in contradiction with the fact that $\frac{N}{\omega} = N$. Hence $\psi(\oslash) \neq N$. We conclude that $\oslash$ and N are not isomorphic. □

With Theorem 3.1.3 one shows that there are many neutrices not isomorphic by internal homomorphisms. Indeed, with $\omega \simeq +\infty$ one has that $\pounds\omega$, $\pounds e^{\pounds\omega}$, $\pounds e^{e^{\pounds\omega}}$, ... are all non-isomorphic neutrices [130, 134].

In the context of IST (see Chapter 4) we have standard sets of all cardinalities and then all neutrices are external sets of the form $\bigcup_{\mathrm{st}(x)\in X}[-a_x, a_x]$ or $\bigcap_{\mathrm{st}(x)\in X}[-a_x, a_x]$, where X is standard and $a : X \to \mathbb{R}$ is internal. Different cardinalities lead obviously to non-isomorphic neutrices.

Though neutrices have an interpretation as external sets, the collection of all neutrices cannot be an external set. We denote this "class" by $\mathcal{N}$.

Addition and multiplication on $\mathcal{N}$ are defined by the Minkowski operations in analogy with Definition 1.4.2. So, if A, B are neutrices we define their *sum* by

$$A + B = \{a + b : (a, b) \in A \times B\},$$

and their *product* by

$$AB = \{ab : (a, b) \in A \times B\}.$$

We also define their *quotient* by

$$A : B = \{c \in \mathbb{R} : cB \subseteq A\}.$$

Let A, B be neutrices. It was shown in Theorem 1.4.1 that $-A = A$ and $A + B = \max(A, B)$ where the maximum is taken in the sense of inclusion. The value of the product, hence also of the quotient, is less obvious (see Chapter 4). We prove here that AB and $A : B$ are neutrices.

Proposition 3.1.4. *The product and the quotient of two neutrices is again a neutrix.*

Proof. Let A, B be neutrices. Then $2AB = (2A)B = AB$ and $-(AB) = (-A)B = AB$. Let $c \in A : B$. Then $(2c)B = c(2B) = cB \subseteq A$ and $(-c)B = -(cB) \subseteq A$. Hence $2c$ and $-c \in A : B$. We conclude that $2(A : B) = -(A : B) = A : B$ and the result follows from Proposition 3.1.1. □

Definition 3.1.2. *An* external number α *is the algebraic sum of a real number* a *with a neutrix* A.

The collection of all external numbers is also a class. We will denote it by $\mathbb{E}$. If $\alpha = a + A$ and $\beta = b + B$ are two such external numbers, the Minkowski sum and product are given by

$$\begin{aligned} \alpha + \beta &= a + b + A + B \\ \alpha\beta &= ab + aB + bA + AB. \end{aligned}$$

Notice that by Part 2 of Theorem 1.4.1

$$\alpha + \beta = a + b + \max(A, B)$$

and

$$\alpha\beta = ab + \max(aB, bA, AB). \tag{3.1}$$

We recall from Theorem 1.4.2 that

$$-\alpha = -a + A.$$

An external number which is not a neutrix was called *zeroless* in Definition 1.4.3. If α is zeroless, by Theorem 1.4.2 we have

$$\frac{1}{\alpha} = \frac{1}{a + A} = \frac{1}{a} + \frac{A}{a^2},$$

which is again an external number.

Recall that in Theorem 1.4.2 it was shown that

$$\frac{A}{a} = \frac{A}{\alpha} \subseteq \oslash. \tag{3.2}$$

So, by Theorem 1.4.1 if $a \notin A$

$$\frac{1}{1 + \frac{A}{a}} = 1 + \frac{A}{a}.$$

In the case where α, β are zeroless we may neglect the last term in (3.1).

Lemma 3.1.5. *Let* $\alpha = a + A$ *and* $\beta = b + B$ *be zeroless external numbers. Then* $\alpha\beta = ab + \max(aB, bA)$.

Proof. Since $B \subseteq \oslash b$ by (3.2), $\max(BA, bA) = bA$. Hence

$$\alpha\beta = ab + \max(aB, bA, BA) = ab + \max(aB, bA). \qquad \square$$

As a consequence, if $\alpha = a + A$ is zeroless

$$\left(1 + \frac{A}{a}\right)\left(1 + \frac{A}{a}\right) = \left(1 + \frac{A}{a}\right).$$

We present two simple cases of the distributivity property that will be dealt with in full generality in Section 3.3.

Lemma 3.1.6. *Let $\alpha = a + A$ be zeroless. Then $\alpha B = aB + AB$ for all $B \in \mathcal{N}$.*

Proof. See Exercise 3.1.2. □

Lemma 3.1.7. *Let $\alpha = a + A$ and $\beta = b + B$ be zeroless. Then $\alpha\beta = \alpha b + \alpha B$.*

Proof. Using Lemmas 3.1.6 and 3.1.5, we derive

$$\alpha b + \alpha B = (a + A)b + aB + AB = ab + bA + aB + AB = \alpha\beta. \qquad \square$$

Definition 3.1.3. *If $\alpha = a + A$ is an external number, then A is called the* neutrix part *of α and is denoted $N(\alpha)$. The number a is called a* representative *of α.*

The next proposition shows that every element of α can be taken as representative.

Proposition 3.1.8. *Let $\alpha = a + A$ be an external number. Then for all $y \in \alpha$ it holds that $\alpha = y + A$. In particular, with $y = 0$ we have $\alpha = A$.*

Proof. Let $y \in \alpha$. Then there is $x \in A$ such that $y = a + x$ and hence, $y + A = a + x + A = a + A = \alpha$. □

Corollary 3.1.9. *The only external numbers containing 0 are neutrices.*

Proposition 3.1.10. *The neutrix part of an external number is the set of all real numbers that leave it invariant by translation.*

Proof. Let $\alpha = a + A$ be an external number and let x be a real number. Then $x + \alpha = \alpha \Leftrightarrow x + A = A \Leftrightarrow x \in A$. This means that the neutrix part of an external number is the set of all real numbers that leave it invariant by translation. □

Definition 3.1.4. *Let A be a neutrix and α be an external number. We say that α is* appreciable *with respect to A if $\alpha A = A$, that α is an* absorber *of A if $\alpha A \subset A$ and that α is an* exploder *of A if $A \subset \alpha A$.*

External numbers which are appreciable with respect to $\oslash$ or $\pounds$ are simply appreciable. Sometimes infinitesimals are not absorbers and infinitely large numbers are not exploders. This is the case of the neutrices m_ε and M_ε for they are stable by multiplication by ε and $\frac{1}{\varepsilon}$.

Proposition 3.1.11. *Let $A \subseteq \oslash, A \neq 0$ be a neutrix.*

1. *Every element t of A is an absorber of A.*
2. *Every neutrix $B \subset A$ is an absorber of A.*

Proof. 1. The property is obvious for $t = 0$. Let $t \in A, t \neq 0$. Because $|t| < 1$ one has $tA \subseteq A$. Suppose by contradiction that $tA = A$. Then there exists $x \in A$ such that $tx = t$. Hence $1 \in A$, in contradiction with the fact that $A \subseteq \oslash$. Hence $tA \subset A$.

2. By Part 1, observe that there exists $t \in \mathbb{R}$ such that $B < t \in A$. □

Given two external numbers, they are either disjoint or one contains the other [76, Proposition 7.4.1].

Proposition 3.1.12. *Let α and β be two external numbers. Then*

$$\alpha \cap \beta = \emptyset \vee \alpha \subseteq \beta \vee \beta \subseteq \alpha.$$

Proof. See Exercise 3.1.3. □

In the set of external numbers we can define the following order relation.

Definition 3.1.5 ([76, 75])**.** *Let $\alpha = a + A, \beta = b + B \in \mathbb{E}$. We define (with abuse of notation)*

1. $\alpha \geq \beta \Leftrightarrow \forall x \in \alpha \exists y \in \beta (x \geq y)$.
2. $\alpha > \beta \Leftrightarrow \forall x \in \alpha \forall y \in \beta (x > y)$.
3. $\alpha \leq \beta \Leftrightarrow \forall x \in \alpha \exists y \in \beta (x \leq y)$.
4. $\alpha < \beta \Leftrightarrow \forall x \in \alpha \forall y \in \beta (x < y)$.

Let ε be an infinitesimal. We have $1 + \varepsilon \pounds > \oslash$, $\varepsilon \leq \oslash$ and $\varepsilon \geq \oslash$. In fact, let $\alpha, \beta \in \mathbb{E}$. If $\alpha \subseteq \beta$ then both $\alpha \leq \beta$ and $\alpha \geq \beta$. We note that $\beta \leq \alpha$ is not equivalent to $\alpha \geq \beta$. For example, we have $\oslash \geq \pounds$, yet $\pounds \not\leq \oslash$. Moreover, we have $\oslash \leq \pounds$. Indeed, $\alpha \leq \beta$ and $\alpha \geq \beta$ can occur simultaneously without equality if $\alpha \subset \beta$. It follows from the definition that $\alpha \leq \beta$ if and only if $\alpha \geq -\beta$. However, $\alpha \not\leq \beta$ does not imply that $\alpha > \beta$ (nor $\beta < \alpha$). Indeed, it is enough to take for example $\alpha = \pounds$ and $\beta = \oslash$. Clearly $\alpha < \beta$ if and only if $\beta > \alpha$.

The following characterizations of the order relations of Definition 3.1.5 are an easy consequence of Proposition 3.1.12.

Proposition 3.1.13. *Let $\alpha, \beta \in \mathbb{E}$. Then*

1. *$\alpha < \beta$ if and only if $\alpha \leq \beta \wedge \alpha \cap \beta = \emptyset$.*
2. *$\alpha > \beta$ if and only if $\alpha \geq \beta \wedge \alpha \cap \beta = \emptyset$.*
3. *$\alpha \leq \beta$ if and only if $\alpha < \beta \vee \alpha \subseteq \beta$.*
4. *$\alpha \geq \beta$ if and only if $\alpha > \beta \vee \alpha \subseteq \beta$.*

The following two results give conditions that allow us to "change" the order relation.

Proposition 3.1.14. *Let M be a neutrix and let α and β be two external numbers such that $N(\alpha) \subseteq M$ and $N(\beta) \subseteq M$. Then $\alpha + M \not\leq \beta + M$ implies that $\alpha + M > \beta + M$.*

Proof. Observe that $N(\alpha+M)=N(\beta+M)=M$. If $(\alpha+M)\cap(\beta+M)\neq\emptyset$, there exists $x\in(\alpha+M)\cap(\beta+M)$. It follows that $\alpha+M=\beta+M=x+M$, in contradiction with the hypothesis. So $(\alpha+M)\cap(\beta+M)=\emptyset$. Since $\alpha+M<\beta+M$ cannot be, we conclude that $\alpha+M>\beta+M$. □

Corollary 3.1.15. *Let α,β be two external numbers such that $N(\alpha)=N(\beta)$. If $\alpha\not\leq\beta$ then $\alpha>\beta$.*

Proof. It follows from Proposition 3.1.14 with $M=N(\alpha)=N(\beta)$. □

Proposition 3.1.16. *Let $\alpha=a+A$ be an external number and let N be a neutrix. If $\alpha+N=N$, then $\alpha\leq N$ and $-\alpha\leq N$.*

Proof. Assume that $\alpha+N=N$. Then $\alpha\subseteq N$. Hence also $-\alpha\subseteq N$ and we conclude that $\alpha\leq N$ and $-\alpha\leq N$. □

Proposition 3.1.17. *Let A be a neutrix and let β and γ be external numbers such that $N(\beta)\leq\beta\leq\gamma$. Then $A\beta\subseteq A\gamma$.*

Proof. Suppose that $\beta\leq\gamma$. Then there exist $x\in\beta$ and $y\in\gamma$ such that $0\leq x\leq y$. Let $a\in A$. Then $|a|\,x\leq|a|\,z\in A\gamma$. Hence $A\beta\subseteq A\gamma$. □

Theorem 3.1.18. *The relation $\leq$ is a total order relation compatible with addition and multiplication in the following way:*

1. $\forall\alpha,\beta,\gamma\in\mathbb{E}(\alpha\leq\beta\Rightarrow\alpha+\gamma\leq\beta+\gamma)$.
2. $\forall\alpha,\beta\in\mathbb{E}(N(\alpha)\leq\alpha\wedge N(\beta)\leq\beta\Rightarrow N(\alpha\beta)\leq\alpha\beta)$.
3. $\forall\alpha,\beta,\gamma\in\mathbb{E}(N(\alpha)\leq\alpha\leq\beta\wedge N(\gamma)\leq\gamma\Rightarrow\alpha\gamma\leq\beta\gamma)$.
4. Whenever $\alpha,\beta,\gamma\in\mathbb{E}$, if $N(\alpha)<\alpha$ and $\beta\leq\gamma$, then $\alpha\beta\leq\alpha\gamma$;
5. Whenever $\alpha,\beta,\gamma\in\mathbb{E}$, if $N(\beta)\leq\beta\leq\gamma$, then $N(\alpha)\beta\leq N(\alpha)\gamma$.

Proof. Let α,β and γ be arbitrary external numbers.

We prove first that the relation $\leq$ is a total order relation. Let $x\in\alpha$. Because $x\leq x$ one has $\alpha\leq\alpha$, so the relation is reflexive. Suppose that $\alpha\leq\beta$ and $\beta\leq\gamma$. Then there is $y\in\beta$ and $z\in\gamma$ such that $x\leq y$ and $y\leq z$. Hence $x\leq z$ and the relation is transitive. Suppose now that $\alpha\leq\beta$ and $\beta\leq\alpha$. Then $\alpha\subseteq\beta$ and $\beta\subseteq\alpha$. Hence by Proposition 3.1.12, one has $\alpha=\beta$ and the relation is antisymmetric. To prove that the relation is total suppose that $\alpha\not\leq\beta$. Then there is $x\in\alpha$ such that $x>y$, for all $y\in\beta$. Hence $\beta\leq\alpha$. We conclude that the relation $\leq$ is a total order relation.

1. Suppose that $\alpha\leq\beta$. Let $a\in\alpha$. Then there is $b\in\beta$ such that $a\leq b$. Let $x\in\alpha+\gamma$. Then there is $c\in\gamma$ such that $x=a+c$. Hence $a+c\leq b+c\in\beta+\gamma$ and one concludes that $\alpha+\gamma\leq\beta+\gamma$.

2. Suppose that $N(\alpha)\leq\alpha$ and $N(\beta)\leq\beta$. If $\alpha\beta$ is not zeroless then $\alpha\beta=N(\alpha\beta)$. If $\alpha\beta$ is zeroless by Lemma 3.1.5 one has $\alpha\beta=ab+\max(aB,bA)$.

Let $x \in N(\alpha\beta)$. Then $|x| \in \max(aB, bA)$. If $\max(aB, bA) = aB$ there is $b \in B$ such that $|x| = ab$. Because $N(\alpha) \leq \alpha$, there are positive numbers $a' \in N(\alpha)$ and $a'' \in \alpha$ such that $a \leq a' \leq a''$. Also, there is $b' \in \beta$ such that $b \leq b'$, because $N(\beta) \leq \beta$. Then $x \leq |x| = ab \leq a''b' \in \alpha\beta$ and $N(\alpha\beta) \leq \alpha\beta$. If $\max(aB, bA) = bA$ the proof is analogous.

3. Suppose that $N(\alpha) \leq \alpha$, $\alpha \leq \beta$ and $N(\gamma) \leq \gamma$. Let $x \in \alpha\gamma$. Then, there is $a \in \alpha$ and $c \in \gamma$ such that $x = ac$. Moreover, a and c may be assumed to be positive because $N(\alpha) \leq \alpha$ and $N(\gamma) \leq \gamma$. Because $\alpha \leq \beta$ there is $b \in \beta$ such that $a \leq b$. Then $ac \leq bc \in \beta\gamma$ and $\alpha\gamma \leq \beta\gamma$.

4. If $N(\alpha) < \alpha$ and $x \in \alpha$ then $0 < x$. Let $y \in \beta$. Because $\beta \leq \gamma$ there exists $z \in \gamma$ such that $y \leq z$. Then $xy \leq xz$. Hence $\alpha\beta \leq \alpha\gamma$.

5. This is a direct consequence of Proposition 3.1.17. □

Definition 3.1.6. *Let $\alpha = a + A$ be an external number. The* absolute value *of α is defined (with abuse of notation) by $|\alpha| := |a| + A$.*

Let $\alpha = a + A$. Clearly, $|\alpha| = \alpha$, if $a > 0$ and $|\alpha| = -a + A = -\alpha$, if $a < 0$. If N is a neutrix, then $|N| = N$.

We finish by studying some properties of the absolute value.

Proposition 3.1.19. *If $\alpha = a + A \in \mathbb{E}$ is zeroless, then $|\alpha| = \{|x| : x \in \alpha\}$.*

Proof. Let $\xi = \{|x| : x \in \alpha\}$. We give the proof for the case where $a > A$, the case where $a < A$ is analogous. Let $x \in |\alpha|$. Then $x = |a| + u$, for some $u \in A$. Since $a > A$, we have $a + u \in \alpha$ and $x = |a| + u = a + u = |a + u|$ and we conclude that $x \in \xi$. Hence $|\alpha| \subseteq \xi$. Conversely, let $y \in \xi$. Since $a > A$ we have $y = |a + v| = a + v = |a| + v$, for some $v \in A$. So $y \in |\alpha|$, and therefore $\xi \subseteq |\alpha|$. Hence $\xi = |\alpha|$. □

Observe that Proposition 3.1.19 fails to be true in the case where α is a neutrix.

Corollary 3.1.20. *Let $\alpha = a + A$ be a zeroless external number. The definition of the absolute value of α does not depend on the choice of the representative a of α.*

Proof. Assume that $\alpha = b + A$. Then $|\alpha| = |b| + A = \{|x| : x \in \alpha\} = |a| + A$. □

Let $\alpha = a + A, \beta = b + B \in \mathbb{E}$. The following triangular inequalities follow directly from the definition of absolute value. We will use them implicitly throughout the book.

$$|\alpha + \beta| \leq |\alpha| + |\beta|. \tag{3.3}$$

$$||\alpha| - |\beta|| \leq |\alpha| - |\beta|. \tag{3.4}$$

Proposition 3.1.21. *Let $\alpha \in \mathbb{E}$ and β be a zeroless, positive external number. Then $|\alpha| < \beta$ if and only if $-\beta < \alpha < \beta$.*

Proof. We start with the direct implication. Assume firstly that α is a neutrix. Then $\alpha = |\alpha| < \beta$. Suppose that $-\beta \not< \alpha$. Then there exists $u \in \alpha$ and $v \in -\beta$ such that $u \leq v$. Observe that $-\beta$ is negative, so $v \leq 0$. Since α is a neutrix and $u, 0 \in \alpha$ we have that $v \in [u, 0] \subseteq \alpha$. It follows that $-v \in \alpha$. Then $-v \in \alpha \cap \beta$, a contradiction since $|\alpha| < \beta$ implies that $\alpha \cap \beta = \emptyset$. Hence $-\beta < \alpha < \beta$. Assume secondly that α is zeroless. Let $u \in \alpha$ and $v \in \beta$. Then $|u| \in |\alpha|$ by Proposition 3.1.19. So $|u| < v$; hence $-\beta < \alpha$. Also $\alpha \leq |\alpha| < \beta$. Thus, $-\beta < \alpha < \beta$.

To prove the converse implication, assume that $-\beta < \alpha < \beta$. If α is a neutrix then $|\alpha| = \alpha < \beta$. If α is zeroless, let $y \in |\alpha|$ and $v \in \beta$. Then $y = |u|$ for some $u \in \alpha$ by Proposition 3.1.19. Since $-\beta < \alpha < \beta$, it holds that $-v < u < v$, so $y = |u| < v$. We conclude that $|\alpha| < \beta$. □

Exercises

Exercise 3.1.1. *Let $\varepsilon \simeq 0$, $\varepsilon > 0$. Show that*

1. $m_\varepsilon \subset M_\varepsilon \subset \oslash\varepsilon$.
2. $M_\varepsilon = £\varepsilon^{\not\propto}$.
3. $m_\varepsilon = £e^{-\frac{@}{\varepsilon}}$.
4. M_ε *and* m_ε *are neutrices.*
5. $\varepsilon M_\varepsilon = M_\varepsilon$ *and* $\varepsilon m_\varepsilon = m_\varepsilon$.

Exercise 3.1.2. *Prove Lemma 3.1.6.*

Exercise 3.1.3. *Prove Proposition 3.1.12.*

Exercise 3.1.4. *Let $\varepsilon \simeq 0, \varepsilon > 0$. Determine whether the following are true or false.*

1. $1 + \oslash \leq £$.
2. $\frac{1+\oslash}{\log(\varepsilon)} \leq \oslash$.
3. $\frac{1+\oslash}{\log(\varepsilon)} \leq -\frac{1}{\varepsilon} + £$.
4. $\oslash \leq £$.
5. $£ \leq \oslash$.
6. $\varepsilon + \varepsilon^2 £ < 2\varepsilon$.

Exercise 3.1.5. *Let $\alpha = a + N(\alpha), \beta = b + N(\beta), \gamma = c + N(\gamma) \in \mathbb{E}$. Show that*

1. *If* $\alpha - \beta < \gamma$, *then* $\alpha + N(\beta) < \beta + \gamma$.
2. *If* $\gamma < \alpha - \beta$, *then* $\gamma + \beta < \alpha + N(\beta)$. *As a consequence,* $\gamma + \beta < \alpha$.
3. *If* $\alpha - \beta < \gamma$, *then* $\alpha < \beta + \gamma$.

3.2 Algebraic properties for addition and multiplication

The external numbers when equipped with addition and zeroless external numbers when equipped with multiplication are regular commutative semigroups. The neutral and unity elements appear in the form of external functions.

We present here algebraic proofs of some properties of addition and multiplication as given in [41] and which were originally proved in [75] by set theoretical arguments.

3.2.1 External numbers and regular semigroups

We recall that a *semigroup* is a structure $(S,*)$ such that S is non-empty and $*$ is a binary operation that satisfies the following axiom

$$\forall x,y,z \in S\,(x*(y*z)=(x*y)*z).$$

A *regular* semigroup is a semigroup S such that every element is *regular*; that is, for every $a \in S$ there is $x \in S$ such that $axa = a$.

Theorem 3.2.1. *The structures* $(\mathbb{E},+)$ *and* $(\mathbb{E}\backslash\mathcal{N},\cdot)$ *are commutative regular semigroups.*

Proof. Let $\alpha = a+A$, $\beta = b+B$ and $\gamma = c+C$ be arbitrary external numbers. Firstly we prove that both operations are commutative. In fact, one has

$$\begin{aligned}\alpha+\beta &= (a+A)+(b+B) = a+b+\max(A,B)\\ &= b+a+\max(B,A) = (b+B)+(a+A) = \beta+\alpha\end{aligned}$$

and

$$\begin{aligned}\alpha\beta &= (a+A)(b+B) = ab+\max(aB,bA,AB)\\ &= ba+\max(bA,aB,BA) = (b+B)(a+A) = \beta\alpha.\end{aligned}$$

Secondly we prove that both operations are associative. Indeed,

$$\begin{aligned}(\alpha+\beta)+\gamma &= ((a+A)+(b+B))+(c+C)\\ &= (a+b+\max(A,B))+(c+C)\\ &= (a+b)+c+\max(\max(A,B),C)\\ &= a+(b+c)+\max(A,\max(B,C))\\ &= (a+A)+(b+c+\max(B,C)) = \alpha+(\beta+\gamma)\end{aligned}$$

and

$$\begin{aligned}(\alpha\beta)\gamma &= ((a+A)(b+B))(c+C)\\ &= (ab+\max(aB,bA,AB))(c+C)\\ &= (ab)c+\max(abC, c\max(aB,bA,AB), \max(aB,bA,AB)C)\\ &= (ab)c+\max(abC,acB,bcA,cAB,aBC,bAC,ABC)\\ &= a(bc)+\max(a\max(bC,cB,BC),bcA,A\max(bC,cB,BC))\\ &= (a+A)(bc+\max(bC,cB,BC))\\ &= (a+A)((b+B)(c+C))\\ &= \alpha(\beta\gamma).\end{aligned}$$

Finally, we prove the regularity properties. Let $\alpha = a + A$ be an arbitrary external number. Put $\delta = (-a + A)$ and $\zeta = \left(\frac{1}{a} + \frac{A}{a^2}\right)$. Then

$$\alpha+\delta+\alpha = (a+A)+(-a+A)+(a+A) = (a+A) = \alpha.$$

If α is zeroless, then necessarily $a \neq 0$, so applying Lemma 3.1.5

$$\begin{aligned}\alpha\zeta\alpha &= (a+A)\left(\frac{1}{a}+\frac{A}{a^2}\right)(a+A) = \left(1+\frac{aA}{a^2}+\frac{A}{a}\right)(a+A)\\ &= \left(1+\frac{A}{a}\right)(a+A) = a+A+\frac{aA}{a} = a+A = \alpha.\end{aligned}$$

Hence $(\mathbb{E},+)$ and $(\mathbb{E}\setminus\mathcal{N},\cdot)$ are commutative regular semigroups. □

3.2.2 Properties of neutral and inverse elements

Uniqueness for identity and inverse elements holds neither for addition nor for multiplication, as seen by the following examples.

Example 3.2.1. Let $\alpha = 1+\oslash$, $\beta = 1+\varepsilon\oslash$ and $\varepsilon \simeq 0$.

1. $\alpha+\oslash = 1+\oslash+\oslash = 1+\oslash = \alpha$ and $\alpha+\varepsilon\oslash = 1+\oslash+\varepsilon\oslash = 1+\oslash = \alpha$.
2. $\alpha\alpha = (1+\oslash)(1+\oslash) = 1+\oslash = \alpha$ and $\alpha\beta = (1+\oslash)(1+\varepsilon\oslash) = 1+\oslash+\varepsilon\oslash = 1+\oslash = \alpha$.

We show the existence, for both addition and multiplication, of functions which generalize the concept of neutral element. For each external number α these functions give the maximal elements that leave α invariant. Thus they are individualized neutral elements, and as such they are unique.

Proposition 3.2.2. *There is a unique function $e : \mathbb{E} \to \mathbb{E}$ such that (i) $\alpha + e(\alpha) = \alpha$ for all $\alpha \in \mathbb{E}$ and (ii) if $f : \mathbb{E} \to \mathbb{E}$ is such that $\alpha + f(\alpha) = \alpha$ for all $\alpha \in \mathbb{E}$, then $e(\alpha) + f(\alpha) = e(\alpha)$. In fact, $e(\alpha) = N(\alpha)$.*

Proof. We prove that $e : \mathbb{E} \to \mathbb{E}$ defined by $e(\alpha) = N(\alpha)$ is the required function. Let $\alpha = a + A$ be an arbitrary external number. Then $\alpha + e(\alpha) = \alpha + N(\alpha) = (a + A) + A = a + A = \alpha$. Hence, for all $f : \mathbb{E} \to \mathbb{E}$ such that $\alpha + f(\alpha) = \alpha$ it holds that $e(\alpha) + f(\alpha) = A + f(\alpha) = \alpha - \alpha + f(\alpha) = -\alpha + \alpha = e(\alpha)$. □

Proposition 3.2.3. *There is a unique function $u : \mathbb{E} \setminus \mathcal{N} \to \mathbb{E}$ such that (i) $\alpha u(\alpha) = \alpha$ for all $\alpha \in \mathbb{E} \setminus \mathcal{N}$ and (ii) if $v : \mathbb{E} \setminus \mathcal{N} \to \mathbb{E}$ is such that $\alpha v(\alpha) = \alpha$ for all $\alpha \in \mathbb{E} \setminus \mathcal{N}$, then $u(\alpha)v(\alpha) = u(\alpha)$. In fact, $u(\alpha) = 1 + \frac{N(\alpha)}{\alpha}$.*

Proof. We prove that $u : \mathbb{E}\setminus\mathcal{N} \to \mathbb{E}$ defined by $u(\alpha) = 1 + \frac{N(\alpha)}{\alpha}$ is the required function. Let $\alpha = a + A$ be an arbitrary zeroless external number. Then using (3.2) and Lemma 3.1.5,

$$\begin{aligned}
\alpha u(\alpha) &= \alpha\left(1 + \frac{N(\alpha)}{\alpha}\right) = (a + A)\left(1 + \frac{A}{a}\right) \\
&= a + \max\left(a\frac{A}{a}, A\right) = a + A = \alpha.
\end{aligned}$$

Let $v : \mathbb{E}\setminus\mathcal{N} \to \mathbb{E}$ be such that $\alpha v(\alpha) = \alpha$ for all $\alpha \in \mathbb{E}\setminus\mathcal{N}$. Then applying Lemma 3.1.7

$$\begin{aligned}
u(\alpha)v(\alpha) &= \left(1 + \frac{N(\alpha)}{\alpha}\right) v(\alpha) = \left(1 + \frac{A}{a}\right) v(\alpha) \\
&= \frac{(a + A)v(\alpha)}{a} = \frac{a + A}{a} = 1 + \frac{A}{a} = u(\alpha).
\end{aligned}$$ □

Corollary 3.2.4. *Let α be a zeroless external number. Then $u(\alpha) \neq e(\alpha)$.*

We prove the existence, again for both addition and multiplication, of functions that give individualized symmetrical elements.

Proposition 3.2.5. *There is a unique function $s : \mathbb{E} \to \mathbb{E}$ such that $\alpha + s(\alpha) = e(\alpha)$ and $e(s(\alpha)) = e(\alpha)$ for all $\alpha \in \mathbb{E}$. In fact, $s(\alpha) = -\alpha$.*

Proof. Let $\alpha = a + A$ be an arbitrary external number. We prove that $s : \mathbb{E} \to \mathbb{E}$ defined by $s(\alpha) = -\alpha$ is the required function. One has $\alpha + s(\alpha) = (a + A) + (-a + A) = A = e(\alpha)$ and $e(s(\alpha)) = e(-a + A) = A = e(\alpha)$. Suppose that $t : \mathbb{E} \to \mathbb{E}$ is such that $\alpha + t(\alpha) = e(\alpha)$ and $e(t(\alpha)) = e(\alpha)$. Then $e(t(\alpha)) = e(\alpha) = e(s(\alpha))$ and

$$\begin{aligned}
t(\alpha) &= t(\alpha) + e(t(\alpha)) = t(\alpha) + e(\alpha) = t(\alpha) + \alpha + s(\alpha) \\
&= e(\alpha) + s(\alpha) = e(s(\alpha)) + s(\alpha) = s(\alpha).
\end{aligned}$$ □

Proposition 3.2.6. *There is a unique function $d : \mathbb{E} \setminus \mathcal{N} \to \mathbb{E}$ such that $\alpha d(\alpha) = u(\alpha)$ and $u(d(\alpha)) = u(\alpha)$, for all $\alpha \in \mathbb{E} \setminus \mathcal{N}$. In fact $d(\alpha) = \frac{1}{\alpha} = \frac{1}{a} + \frac{A}{a^2}$.*

Proof. Let $\alpha = a + A$ be an arbitrary external number. We prove that $d : \mathbb{E} \setminus \mathcal{N} \to \mathbb{E}$, defined by

$$d(\alpha) = \frac{1}{\alpha} = \frac{1}{a}\left(\frac{1}{1+\frac{A}{a}}\right) = \frac{1}{a}\left(1 + \frac{A}{a}\right) = \frac{1}{a} + \frac{A}{a^2}$$

verifies the required conditions. One has

$$\alpha d(\alpha) = (a + A)\left(\frac{1}{a} + \frac{A}{a^2}\right) = 1 + \frac{A}{a} = u(\alpha)$$

and, by Proposition 3.2.3 and (3.2),

$$u(d(\alpha)) = u\left(\frac{1}{a} + \frac{A}{a^2}\right) = 1 + \frac{\frac{A}{a^2}}{\frac{1}{a}} = 1 + \frac{A}{a} = u(\alpha).$$

Uniqueness is shown in the same way as in the proof of Proposition 3.2.5. □

Since neutrices are ordered by inclusion one always has $e(\alpha + \beta) = e(\alpha)$ or $e(\alpha + \beta) = e(\beta)$. An analogous property holds for multiplication.

Proposition 3.2.7. *Let α and β be zeroless external numbers. Then $u(\alpha\beta) = u(\alpha)$ or $u(\alpha\beta) = u(\beta)$.*

Proof. Let $\alpha = a + A$ and $\beta = b + B$ be zeroless. Then

$$\begin{aligned} u(\alpha\beta) &= u(ab + aB + bA) = 1 + \frac{aB + bA}{ab} \\ &= 1 + \max\left(\frac{B}{b}, \frac{A}{a}\right) = 1 + \max\left(\frac{N(\beta)}{\beta}, \frac{N(\alpha)}{\alpha}\right). \end{aligned}$$

Hence $u(\alpha\beta) = u(\alpha)$ or $u(\alpha\beta) = u(\beta)$. □

The fact that addition and multiplication have the above properties in common lead to the introduction of an algebraic structure which was called *assembly* in [41] (see also [44]).

We finish with a connection between the neutral and inverse functions of addition and multiplication.

Proposition 3.2.8. *The functions e and s have the following properties with respect to multiplication. For all $\alpha, \beta \in \mathbb{E}$*

1. $e(\alpha\beta) = \beta e(\alpha) + \alpha e(\beta)$.

2. $s(\alpha\beta) = s(\alpha)\beta = \alpha s(\beta)$.

3. $\alpha\beta = e(\alpha\beta) \Leftrightarrow \alpha = e(\alpha) \vee \beta = e(\beta)$.

Proof. Let $\alpha = a + A$ and $\beta = b + B$ be arbitrary external numbers. By Proposition 3.2.2 one has $e(\alpha) = N(\alpha) = A$ and $e(\beta) = N(\beta) = B$.

1. One has $e(\alpha\beta) = e((a+A)(b+B)) = e(ab + aB + bA + AB) = aB + bA + AB$ and $\beta e(\alpha) + \alpha e(\beta) = (b+B)A + (a+A)B = bA + AB + aB + AB = aB + bA + AB$. Therefore $e(\alpha\beta) = \beta e(\alpha) + \alpha e(\beta)$.

2. One has $s(\alpha\beta) = s(ab + aB + bA + AB) = -ab + aB + bA + AB = (-a + A)(b + B) = s(\alpha)\beta$.

3. To prove the direct implication we assume that $\alpha\beta = e(\alpha\beta)$. Then

$$ab + aB + bA + AB = aB + bA + AB.$$

This implies that $ab \in aB$, $ab \in bA$ or $ab \in AB$. Suppose that $\alpha \neq e(\alpha)$ and $\beta \neq e(\beta)$. Then $a \notin A$ and $b \notin B$. Hence $ab \notin aB$, $ab \notin bA$ and $ab \notin AB$, a contradiction. We conclude that

$$\alpha\beta = e(\alpha\beta) \Rightarrow \alpha = e(\alpha) \vee \beta = e(\beta). \tag{3.5}$$

Assume now that $\alpha = e(\alpha)$ or $\beta = e(\beta)$. If $\alpha = e(\alpha)$, by Lemma 3.1.6 one has $\alpha\beta = A\beta = Ab + AB = e(\alpha\beta)$. The other case is analogous. Hence

$$\alpha = e(\alpha) \vee \beta = e(\beta) \Rightarrow \alpha\beta = e(\alpha\beta). \tag{3.6}$$

Combining (3.5) and (3.6), we obtain Part 3. □

The results in this section justify interpreting neutrices as generalized zeros. For example, Part 3 of Proposition 3.2.8 establishes in a sense that zero divisors can only be neutrices. The interpretation is further enhanced by Exercise 3.2.1, expressing that neutral elements for addition are invariant for neutral elements for multiplication.

Exercises

Exercise 3.2.1. *Show that for all* $\alpha \in \mathbb{E}$ *and* $\beta \in \mathbb{E} \setminus \mathcal{N}$ *it holds that* $e(\alpha)u(\beta) = e(\alpha)$.

Exercise 3.2.2. *Show that for all* $\alpha \in \mathbb{E} \setminus \mathcal{N}$ *it holds that* $e(u(\alpha)) = e(\alpha)d(\alpha)$.

3.3 Distributivity

Being intervals of real numbers, for external numbers multiplication is *subdistributive* with respect to addition [90]; i.e. for all external numbers α, β and γ

$$\alpha(\beta + \gamma) \subseteq \alpha\beta + \alpha\gamma. \tag{3.7}$$

However, as already observed in Chapter 1, proper distributivity does not always hold. The following theorem indicates the price to pay. This is expressed by adding two neutrices which, in general, correspond to increasing the imprecision of the expression $\alpha(\beta+\gamma)$.

Theorem 3.3.1. *Let α, β and γ be external numbers such that $\alpha = a + A$. Then $\alpha\beta + \alpha\gamma = \alpha(\beta+\gamma) + A\beta + A\gamma$.*

The theorem will be proved in Section 3.3.2. Theorem 3.3.4 below gives a complete characterization of distributivity. In fact Theorem 3.3.4 and Theorem 3.3.1 are equivalent, as shown in [41]. In Chapter 11 we give an axiomatic system for the external numbers and will take Theorem 3.3.1 as one of the axioms.

Before stating Theorem 3.3.4 we need the notion of relative uncertainty of Definition 2.4.2 and a new notion of oppositeness.

We recall that the *relative uncertainty* of α is the neutrix $\mathcal{R}(\alpha) = \frac{A}{a} \left(= \frac{A}{\alpha}\right)$, if α is zeroless and $\mathbb{R}$, otherwise.

Remark 3.3.2. Let $\alpha = a + A$ be a zeroless external number. Then $\mathcal{R}(\alpha) \subseteq \oslash$ by (3.2). Moreover $\alpha = a(1+\mathcal{R}(\alpha))$, because Lemma 3.1.7 implies that $a(1+\mathcal{R}(\alpha)) = a + a\mathcal{R}(\alpha) = a + A = \alpha$.

Definition 3.3.1. *Let α and β be external numbers. We say that α is* (asymptotically) more precise *than β if $\mathcal{R}(\alpha) \subseteq \mathcal{R}(\beta)$. If $\mathcal{R}(\alpha) = \mathcal{R}(\beta)$ we may say that α is* as precise as *β. Also, if $\alpha \in \mathbb{R}$ we say that α is* precise.

Observe that if $\alpha \in \mathbb{R}\backslash\{0\}$, then $\mathcal{R}(\alpha) = \{0\}$, and so real non-zero numbers are more precise than any external number.

Definition 3.3.2. *Let A be a neutrix and β and γ be external numbers. Then β and γ are called* opposite with respect to A *if $(\beta+\gamma)A \subset \max(|\beta|, |\gamma|)A$.*

If b and c are two opposite real numbers, i.e. such that $b = -c$, then b and c are opposite with respect to all neutrices $N \neq 0$. Let $\omega \simeq \infty$. We have that $\omega + 1$ and $-\omega$ are not nearly equal but they are opposite with respect to $\oslash$ since $((\omega+1)-\omega)\oslash = \oslash \subset \omega\oslash = (\omega+1)\oslash$.

If two numbers are opposite with respect to a given neutrix, neither of them can be a neutrix. To see this observe first that $\beta + \gamma = \max(|\beta|, |\gamma|)$ if at least one of them, say β is a neutrix and $\gamma \subseteq \beta$. In the remaining case we may suppose that β is a neutrix and $\beta < |\gamma|$. Clearly $\gamma A \subseteq (\beta+\gamma)A$. Since $\frac{\beta}{\gamma} \subseteq \oslash$, it follows from Lemma 3.1.7 that $\beta + \gamma \subseteq \oslash\gamma + \gamma = (1+\oslash)\gamma$, so $(\beta+\gamma)A \subseteq (1+\oslash)\gamma A = \gamma A$. Hence $(\beta+\gamma)A = \gamma A = \max(|\beta|, |\gamma|)A$.

Two external numbers β and γ of the same sign are never opposite with respect to a given neutrix. In fact, if β and γ are opposite with respect to a given neutrix A then $\frac{\beta}{\gamma} \subseteq -1 + \oslash$. Indeed, if $-1 + d \in \frac{\beta}{\gamma}$ with $d \not\simeq 0$,

$$(\beta+\gamma)A = \begin{cases} \beta A = \max(|\beta|, |\gamma|)A, & |d| \simeq \infty \\ (-1+d)\gamma A = \gamma A = \beta A, & -1+d \text{ appreciable} \\ \gamma A = \max(|\beta|, |\gamma|)A, & -1+d \simeq 0. \end{cases}$$

The latter observation enables a characterization in terms of absorbers: If β and γ are opposite with respect to A, both $(\beta+\gamma)A \subset \beta A$ and $(\beta+\gamma)A \subset \gamma A$; hence $\frac{\beta+\gamma}{\gamma}$ and $\frac{\beta+\gamma}{\beta}$ are absorbers of A. The notion of oppositeness is directly related to distributivity. Indeed, if β and γ are opposite with respect to A, then

$$(\beta+\gamma)A \subset \max(|\beta|, |\gamma|)A = \max(\beta A, \gamma A) = \beta A + \gamma A,$$

and if β and γ are not opposite with respect to A, then

$$(\beta+\gamma)A = \max(|\beta|, |\gamma|)A = \max(\beta A, \gamma A) = \beta A + \gamma A. \tag{3.8}$$

The following lemma permits us to deal with oppositeness with respect to linear combinations of neutrices.

Lemma 3.3.3. *Let α, β and γ be external numbers. Let M and N be neutrices.*

1. *If α and β are not opposite with respect to M nor N, then α and β are not opposite with respect to $M+N$.*

2. *If α and β are not opposite with respect to M, then α and β are not opposite with respect to γM.*

Proof. Suppose that α and β are not opposite with respect to M nor N.

1. As a consequence of (3.8) one has

$$\begin{aligned}(\alpha+\beta)(M+N) &= (\alpha+\beta)M + (\alpha+\beta)N \\ &= \alpha M + \beta M + \alpha N + \beta N \\ &= \alpha(M+N) + \beta(M+N).\end{aligned}$$

2. See Exercise 3.3.2. □

The following theorem gives a characterization of the validity of the distributive law.

Theorem 3.3.4 (Distributivity criterion). *Let α, β and γ be external numbers. Then $\alpha(\beta+\gamma) = \alpha\beta + \alpha\gamma$ if and only if (i) α is more precise than β or γ, or (ii) β and γ are not opposite with respect to $N(\alpha)$.*

If at least one of the external numbers α, β, γ is a neutrix some simplification in the criterion is possible. Indeed, we have the following special cases.

Theorem 3.3.5. *Let α, β and γ be external numbers.*

1. *If $\alpha \in \mathcal{N}$ and neither $\beta \in \mathcal{N}$, nor $\gamma \in \mathcal{N}$, then $\alpha(\beta+\gamma) = \alpha\beta + \alpha\gamma$ if and only if β and γ are not opposite with respect to α.*

2. *If $\beta \in \mathcal{N}$, or $\gamma \in \mathcal{N}$, then $\alpha(\beta+\gamma) = \alpha\beta + \alpha\gamma$.*

A second important special case, which generalizes Theorem 1.4.3, concerns external numbers of the same sign. This is in line with the fact discussed above that two external numbers of the same sign are never opposite with respect to a given neutrix.

Theorem 3.3.6. *If α is an external number and β and γ are external numbers of the same sign, then $\alpha(\beta+\gamma)=\alpha\beta+\alpha\gamma$.*

We also mention the special cases given by Lemma 3.1.6 and Lemma 3.1.7. Let us illustrate Theorems 3.3.4-3.3.6 with some examples.

Example 3.3.1. 1. Let $\alpha = 1+\oslash$, $\beta = 1+\varepsilon$ with $\varepsilon \simeq 0$, and $\gamma = 1$. Then $\alpha(\beta+\gamma) = \alpha\beta+\alpha\gamma$ by Theorem 3.3.6. The equality follows also from Theorem 3.3.4 (ii). On the one hand,

$$\begin{aligned}\alpha(\beta+\gamma) &= (1+\oslash)(1+\varepsilon+1) = (1+\oslash)(2+\varepsilon)\\ &= 2+\varepsilon+(2+\varepsilon)\oslash \qquad \text{by Theorem 3.3.4 (ii)}\\ &= 2+\varepsilon+\oslash \qquad \text{because } 2+\varepsilon \text{ is appreciable}\\ &= 2+\oslash.\end{aligned}$$

On the other hand,

$$\begin{aligned}\alpha\beta+\alpha\gamma &= (1+\oslash)(1+\varepsilon)+(1+\oslash)1\\ &= 1+\oslash+(1+\oslash)\varepsilon+1+\oslash \qquad \text{by Theorem 3.3.4 (ii)}\\ &= 1+\oslash+\varepsilon+\varepsilon\oslash+1+\oslash \qquad \text{by Theorem 3.3.4 (ii)}\\ &= 2+\oslash.\end{aligned}$$

2. Let $\alpha = 1+\oslash$, $\beta = 1+\varepsilon$ with $\varepsilon \simeq 0$, and $\gamma = -1$. Then

$$\alpha(\beta+\gamma) = (1+\oslash)(1+\varepsilon-1) = (1+\oslash)\varepsilon = \varepsilon+\varepsilon\oslash$$

and

$$\begin{aligned}\alpha\beta+\alpha\gamma &= (1+\oslash)(1+\varepsilon)+(1+\oslash)(-1)\\ &= 1+\oslash+(1+\oslash)\varepsilon-1+\oslash \qquad \text{by Theorem 3.3.4 (ii)}\\ &= \oslash+\varepsilon+\oslash\varepsilon \qquad \text{by Theorem 3.3.4 (ii)}\\ &= \oslash.\end{aligned}$$

Because $\varepsilon\oslash \subset \oslash$, subdistributivity holds, but distributivity does not. This is in line with the fact that α is less precise than both β and γ and $1+\varepsilon$ and -1 are opposite with respect to $\oslash$. If we change β into $\beta' = 1+\oslash$, then α is as precise as β' and one verifies that

$$\alpha(\beta'+\gamma) = \oslash = \alpha\beta'+\alpha\gamma.$$

If ε is appreciable, the numbers $1+\varepsilon$ and -1 are no longer opposite with respect to $\oslash$ and then

$$\alpha(\beta+\gamma) = \varepsilon+\varepsilon\oslash = \varepsilon+\oslash = \oslash+\varepsilon+\oslash\varepsilon = \alpha\beta+\alpha\gamma.$$

3. If $\alpha = \oslash$, $\beta = \omega + \pounds$, $\gamma = -\omega + \oslash$, with $\omega \simeq +\infty$, distributivity does not hold by Part 1 of Theorem 3.3.5, for β and γ are opposite with respect to $\oslash$. Indeed, one shows with the aid of Part 2 of Theorem 3.3.5 that

$$\begin{aligned}\alpha(\beta+\gamma) &= \oslash\pounds = \oslash \subset \oslash\omega \\ &= \oslash\omega + \oslash\pounds = \oslash(\omega+\pounds) \\ &= \alpha\max(|\beta|, |\gamma|).\end{aligned}$$

4. If $\alpha = \sqrt{\omega} + \pounds$, $\beta = \omega + \pounds$, $\gamma = \pounds$,with $\omega \simeq +\infty$, distributivity holds. Indeed, using Theorem 3.3.4 (i) and Part 2 of Theorem 3.3.5

$$\begin{aligned}\alpha(\beta+\gamma) &= (\sqrt{\omega}+\pounds)(\omega+\pounds) \\ &= (\sqrt{\omega}+\pounds)\omega + (\sqrt{\omega}+\pounds)\pounds \\ &= \omega\sqrt{\omega} + \pounds\omega + \sqrt{\omega}\pounds + \pounds\pounds \\ &= \omega\sqrt{\omega} + \pounds\omega.\end{aligned}$$

   Note that we have in fact calculated $\alpha\beta$, and that $\alpha\gamma = (\sqrt{\omega}+\pounds)\pounds = \sqrt{\omega}\pounds + \pounds\pounds = \sqrt{\omega}\pounds$ is contained in $N(\alpha\beta)$. Hence $\alpha\beta + \alpha\gamma = \alpha\beta = \alpha(\beta+\gamma)$.

Some of these calculations may be obtained directly by the definition of multiplication for external numbers. As it turns out this rule plays a substantial part in the proof of Theorem 3.3.4. In order to prove Theorem 3.3.4 we will consider two separate cases. The first case deals with distributivity when one of the numbers is a neutrix (Section 3.3.1) and the second case deals with distributivity when neither of the external numbers is a neutrix (Section 3.3.2).

3.3.1 Distributivity with neutrices

We prove Theorem 3.3.4 in the case where at least one of the external numbers α, β, γ is a neutrix. We consider first the case where β or γ is a neutrix and then the case where α is a neutrix.

Proof of Theorem 3.3.4 in the case where β or γ is a neutrix: The criterion

$$\mathcal{R}(\alpha) \leq \max(\mathcal{R}(\beta), \mathcal{R}(\gamma)) \vee (\beta+\gamma)A = A\max(|\beta|, |\gamma|)$$

is trivially satisfied because either $\mathcal{R}(\beta) = \mathbb{R}$ or $\mathcal{R}(\gamma) = \mathbb{R}$. Conversely, suppose without loss of generality that $\gamma = C \in \mathcal{N}$. Because aB, aC, bA, AB, AC are neutrices one has

$$\begin{aligned}\alpha(\beta+\gamma) &= (a+A)(b+\max(B,C)) \\ &= ab + a\max(B,C) + bA + A\max(B,C) \\ &= ab + \max(aB, aC, bA, AB, AC)\end{aligned}$$

and

$$\alpha\beta + \alpha\gamma = ab + \max(aB, bA, AB) + \max(aC, AC)$$
$$= ab + \max(aB, aC, bA, AB, AC).$$

Hence $\alpha(\beta + \gamma) = \alpha\beta + \alpha\gamma$. □

We have the following trivial corollary.

Corollary 3.3.7. *1. Let $\alpha \in \mathbb{E}$ and $B, C \in \mathcal{N}$. Then $\alpha(B+C) = \alpha B + \alpha C$.*

2. If A, B, $C \in \mathcal{N}$, then $A(B + C) = AB + AC$.

3. Let A, $C \in \mathcal{N}$ and let β be zeroless. Then $A(\beta + C) = A\beta + AC$.

Proof of Theorem 3.3.4 in the case where α is a neutrix: Let $\alpha = A$ be an arbitrary neutrix. Without loss of generality we may assume that $|\beta| \geq |\gamma|$. Firstly suppose that $A(\beta + \gamma) = A\beta + A\gamma$. Then $A\beta + A\gamma = A\beta = A\max(|\beta|, |\gamma|)$. Hence β and γ are not opposite with respect to A.

Assume now that A is more precise than β or γ, or that β and γ are not opposite with respect to A. In the first case either β or γ have to be a neutrix because $\mathcal{R}(A) = \mathbb{R}$. This case is contained in Part 2 and Part 3 of Corollary 3.3.7. In the second case $A(\beta + \gamma) = A\max(|\beta|, |\gamma|) = A\beta = A\beta + A\gamma$. □

3.3.2 Distributivity with zeroless external numbers

We now turn to the cases where α, β and γ are zeroless external numbers. Unless stated otherwise, we always write $\alpha = a + A$, $\beta = b + B$ and $\gamma = c + C$, where a, b and c are real numbers and A, B and C are neutrices, with $A < |a|$, $B < |b|$ and $C < |c|$. By Remark 3.3.2 this means that $\alpha = a(1 + \mathcal{R}(\alpha))$, $\beta = b(1 + \mathcal{R}(\beta))$ and $\gamma = c(1 + \mathcal{R}(\gamma))$.

We may assume without loss of generality that $|\beta| \geq |\gamma|$. Then we may also suppose that $0 < \left|\frac{c}{b}\right| \leq 1$.

We deal first with the case where $a = 1$, $b = 1$ and $0 < |c| \leq 1$; then $A \leq \oslash$ and $B \leq \oslash$ by (3.2). The general case is obtained by rescaling. First, we need to give direct proofs of the distributive law in some relatively easy special cases.

Lemma 3.3.8. *Let $a \in \mathbb{R}$ and β, γ be zeroless external numbers. Then $a(\beta + \gamma) = a\beta + a\gamma$.*

Proof. By Lemma 3.1.7 and Part 1 of Corollary 3.3.7 one has

$$\begin{aligned} a(\beta + \gamma) &= a((b + c) + \max(B, C)) \\ &= a(b + c) + a(B + C) \\ &= ab + ac + aB + aC \\ &= a(b + B) + a(c + C) \\ &= a\beta + a\gamma. \end{aligned}$$

□

Proof of Theorem 3.3.1. Using Lemma 3.1.7 and Lemma 3.3.8 one has

$$\begin{aligned}\alpha(\beta+\gamma)+A\beta+A\gamma &= (a+A)(\beta+\gamma)+A\beta+A\gamma\\ &= a(\beta+\gamma)+A(\beta+\gamma)+A\beta+A\gamma\\ &= a\beta+a\gamma+A(\beta+\gamma)+A\beta+A\gamma.\end{aligned}$$

Because $A\beta$ and $A\gamma$ are neutrices, it follows from formula (3.7) that

$$a(\beta+\gamma)+A\beta+A\gamma = a\beta+a\gamma+A\beta+A\gamma$$

Hence $\alpha(\beta+\gamma)+A\beta+A\gamma=\alpha\beta+\alpha\gamma$ by Lemma 3.1.7. □

Proposition 3.3.9. *Let α,β,γ be external numbers such that β and γ are not opposite with respect to $\pounds$; then $\alpha(\beta+\gamma)=\alpha\beta+\alpha\gamma$.*

Proof. We assume without loss of generality that α and β are positive. Because $A\pounds=A$, by Part 2 of Lemma 3.3.3 and formula (3.8) it holds that $(\beta+\gamma)A=\beta A+\gamma A$. Notice that $\beta+\gamma$ is zeroless. If not, both $\pounds(\beta+\gamma)=\pounds(B+C)=B+C$ and $\pounds(\beta+\gamma)=\pounds\beta\geq b$, with $b>B$ and $b\geq |c|>C$, a contradiction. Then by Lemma 3.1.7 and Lemma 3.3.8 one has

$$\begin{aligned}\alpha(\beta+\gamma) &= a(\beta+\gamma)+A(\beta+\gamma)\\ &= a\beta+a\gamma+A\beta+A\gamma\\ &= \alpha\beta+\alpha\gamma.\end{aligned}$$

□

Lemma 3.3.10. *Assume that $a=1$, $b=1$ and $0<|c|\leq 1$. Then*

1. *If $\beta+\gamma\in\mathcal{N}$,*

$$\begin{aligned}\alpha(\beta+\gamma)=\alpha\beta+\alpha\gamma &\Leftrightarrow A\leq \max(B,C)\\ &\Leftrightarrow \alpha\beta=\beta\vee\alpha\gamma=\gamma.\end{aligned}$$

2. *If $\beta+\gamma\notin\mathcal{N}$,*

$$\begin{aligned}\alpha(\beta+\gamma)=\alpha\beta+\alpha\gamma &\Leftrightarrow A\leq\max(B,C)\vee(1+c)A=A\\ &\Leftrightarrow \alpha\beta=\beta\vee\alpha\gamma=\gamma\vee(\beta+\gamma)A=A.\end{aligned}$$

Proof. Observe first that by Lemma 3.1.5 and Lemma 3.1.6, one has

$$\alpha\beta=(1+A)(1+B)=1+A+B \tag{3.9}$$

and

$$\alpha\gamma=(1+A)(c+C)=c+cA+C. \tag{3.10}$$

Hence

$$\begin{aligned}\alpha\beta+\alpha\gamma &= 1+A+B+c+cA+C\\ &= 1+c+A+B+C=1+c+\max(A,B,C).\end{aligned} \tag{3.11}$$

1. As for the first equivalence, by hypothesis and Lemma 3.1.7, one has

$$\alpha(\beta+\gamma) = (1+A)(B+C) = \max(B,C). \tag{3.12}$$

Hence, by (3.11) and (3.12) we conclude that $\alpha(\beta+\gamma) = \alpha\beta+\alpha\gamma$ if and only if $A \leq \max(B,C)$.

To prove the second equivalence, assume first that $\alpha\beta = \beta$ or $\alpha\gamma = \gamma$. In the first case, $1+A+B = 1+B$ by (3.9), which implies $A \leq B$. In the second case, by (3.10)

$$c + cA + C = c + C. \tag{3.13}$$

Because β and γ are zeroless, one has $B, C \subseteq \oslash$. Now $\beta+\gamma = 1+c+B+C \subseteq 1+c+\oslash \in \mathcal{N}$, which implies that $c \simeq -1$; hence $cA = A$. Then we derive from (3.13) that $c+A+C = c+C$, so $A \leq C$. We conclude that $A \leq \max(B,C)$. Conversely, assume that $A \leq \max(B,C)$. Hence $A \leq B$ or $A \leq C$. Then clearly $1+A+B = 1+B$ or $c+A+C = c+C$ and then $\alpha\beta = \beta$ or $\alpha\gamma = \gamma$ by (3.9) and (3.10). Hence $A \leq \max(B,C) \Leftrightarrow \alpha\beta = \beta \vee \alpha\gamma = \gamma$.

2. One has $\beta+\gamma \neq \max(B,C)$. Hence

$$(\beta+\gamma)A = (1+B+c+C)A = (1+c)A.$$

We prove first that

$$\alpha(\beta+\gamma) = \alpha\beta+\alpha\gamma \Leftrightarrow A \leq \max(B,C) \vee (1+c)A = A. \tag{3.14}$$

Lemma 3.1.5 yields $\alpha(\beta+\gamma) = (1+A)(1+B+c+C) = 1+c+B+C+(1+c)A$.

If $A \leq \max(B,C)$, because $(1+c)A \subseteq A$ both $\alpha\beta+\alpha\gamma = 1+c+\max(B,C)$ and $\alpha(\beta+\gamma) = 1+c+\max(B,C)$. If $(1+c)A = A$, we conclude from (3.11) that $\alpha\beta+\alpha\gamma = \alpha(\beta+\gamma)$. If $\alpha\beta+\alpha\gamma = \alpha(\beta+\gamma)$, by (3.11) it holds that $A+B+C = B+C+(1+c)A$, so $A \leq \max(B,C)$ or $A = (1+c)A$. Hence (3.14) holds. Finally we prove that

$$A \leq \max(B,C) \vee (1+c)A = A \Leftrightarrow \alpha\beta = \beta \vee \alpha\gamma = \gamma \vee (\beta+\gamma)A = A. \tag{3.15}$$

By (3.9) and (3.10) one has $\alpha\beta = \beta \Leftrightarrow A \leq B$ and $\alpha\gamma = \gamma \Leftrightarrow C + cA = C$. Assume that $A \leq \max(B,C)$ or $(1+c)A = A$. If $A \leq B$, then $\alpha\beta = \beta$. If $A \leq C$, then $C + cA = C$; hence $\alpha\gamma = \gamma$. If $(1+c)A = A$, then $(\beta+\gamma)A = (1+c+B+C)A = (1+c)A + BA + CA = A$. Assume now that $\alpha\beta = \beta$ or $\alpha\gamma = \gamma$, or $(\beta+\gamma)A = A$. If $\alpha\beta = \beta$, then $A \leq B \leq \max(B,C)$. If $\alpha\gamma = \gamma$, then $cA \leq C$. If $cA = A$ one has $A \leq C \leq \max(B,C)$. If $cA < A$ then $1+c \simeq 1$; hence $(1+c)A = A$. If $(\beta+\gamma)A = A$, then $(1+c)A = A$, because $\beta+\gamma$ is zeroless. Hence (3.15) holds. □

We are now able to characterize distributivity for zeroless external numbers:

Theorem 3.3.11. *Let α, β, γ be zeroless external numbers.*

1. *If $\beta + \gamma \in \mathcal{N}$,*

$$\alpha(\beta+\gamma) = \alpha\beta + \alpha\gamma \Leftrightarrow \mathcal{R}(\alpha) \leq \max(\mathcal{R}(\beta), \mathcal{R}(\gamma)).$$

2. *If $\beta + \gamma \notin \mathcal{N}$,*

$$\begin{aligned} \alpha(\beta+\gamma) &= \alpha\beta + \alpha\gamma \\ &\Leftrightarrow \mathcal{R}(\alpha) \leq \max(\mathcal{R}(\beta), \mathcal{R}(\gamma)) \vee (\beta+\gamma)A = A\max(|\beta|, |\gamma|) \end{aligned}$$

Proof. First, we put the products in a convenient form. By Lemma 3.1.6 and by Lemma 3.3.8 one has

$$\begin{aligned} \alpha(\beta+\gamma) &= a(1+\mathcal{R}(\alpha))(b(1+\mathcal{R}(\beta)) + c(1+\mathcal{R}(\gamma))) \\ &= ab(1+\mathcal{R}(\alpha))\left(1+\mathcal{R}(\beta)+\frac{c}{b}+\frac{c}{b}\mathcal{R}(\gamma)\right) \end{aligned} \tag{3.16}$$

and

$$\begin{aligned} \alpha\beta + \alpha\gamma &= a(1+\mathcal{R}(\alpha))b(1+\mathcal{R}(\beta)) + a(1+\mathcal{R}(\alpha))c(1+\mathcal{R}(\gamma)) \\ &= ab\left((1+\mathcal{R}(\alpha))(1+\mathcal{R}(\beta)) + (1+\mathcal{R}(\alpha))\left(\frac{c}{b}+\frac{c}{b}\mathcal{R}(\gamma)\right)\right). \end{aligned} \tag{3.17}$$

From (3.16) and (3.17) distributivity is equivalent to

$$\begin{aligned} &(1+\mathcal{R}(\alpha))\left(1+\mathcal{R}(\beta)+\frac{c}{b}+\frac{c}{b}\mathcal{R}(\gamma)\right) \\ &= (1+\mathcal{R}(\alpha))(1+\mathcal{R}(\beta)) + (1+\mathcal{R}(\alpha))\left(\frac{c}{b}+\frac{c}{b}\mathcal{R}(\gamma)\right). \end{aligned} \tag{3.18}$$

Since by assumption $|\gamma| \leq |\beta|$ and $|c| \leq |b|$, we are able to apply Lemma 3.3.10. To prove Part 1, suppose that $\beta + \gamma \in \mathcal{N}$. Then $b + B + c + C = B + C \subseteq \oslash b + \oslash c = \oslash b$, so $1 + \frac{c}{b} + \oslash \subseteq \oslash$. Hence $\frac{c}{b} \simeq -1$. This implies that $\frac{c}{b}\mathcal{R}(\gamma) = \mathcal{R}(\gamma)$. Then

$$\mathcal{R}(\alpha) \leq \max\left(\mathcal{R}(\beta), \frac{c}{b}\mathcal{R}(\gamma)\right) \Leftrightarrow \mathcal{R}(\alpha) \leq \max(\mathcal{R}(\beta), \mathcal{R}(\gamma)).$$

Hence, by Part 1 of Lemma 3.3.10,

$$\begin{aligned} \alpha(\beta+\gamma) = \alpha\beta + \alpha\gamma &\Leftrightarrow \mathcal{R}(\alpha) \leq \max\left(\mathcal{R}(\beta), \frac{c}{b}\mathcal{R}(\gamma)\right) \\ &\Leftrightarrow \mathcal{R}(\alpha) \leq \max(\mathcal{R}(\beta), \mathcal{R}(\gamma)). \end{aligned}$$

To prove Part 2, suppose that $\beta + \gamma \notin \mathcal{N}$. Then, by Part 2 of Lemma 3.3.10 and (3.18),

$$\begin{aligned} \alpha(\beta+\gamma) = \alpha\beta + \alpha\gamma &\Leftrightarrow (1+\mathcal{R}(\alpha))(1+\mathcal{R}(\beta)) = 1+\mathcal{R}(\beta) \\ &\vee (1+\mathcal{R}(\alpha))\left(\frac{c}{b}+\frac{c}{b}\mathcal{R}(\gamma)\right) = \frac{c}{b}+\frac{c}{b}\mathcal{R}(\gamma) \\ &\vee \left(1+\mathcal{R}(\beta)+\frac{c}{b}+\frac{c}{b}\mathcal{R}(\gamma)\right)A = A. \end{aligned} \tag{3.19}$$

First, we prove the direct implication of Part 2. With respect to (3.19) there are three cases to consider: (i) $(1+\mathcal{R}(\alpha))(1+\mathcal{R}(\beta)) = 1+\mathcal{R}(\beta)$, (ii) $(1+\mathcal{R}(\alpha))\left(\frac{c}{b}+\frac{c}{b}\mathcal{R}(\gamma)\right) = \frac{c}{b}+\frac{c}{b}\mathcal{R}(\gamma)$ and (iii) $\left(1+\mathcal{R}(\beta)+\frac{c}{b}+\frac{c}{b}\mathcal{R}(\gamma)\right)A = A$.

(i) By Lemma 3.1.5

$$1+\mathcal{R}(\alpha)+\mathcal{R}(\beta) = 1+\mathcal{R}(\beta); \tag{3.20}$$

hence $\mathcal{R}(\alpha) \leq \mathcal{R}(\beta)$. This implies that $\mathcal{R}(\alpha) \leq \max(\mathcal{R}(\beta), \mathcal{R}(\gamma))$.

(ii) Similarly to (3.20) one has

$$\frac{c}{b}(1+\mathcal{R}(\alpha)+\mathcal{R}(\gamma)) = \frac{c}{b}(1+\mathcal{R}(\gamma)),$$

implying that $\mathcal{R}(\alpha) \leq \mathcal{R}(\gamma)$. Hence $\mathcal{R}(\alpha) \leq \max(\mathcal{R}(\beta), \mathcal{R}(\gamma))$.

(iii) By Lemma 3.1.7 one has $(b+B+c+C)A = bA$, and therefore

$$(\beta+\gamma)A = \beta A = A\max(|\beta|, |\gamma|).$$

Combining the three cases, we conclude that

$$\begin{aligned} &\alpha(\beta+\gamma) = \alpha\beta+\alpha\gamma \\ &\qquad \Rightarrow \mathcal{R}(\alpha) \leq \max(\mathcal{R}(\beta), \mathcal{R}(\gamma)) \vee (\beta+\gamma)A = A\max(|\beta|, |\gamma|). \end{aligned} \tag{3.21}$$

To prove the reverse implication we consider two cases: (a) $(\beta+\gamma)A = A\max(|\beta|, |\gamma|)$ and (b) $\mathcal{R}(\alpha) \leq \max(\mathcal{R}(\beta), \mathcal{R}(\gamma))$.

(a) One has $(\beta+\gamma)A = |\beta|\,A$. Then $(b+B+c+C)\,A = (b+B)A$. This implies that $\left(1+\mathcal{R}(\beta)+\frac{c}{b}+\frac{c}{b}\mathcal{R}(\gamma)\right)A = A$. Then by (3.19) we conclude that

$$(\beta+\gamma)A = A\max(|\beta|, |\gamma|) \Rightarrow \alpha(\beta+\gamma) = \alpha\beta+\alpha\gamma. \tag{3.22}$$

(b) If $\mathcal{R}(\alpha) \leq \max(\mathcal{R}(\beta), \mathcal{R}(\gamma))$, then $\mathcal{R}(\alpha) \leq \mathcal{R}(\beta)$ or $\frac{c}{b}\mathcal{R}(\alpha) \leq \frac{c}{b}\mathcal{R}(\gamma)$ and by Lemma 3.1.5 one has $(1+\mathcal{R}(\alpha))(1+\mathcal{R}(\beta)) = 1+\mathcal{R}(\beta)$ or $(1+\mathcal{R}(\alpha))\left(\frac{c}{b}+\frac{c}{b}\mathcal{R}(\gamma)\right) = \frac{c}{b}+\frac{c}{b}\mathcal{R}(\gamma)$. Then by (3.19) we conclude that

$$\mathcal{R}(\alpha) \leq \max(\mathcal{R}(\beta), \mathcal{R}(\gamma)) \Rightarrow \alpha(\beta+\gamma) = \alpha\beta+\alpha\gamma. \tag{3.23}$$

From (3.22) and (3.23) we obtain

$$\begin{aligned} &\mathcal{R}(\alpha) \leq \max(\mathcal{R}(\beta), \mathcal{R}(\gamma)) \vee (\beta+\gamma)A = A\max(|\beta|, |\gamma|) \\ &\qquad \Rightarrow \alpha(\beta+\gamma) = \alpha\beta+\alpha\gamma. \end{aligned} \tag{3.24}$$

We conclude by combining (3.21) and (3.24). □

To complete the proof of Theorem 3.3.4 for the case of zeroless numbers we need one final lemma.

Lemma 3.3.12. *Let α, β, γ be zeroless external numbers. If $\beta+\gamma \in \mathcal{N}$ and $\alpha(\beta+\gamma) \neq \alpha\beta+\alpha\gamma$, then $(\beta+\gamma)A \subset A\max(|\beta|, |\gamma|)$.*

Proof. Suppose that $\alpha(\beta+\gamma) \neq \alpha\beta+\alpha\gamma$. Then, since $\beta+\gamma = B+C$, by Theorem 3.3.11 we obtain $\mathcal{R}(\alpha) > \max(\mathcal{R}(\beta), \mathcal{R}(\gamma))$. It follows from Proposition 3.1.11 that $\max(\mathcal{R}(\beta), \mathcal{R}(\gamma))\mathcal{R}(\alpha) \subset \mathcal{R}(\alpha)$. This implies that $\left(\frac{B}{b}+\frac{C}{c}\right)\frac{A}{a} \subset \frac{A}{a}$. Because $c \leq b$, we obtain $\left(\frac{B}{b}+\frac{C}{b}\right)A \subset A$. Because $\beta+\gamma = B+C$, we have that $(\beta+\gamma)A \subset A\beta$ and we deduce that $(\beta+\gamma)A \subset A\max(|\beta|, |\gamma|)$. □

Proof of Theorem 3.3.4 for zeroless external numbers: If follows immediately from Theorem 3.3.11 that if $\alpha(\beta+\gamma) = \alpha\beta+\alpha\gamma$, then

$$\mathcal{R}(\alpha) \leq \max(\mathcal{R}(\beta), \mathcal{R}(\gamma)) \vee (\beta+\gamma)A = A\max(|\beta|, |\gamma|).$$

If $\beta+\gamma \notin \mathcal{N}$, the reverse implication follows from Part 2 of Theorem 3.3.11. If $\beta+\gamma \in \mathcal{N}$, the reverse implication follows from Part 1 of Theorem 3.3.11 and Lemma 3.3.12. □

3.3.3 Application: Binomial formulas

As an application we study the effect of the distributive law on some binomial forms. Let $\alpha = a+A$, $\beta = b+B$ and $\gamma = c+C$ with a, b and c real numbers and A, B and C neutrices. We are interested in the domains of validity of the following formulas:

$$\alpha(\alpha+\beta) = \alpha^2+\alpha\beta. \tag{3.25}$$

$$(\alpha-\beta)(\alpha+\beta) = \alpha^2-\beta^2. \tag{3.26}$$

$$(\alpha+\beta)^2 = \alpha^2+2\alpha\beta+\beta^2. \tag{3.27}$$

Also we extend the equality (3.27) to a Binomial Theorem for external numbers.

Theorem 3.3.13. *Let $\alpha = a+A$ and $\beta = b+B$ be external numbers. If α and β are neither opposite with respect to A nor to B, for standard $n \in \mathbb{N}, n \geq 1$*

$$(\alpha+\beta)^n = \sum_{k=0}^{n} \binom{n}{k} \alpha^{n-k}\beta^k. \tag{3.28}$$

We start by noting that (3.25) is always valid. The reason for this is because α is as precise as itself.

We now investigate the validity of the equality (3.26). If α and β are neutrices the equality (3.26) may be verified directly very easily. In the remaining case, we may assume without loss of generality that α and β are both non-negative. Then by Theorem 3.3.6

$$(\alpha-\beta)(\alpha+\beta) = (\alpha-\beta)\alpha+(\alpha-\beta)\beta.$$

Hence (3.25) yields

$$(\alpha-\beta)\alpha+(\alpha-\beta)\beta=\alpha^2-\alpha\beta+\alpha\beta-\beta^2=\alpha^2-\beta^2+N(\alpha\beta).$$

Hence the inclusion $(\alpha-\beta)(\alpha+\beta)\supseteq\alpha^2-\beta^2$ always holds. Observe that $N(\alpha^2-\beta^2)=\alpha A+\beta B$ and $N(\alpha\beta)=\alpha B+\beta A$. Hence (3.26) holds if $\alpha B+\beta A\subseteq\alpha A+\beta B$, say, if $B\leq A$ and $\beta\leq\alpha$.

We show that (3.27) holds if α and β are neither opposite with respect to A nor to B. Indeed, by Part 1 of Lemma 3.3.3 the numbers α and β are not opposite with respect to $A+B=N(\alpha+\beta)$. Then by Theorem 3.3.4 and (3.25)

$$\begin{aligned}(\alpha+\beta)^2&=(\alpha+\beta)(\alpha+\beta)=\alpha(\alpha+\beta)+\beta(\alpha+\beta)\\&=\alpha^2+\alpha\beta+\beta\alpha+\beta^2=\alpha^2+2\alpha\beta+\beta^2.\end{aligned}$$

To prove Theorem 3.3.13 we need the following properties of the relative uncertainty.

Lemma 3.3.14. *Let $\alpha=a+A$ and $\beta=b+B$ be external numbers. Then*

1. $\mathcal{R}(\alpha\beta)=\mathcal{R}(\alpha)+\mathcal{R}(\beta)$
2. If $k\in\mathbb{N}$ is standard, $\mathcal{R}(\alpha^k)=\mathcal{R}(\alpha)$.

Proof. See Exercise 3.3.5. □

Proof of Theorem 3.3.13: The proof is by External Induction. If $n=1$, formula (3.28) is clearly true. Suppose that (3.28) is true for standard n. Then

$$(\alpha+\beta)^{n+1}=(\alpha+\beta)(\alpha+\beta)^n=(\alpha+\beta)\sum_{k=0}^{n}\binom{n}{k}\alpha^{n-k}\beta^k.$$

The neutrix $C:=N\left(\sum_{k=0}^{n}\binom{n}{k}\alpha^{n-k}\beta^k\right)$ is a sum with a standard finite number of multiples of A and B. Hence Lemma 3.3.3 yields that α and β are not opposite with respect to C. By Theorem 3.3.4

$$(\alpha+\beta)\sum_{k=0}^{n}\binom{n}{k}\alpha^{n-k}\beta^k=\alpha\sum_{k=0}^{n}\binom{n}{k}\alpha^{n-k}\beta^k+\beta\sum_{k=0}^{n}\binom{n}{k}\alpha^{n-k}\beta^k.$$

It follows from Lemma 3.3.14 that $\mathcal{R}(\alpha)\leq\mathcal{R}\left(\binom{n}{k}\alpha^{n-k}\beta^k\right)$ for all k such that $0\leq k\leq n-1$, and $\mathcal{R}(\beta)\leq\mathcal{R}\left(\binom{n}{k}\alpha^{n-k}\beta^k\right)$ for all k such that $1\leq k\leq n$. Repeated application of Theorem 3.3.4 yields

$$\alpha\sum_{k=0}^{n}\binom{n}{k}\alpha^{n-k}\beta^k+\beta\sum_{k=0}^{n}\binom{n}{k}\alpha^{n-k}\beta^k=\sum_{k=0}^{n}\binom{n}{k}\alpha^{n+1-k}\beta^k+\sum_{k=0}^{n}\binom{n}{k}\alpha^{n-k}\beta^{k+1}.$$

Because the relative uncertainty of natural numbers is zero, again by Theorem 3.3.4

$$\sum_{k=0}^{n}\binom{n}{k}\alpha^{n+1-k}\beta^k+\sum_{k=0}^{n}\binom{n}{k}\alpha^{n-k}\beta^{k+1}=\sum_{k=0}^{n+1}\binom{n+1}{k}\alpha^{n+1-k}\beta^k.$$ □

Exercises

Exercise 3.3.1. *Show that appreciable real numbers β and γ are opposite with respect to $\oslash$ and $£$ if and only if $\beta \simeq -\gamma$.*

Exercise 3.3.2. *Prove Part 2 of Lemma 3.3.3.*

Exercise 3.3.3. *Let $\varepsilon \simeq 0, \varepsilon > 0$ and $\omega \simeq +\infty$. For each case, calculate $\mathcal{R}(\alpha)$.*

1. $\alpha = 1 + \varepsilon\oslash$.
2. $\alpha = -5 + \oslash$.
3. $-\frac{2}{\varepsilon} + \varepsilon^2 £$.
4. $\omega^2 + \omega £$.
5. $\varepsilon^2 + \frac{\omega}{\varepsilon}$.
6. $\oslash$.

Exercise 3.3.4. *Determine the value of $\mathcal{R}\left(\frac{\alpha}{\beta}\right)$, if $\alpha = 1 + \varepsilon\oslash$ and $\beta = \frac{5}{\varepsilon} + \varepsilon^2 £$, where $\varepsilon \simeq 0, \varepsilon > 0$.*

Exercise 3.3.5. *Prove Lemma 3.3.14.*

Exercise 3.3.6. *Let $\varepsilon \simeq 0, \varepsilon > 0$. Explain why distributivity $\alpha(\beta + \gamma) = \alpha\beta + \alpha\gamma$ holds or fails in each of the following cases.*

1. $\alpha = \oslash$, $\beta = 1 + \varepsilon\oslash$, $\gamma = -1 + \varepsilon\oslash$.
2. $\alpha = 1 + \oslash$, $\beta = -\varepsilon$, $\gamma = 1 + \varepsilon\oslash$.
3. $\alpha \in \mathbb{E} \setminus \mathbb{R}$, $\beta - \gamma \in N(\alpha)$.
4. $\alpha, \beta \in \mathbb{R}$ *and* γ *is an arbitrary external number.*
5. $\alpha, \gamma, \beta \in \mathbb{R}$.

Exercise 3.3.7. *Let a be a real number and A be a neutrix such that $|a| > A$. Show that, for all standard n,*

$$a^n A + a^{n-1} A^2 = a^n A.$$

Exercise 3.3.8. *Let $\alpha = a + A$ be a zeroless external number. Show that for all standard n*

$$\alpha^n = a^n + a^{n-1} A.$$

Exercise 3.3.9. *Let $\alpha_i = a_i + A_i$, $i = 1, ..., n$ be zeroless external numbers. Show that for all standard n*

$$\prod_{i=1}^{n} (a_i + A_i) = \prod_{i=1}^{n} a_i + \sum_{j=1}^{n} \left[\left(\prod_{\substack{K=1 \\ K \neq j}}^{n} a_k \right) A_j \right].$$

Chapter 4

Advanced properties

4.1 Introduction to Internal Set Theory 85
4.1.1 Properties of IST 87
4.1.2 External sets 89
4.2 The nature of halflines, neutrices and external numbers 94
4.3 Generalized Dedekind completeness 96
4.4 Flexible sequences and functions 100
4.4.1 Flexible functions 100
4.4.2 Flexible sequences 104
4.5 Idempotent neutrices and ideals 109
4.5.1 Idempotent neutrices 110
4.5.2 Ideals and the product of neutrices 111

In this chapter we develop more advanced properties of neutrices and external numbers. To this end we need to work in a stronger theory. This theory consists of Nelson's Internal Set Theory [92] to which we add an axiom that enables us to have some external sets, in particular neutrices and external numbers. This is done in Section 4.1. In Section 4.2 we show that, as external sets, neutrices have the lowest possible complexity. In Section 4.3 we show that Dedekind completeness can be reformulated in such a way that it is carried over to external sets of real numbers. In Section 4.5 we characterize this generalized Dedekind completeness algebraically. Firstly it is equivalent to the property that every neutrix is a multiple of a neutrix which is idempotent for multiplication. Secondly, the set $\pounds$ of limited numbers may also be seen as a valuation domain, and then generalized Dedekind completeness is equivalent to its maximality.

4.1 Introduction to Internal Set Theory

We introduce Nelson's Internal Set Theory IST and show that it extends the system ENA^- of Chapter 1. We will also extend Axiom 1.7 on external sets with internal elements, in order to be able to define neutrices and external numbers. We refer to Kanovei and Reeken's Axiomatics HST for a formal treatment of external sets [71] (see Appendix A for a short presentation).

The system IST (Internal Set Theory) was first presented in 1977 [92] and in a sense formulates within first-order language the behaviour of standard and internal sets of a (strong) nonstandard model. This is done by adding the unary predicate "st" to the language of ZFC as well as adding to the axioms of ZFC three new axiom schemes involving the predicate "st": Idealization, Standardization and Transfer. This means that IST is an extension of set theory ZFC, and uses the language $[\in, \mathrm{st}]$, considered already in Chapter 1. We recall that one should read $\mathrm{st}(x)$ as "x is standard"; Nelson suggests interpreting "standard" informally as "fixed". Formulas which do not use the predicate st are called *internal* formulas (or $\in$-formulas) and formulas that use this new predicate are called *external* formulas (or st-$\in$-formulas). A formula φ is *standard* if only standard constants occur in φ. Before formulating the new axioms we introduce some useful abbreviations. We write $\mathrm{fin}(x)$ meaning that x is finite. Let $\varphi(x)$ be a st-$\in$-formula. Then

$\forall^{\mathrm{st}} x\, \varphi(x)$	abbreviates	$\forall x(\mathrm{st}(x) \Rightarrow \varphi(x))$
$\exists^{\mathrm{st}} x\, \varphi(x)$	abbreviates	$\exists x(\mathrm{st}(x) \wedge \varphi(x))$
$\forall^{\mathrm{fin}} x\, \varphi(x)$	abbreviates	$\forall x(\mathrm{fin}(x) \Rightarrow \varphi(x))$
$\exists^{\mathrm{fin}} x\, \varphi(x)$	abbreviates	$\exists x(\mathrm{fin}(x) \wedge \varphi(x))$
$\forall^{\mathrm{stfin}} x\, \varphi(x)$	abbreviates	$\forall x(\mathrm{st}(x) \wedge \mathrm{fin}(x) \Rightarrow \varphi(x))$
$\exists^{\mathrm{stfin}} x\, \varphi(x)$	abbreviates	$\exists x(\mathrm{st}(x) \wedge \mathrm{fin}(x) \wedge \varphi(x))$

The new axiom-schemes of IST are the following:

Transfer:

$$\forall^{\mathrm{st}} x\, A(x,t) \Leftrightarrow \forall x\, A(x,t). \tag{T}$$

The only nonlogical symbol of A must be $\in$ (i.e. A must be internal). The parameter t must take a standard value.

Idealization:

$$\forall^{\mathrm{stfin}} z \exists y \forall x \in z\, B(x,y,t) \Leftrightarrow \exists y \forall^{\mathrm{st}} x\, B(x,y,t). \tag{I}$$

The only nonlogical symbol of B must be $\in$ (i.e. B must be internal). The parameter t may take any value.

Standardization:

$$\forall^{\mathrm{st}} x \exists^{\mathrm{st}} y \forall^{\mathrm{st}} z (z \in y \Leftrightarrow z \in x \wedge C(z,t)). \tag{S}$$

The formula C may contain the nonlogical symbols $\in$ and st (i.e. C can be any formula of the language). The parameter t may take any value.

4.1.1 Properties of IST

We will try to make the axioms more intuitive[1], give some consequences and show that IST implies the axiomatic system ENA^-.

Let us start with Transfer. As a first consequence, many more sets are declared to be standard other than just some integers, in fact so are all well-defined sets, i.e. sets which are unique satisfying some internal property. Indeed, the contrapositive of (T) is

$$\exists x\, A(x,t) \Leftrightarrow \exists^{\mathrm{st}} x\, A(x,t), \tag{T$\exists$}$$

so if x is the unique set satisfying the property $A(x,t)$ it must be equal to the standard set satisfying the property $A(x,t)$. As a consequence 0, 1, $\sqrt{2}$, π, e, $\mathbb{N}$, $\mathbb{R}$, log, sin, as well as every uniquely defined object in usual mathematics, are all standard. Informally, this is in line with our earlier observation that "standard" is a formalization of "concrete, fixed".

When applied to a negation $\neg A$, formula ($\mathrm{T}_\exists$) expresses that if something is not true, one must be able to furnish a concrete counterexample. Also the universal formulation (T) may be interpreted in this sense: Once something is proved for all concrete objects in a set, it must also be true for the possible "ideal" objects of the sets, for counterexamples can only come from concrete objects.

Observe that it is necessary that the parameter t in (T) takes standard ("concrete") values. Indeed, let $\omega \in \mathbb{N}$ be unlimited. Put

$$A(n,\omega) :\equiv n \in \mathbb{N} \wedge \omega \in \mathbb{N} \wedge \mathrm{st}(n) \wedge n < \omega.$$

Then $\forall^{\mathrm{st}} n\, A(n,\omega)$, but $\neg\forall n\, A(n,\omega)$, for ω is a counterexample.

The Idealization axioms express the idea that we can only fix a finite number of objects at a time. One interesting consequence of (I) is a theorem by Nelson [92] which states that there is a finite set F containing all standard sets (see Exercise 4.1.9). This implies that for any set A, there is a finite set F containing all standard elements of A, so all infinite sets have nonstandard elements. To see for example that $\mathbb{N}$ has a nonstandard element ω indeed, take the formula

$$B(n,y) :\equiv n \in \mathbb{N} \wedge y \in \mathbb{N} \wedge n < y.$$

Let $z \subseteq \mathbb{N}$ be a standard finite set. Being finite z has a maximum, say M. Then $B(n,M) \equiv n \in \mathbb{N} \wedge M \in \mathbb{N} \wedge n < M$ holds for all $n \in z$. By Idealization there exists $\omega \in \mathbb{N}$ with $B(n,\omega)$, i.e. $n < \omega$ for all standard $n \in \mathbb{N}$. Hence ω must be nonstandard, in fact unlimited.

Taking $B(x,y) :\equiv y \in X \wedge x \neq y$, one shows in an analogous way that every infinite set, standard or not, has nonstandard elements. Informally, this extends Reeb's observation that not all elements of $\mathbb{N}$ need to be naive: an infinite set in ZFC, having a definition with a finite number of symbols, can

[1]We somewhat follow [91], where Nelson proposes the intuition for (T), (S) and (I).

never be characterized by summing up its elements, leaving "space" for ideal elements.

Standardization implies that standard elements are taken care of. The intuition is that if we have a fixed set, then we can specify a fixed subset of it by giving a criterion for judging whether each fixed element is a member of it or not. Let X be a standard reference set. Let $C(z)$ be a property of some elements of X; it may be external, i.e. it is possible that st is used to formulate the property. Then there exists a standard set y such that its standard elements are exactly the standard elements of X satisfying the property C; concrete leads to concrete. Allowing for external sets in the sense of Axiom 1.7, the set y is called the *standardized set* of $\{\mathrm{st}(z) : z \in X \wedge C(z)\}$; we write

$$y = {}^{s}\{\mathrm{st}(z) : z \in X \wedge C(z)\}.$$

For some examples, let $X = \mathbb{N}$ and $S \subseteq \mathbb{N}$. If $S = \{0, 1, 2\}$, then ${}^{s}S = \{0, 1, 2\}$. Let $\omega \in \mathbb{N}$ be unlimited. If $S = \{0, 1, 2, ..., \omega\}$, we have ${}^{s}S = \mathbb{N}$. If $S = \{n : n \in \mathbb{N} \wedge n > \omega\}$, we have ${}^{s}S = \emptyset$.

A useful consequence of Standardization is the principle of *External Induction,* given in Axiom 1.11, that we recall here for convenience. External Induction states that for any (external or internal) formula Φ, one has

$$(\Phi(0) \wedge \forall^{\mathrm{st}} n(\Phi(n) \Rightarrow \Phi(n+1))) \Rightarrow \forall^{\mathrm{st}} n\, \Phi(n). \tag{EI}$$

Also an interesting consequence of Idealization and Standardization is the following *Saturation Principle* [95, Theorem 5].

Definition 4.1.1. *A formula Φ in the language $\{\in, \mathrm{st}\}$ is called* bounded *if all quantifications involving standard elements are of the form $\forall x(x \in X \Rightarrow \mathrm{st}(x)...)$ or $\exists y(y \in Y \wedge \mathrm{st}(y)...)$, where X and Y are standard, and every other quantification is with respect to an internal subset of a standard set.*

Theorem 4.1.1. *Let Φ be a bounded formula of the language of* IST*. Then*

$$\forall^{\mathrm{st}} x \exists y\, \Phi(x, y) \Rightarrow \exists \tilde{y} \forall^{\mathrm{st}} x\, \Phi(x, \tilde{y}(x)). \tag{4.1}$$

The next theorem shows that IST is indeed stronger than the system ENA$^-$. Since Nelson showed that IST is consistent with ZFC [92], as a consequence one obtains the (relative) consistency of ENA$^-$.

Theorem 4.1.2. *The system* IST *implies* ENA$^-$*.*

Proof. The number 0 is standard, being unique. If $n \in \mathbb{N}$ is standard, its unique successor $n + 1$ is standard. As shown above, the set $\mathbb{N}$ (being infinite) has a nonstandard element. Let A be a formula of the language $\{\in, \mathrm{st}\}$ satisfying $A(0)$ and $\forall^{\mathrm{st}} n \in \mathbb{N}(A(n) \Rightarrow A(n+1))$. Suppose that $B = \{\mathrm{st}(n) : n \in \mathbb{N} \wedge \neg A(n)\}$ is not empty. Let $C = {}^{s}B$, then $C \subseteq \mathbb{N} \neq \emptyset$. It has a (unique) minimal element, say m, which by uniqueness must be standard, and then $m \in B$. Also $m = \min(B)$, hence $\neg A(m)$. If $m = 0$, we have $\neg A(0)$, a contradiction. If $m > 0$, we have $A(m-1)$, hence also $A(m) = A((m-1)+1)$, a contradiction. Hence B must be empty, implying that $\forall^{\text{st}} n \in \mathbb{N}\, A(n)$. We conclude that External Induction holds in $\mathbb{N}$. Hence all axioms of ENA$^-$ are a consequence of the axioms of IST. □

Corollary 4.1.3. *The system* ENA$^-$ *is consistent with* ZFC.

4.1.2 External sets

To be able to define neutrices and external numbers, it is convenient to extend Axiom 1.7 to the bounded formulas of IST.

Axiom 4.1. *Let Φ be a bounded formula using the symbols $\in$ and* st. *Then*

$$\forall z \exists Y (x \in Y \Leftrightarrow (x \in z \wedge \Phi(x))).$$

We use the same convention on external sets and pre-external sets as in Axiom 1.7: we let lower-case letters range over internal sets, and upper-case letters range over external sets. Also we will not be pedantic with respect to the use of upper-case letters.

We extend the notions of galaxy and halo of Definition 1.2.3 as follows.

Definition 4.1.2. *A set G of the form $G = \bigcup_{\text{st}(x)\in X} A_x$, where* $\text{st}(X)$ *and $A : X \to \mathcal{P}(\mathbb{R})$ is an internal mapping is called a* pregalaxy *and a* galaxy *if it is external. A set H of the form $H = \bigcap_{\text{st}(y)\in Y} B_y$, where* $\text{st}(Y)$ *and $B : Y \to \mathcal{P}(\mathbb{R})$ is an internal mapping is called a* prehalo *and a* halo *if it is external.*

It will be shown that all external lower halflines are either halos or galaxies (Theorem 4.2.1). So if we extend the definition of neutrices and external numbers to formulas of IST, they are also halos or galaxies, meaning that within the external sets given by Axiom 4.1 they are of the lowest complexity.

Let X be an uncountable standard set. Since every internal set containing only standard elements must be standard and finite, the set ${}^{\sigma}X$ of the standard elements of X cannot be written as a union of internal sets indexed by standard natural numbers. So ${}^{\sigma}X$ is a galaxy which corresponds to an uncountable union. Example 4.1.1 below exhibits a neutrix which is not reduced to a union on the standard natural numbers.

The Fehrele Principle (formula (1.2.5)) can also be extended to this setting.

Let X, Y be sets and $z \subseteq X \times Y$. We denote by $\Pi_X(z)$ the projection of X into z.

Theorem 4.1.4. *No halo is a galaxy.*

Proof. Let $\text{st}(X)$, and $G = \bigcup_{\text{st}(x)\in X} A_x$, where A is a set-valued internal mapping defined on X. Let $\text{st}(Y)$ and $H = \bigcap_{\text{st}(y)\in Y} B_y$, with B a set-valued internal mapping defined on Y. Assume $G \subseteq H$. Let $z \subseteq X \times Y$ be standard finite. Then there exists an internal set I such that $A_x \subseteq I \subseteq B_y$ for all $(x,y) \in z$. Indeed, put $I = \bigcup_{x \in \Pi_X(z)} A_x$. Then $I \subseteq G \subseteq H \subseteq B_y$ for all standard $y \in \Pi_X(z)$. By Idealization there exists an internal set J such that $A_x \subseteq J \subseteq B_y$ for all standard $(x,y) \in X \times Y$. Then $G \subseteq J \subseteq H$, in fact $G \subset J \subset H$. Hence $G \neq H$. □

It is an interesting feature of IST that all external sets E in the sense of Axiom 4.1 are of the form

$$E = \bigcup_{\mathrm{st}(x) \in X} \bigcap_{\mathrm{st}(y) \in Y} I_{xy}, \tag{4.2}$$

where $\mathrm{st}(X)$ and $\mathrm{st}(Y)$ and $I : X \times Y \to Z$ is an internal mapping into some internal set Z. Such external sets are therefore unions on standard indices of halos; it may be shown that they have an alternative representation as intersections on standard indices of galaxies (see (4.5) below).

Remark 4.1.5. We will always assume that external sets are of the form (4.2) and call them *definable in* IST. By construction these sets are defined by a bounded formula; this means that every variable is bounded by a standard set.

The reduction of complexity is a consequence of Nelson's Reduction Algorithm [92]. To start with, for every bounded formula Φ of IST there exists an equivalent bounded formula Ψ of IST, which is of the form

$$\Psi(z) :\equiv \exists^{\mathrm{st}} x \in X \forall^{\mathrm{st}} y \in Y \, \theta(x,y,z), \tag{4.3}$$

where $\mathrm{st}(X)$ and $\mathrm{st}(Y)$ and θ is an internal formula; also z must have values in some standard set Z. Two properties of IST enable this reduction of complexity. Idealization, and its existential contrapositive, permit us to interchange internal and external quantifiers, until all external quantifiers are to the left, leaving an internal formula to the right of them. Standardization guarantees the existence of a standard choice function, as follows.

Let Φ be a possibly external formula such that $\forall^{\mathrm{st}} x \in X \exists^{\mathrm{st}} y \in Y \, \Phi(x,y)$. Observe that $\mathrm{st}(\mathcal{P}(Y))$. For all standard $x \in X$ we define standard $T = T_x \in \mathcal{P}(Y)$ by

$$T_x = {}^{s}\{\mathrm{st}(y) : y \in Y \wedge \Phi(x,y)\}.$$

Let

$$\widetilde{T} = {}^{s}\{\mathrm{st}(x, T_x) : x \in X\}.$$

For all standard $x \in X$ the set T_x is unique; hence $\widetilde{T}$ is a function when restricted to the standard elements of X, and being standard, by Transfer it is a function $\widetilde{T} : X \to \mathcal{P}(Y)$ indeed. By the Axiom of Choice it admits a choice function, and by Transfer it admits a standard choice function $\widetilde{y} : X \to Y$ that assumes standard values $\widetilde{y}(x)$ for standard arguments x. As a consequence

$$\forall^{\mathrm{st}} x \in X \exists^{\mathrm{st}} y \in Y \, \Phi(x,y) \Rightarrow \exists^{\mathrm{st}} \widetilde{y} \in Y^X \forall^{\mathrm{st}} x \in X \, \Phi(x, \widetilde{y}(x)).$$

If the right-hand side of the latter implications holds, clearly the left-hand side also holds, taking $y = \widetilde{y}(x)$ for all standard $x \in X$. Hence we have the equivalence

$$\forall^{\mathrm{st}} x \in X \exists^{\mathrm{st}} y \in Y\, \Phi(x,y) \Leftrightarrow \exists^{\mathrm{st}} \widetilde{y} \in Y^X \forall^{\mathrm{st}} x \in X\, \Phi(x, \widetilde{y}(x)). \tag{4.4}$$

We are now able to give examples of neutrices which are not reduced to a union, or intersection of intervals indexed by the standard natural numbers.

Example 4.1.1. Let $F := \{f : \mathbb{R}^+ \to \mathbb{R}^+, f \text{ increasing}\}$.

1. Let $\omega \simeq +\infty$. Let $G_\omega := \{x : |x| \leq f(\omega), \text{ for some standard } f \in F\}$. Then G_ω is a galaxy. Clearly G_ω is a pregalaxy. Observe that G_ω is bounded by Idealization, and that $f(\omega) \in G_\omega$ implies that $2f(\omega) \in G_\omega$; hence G_ω does not have a supremum. So G_ω must be external; in fact it is a neutrix. We show that G_ω cannot be of the form $\bigcup_{\mathrm{st}(n) \in \mathbb{N}} [0, u_n(\omega)]$ for an internal sequence (u_n) of real numbers. Indeed, for every standard $n \in \mathbb{N}$ there exists standard $f \in F$ such that $u_n < f(\omega)$. By (4.4) there exists a standard sequence (f_n) such that $u_n < f_n(\omega)$ for all standard $n \in \mathbb{N}$. Du Bois-Reymond's Lemma [60] implies that there exists a function g which grows faster than any function f_n, $n \in \mathbb{N}$. By Transfer g may be supposed standard. It follows again from Transfer that $f_n(\omega) < g(\omega) \in G_\omega$.

2. Let $\varepsilon \simeq 0$, $\varepsilon > 0$. Let

$$\mu_\varepsilon := \left\{x : \forall^{\mathrm{st}} f \in F\, (|x| < f(\varepsilon))\right\}.$$

In a sense, the elements of μ_ε are too small to be attained from ε by standard means. We call μ_ε the *functional microhalo* of ε. It is seen analogously that the prehalo μ_ε is an external neutrix, which cannot be written as an intersection of intervals indexed by standard natural numbers.

Assume that we have a formula with all the external quantifiers to the left. By a procedure based on (4.4) all existential external quantifiers can be put to the left, and all universal external quantifiers can be put to the right. Sequences of existential quantifiers may be replaced by quantifications on Cartesian products, and the same is true for sequences of universal quantifiers. Finally, we obtain a formula of the form (4.3). This implies that an external set E may be reduced to the form (4.2). The equivalence (4.4) permits us to prove that we have also the alternative representation

$$E = \bigcap_{\mathrm{st}(x) \in X} \bigcup_{\mathrm{st}(y) \in y} I_{xy}. \tag{4.5}$$

The Saturation Principle of Theorem 4.1.1 can be seen as a counterpart of formula (4.4) in the case where y is internal instead of standard, since it postulates the existence of an internal choice function:

$$\forall^{\mathrm{st}} x \in X \exists y \in Y\, \Phi(x,y) \Leftrightarrow \exists \widetilde{y} \in Y^X \forall^{\mathrm{st}} x \in X\, \Phi(x, \widetilde{y}(x)). \tag{4.6}$$

In the case where Φ is internal, (4.6) follows from Idealization: for a standard finite subset of X, a finite choice function exists, so (I) gives that a choice function exists which is at least defined for all standard elements of X. If Φ is external, by Nelson's Reduction Algorithm it is equivalent to a formula of the form $\forall^{\mathrm{st}} x \in X \exists^{\mathrm{st}} y \in Y \forall u \in U \exists v \in V\, \psi(x,y,u,v)$. The proof of (4.6) now involves repeated quantifier permutation based on (I) and (4.4); we refer to [95] for the details.

We often use Saturation in the following form.

Definition 4.1.3. *Let A be a set and $\leq$ be a binary relation on A. A set $B \subseteq A$ is said to be* cofinal *in A if $\forall a \in A \exists b \in B (a \leq b)$ and is said to be* coinitial *in A if $\forall a \in A \exists b \in B (b \leq a)$.*

Corollary 4.1.6. *Let $H \subseteq \mathbb{N}$ be a convex prehalo, let X be a standard set and $f : X \to \mathcal{P}(\mathbb{N})$ be a definable mapping such that $f(x) \subseteq H$ for all standard $x \in X$.*

1. *If for all standard $x \in X$ there exists $n \in H$ such that $f(x) < n$, then there exists $h \in H$ such that $f(x) < h$ for all standard $x \in X$.*
2. *If $H = \bigcup_{\mathrm{st}(x) \in X} f_x$, then there exists a standard $x \in X$ such that $f(x)$ is cofinal in H.*

Proof. 1. By the Saturation Principle there exists an internal function $\widetilde{n} : X \to \mathbb{N}$ such that $\widetilde{n}(x) \in H$ and $f(x) < \widetilde{n}(x)$ for all standard $x \in X$. By the Fehrele Principle there exists $h \in H$ such that $f(x) < h$ for all standard $x \in X$.

2. This is a direct consequence of Part 1. □

As for "external" sets of external sets, Kanovei and Reeken [71], gave a satisfactory axiomatic system HST (see Appendix A) for external sets, though there remain some problems to be settled. Generalizing the construction of Example 2 we see that there are neutrices of the form $\bigcup_{\mathrm{st}(x) \in X} [-a(x), a(x)]$ for standard ordered sets X of any cardinality, with internal mappings a which are increasing at least for the standard elements of X. This makes that the neutrices do not form a set, but a class, defined by an unbounded formula. As a consequence the external numbers also form a class, hence also functions, sequences, Cartesian products, etc. In order to avoid such problems, we aim to not exceed naive set formation.

Exercises

Exercise 4.1.1. *Indicate whether the following are standard or nonstandard.*

1. $\lim_{n \to \infty} a_n$*, where (a_n) is a standard convergent sequence.*

2. $\{x \in \mathbb{R}^2 : f(x) = x\}$, *where* $f : [0,1]^2 \to [0,1]^2$ *is a standard continuous function.*

3. $\arctan[0,\omega)$, $\sin[0,\omega)$, *where* $\omega \in \mathbb{R}$ *is positive unlimited.*

4. $\{\mathrm{st}(n) \in \mathbb{N} : n \text{ even}\}$.

Exercise 4.1.2. *Let* (u_n) *be a standard sequence of real numbers converging to a real number* a.

1. *Show that* $u_\omega \simeq a$ *for all unlimited indices* ω.

2. *Deduce that* $u_\omega \in a + \oslash$ *for all unlimited indices* ω.

Exercise 4.1.3. *Show that a standard real function is continuous at a standard point, if and only if it is S-continuous at that point.*

Exercise 4.1.4. *Let* F *denote the set of functions* $f : \mathbb{R}^+ \to \mathbb{R}^+ \setminus \{0\}$. *Let* $f \in F$ *be standard. Show that*

1. $f(x) = O(1)$ *for* $x \to 0$ *if and only if* $f(\oslash) \subset \pounds$.

2. $f(x) = o(1)$ *for* $x \to 0$ *if and only if* $f(\oslash) \subseteq \oslash$.

3. $f(x) = o(x)$ *for* $x \to 0$ *if and only if* $f(\varepsilon) \in \oslash\varepsilon$ *for all* $\varepsilon \simeq 0$, $\varepsilon > 0$.

4. $f(x) = o(x^n)$ *for* $x \to 0$, *for all* $n \in \mathbb{N}$, *if and only if* $f(\varepsilon) \in M_\varepsilon$ *for all* $\varepsilon \simeq 0$, $\varepsilon > 0$.

5. *For all* $a > 0$ *there exist* $b, K > 0$ *such that* $|f(x)| \leq Ke^{\frac{-b}{x}}$ *for all* $x \leq a$ *if and only if* $f(\varepsilon) \in m_\varepsilon$ *for all* $\varepsilon \simeq 0$, $\varepsilon > 0$.

Exercise 4.1.5. *Show that there exists a (necessarily nonstandard) internal function* $\phi : \mathbb{R}^+ \to \mathbb{R}^+ \setminus \{0\}$ *such that* $\phi(\varepsilon) \in \mu_\varepsilon$ *for all* $\varepsilon \simeq 0$, $\varepsilon > 0$.

Exercise 4.1.6. *Consider a differential equation of type* $E : \frac{dy}{dt} = f(t,y)$, *where* f *is of class* C^1. *Let* $x \in \mathbb{R}^2$ *be standard, and* λ *be a solution with initial condition* $\lambda(0) = x$. *Is* λ *a standard solution of* E*? Is* λ *standard if* f *is standard?*

Exercise 4.1.7. *Is there a* $1-1$*-continuous curve passing by every standard point of the plane* $\mathbb{R}^2$*?*

Exercise 4.1.8. *Prove that every infinite set has a nonstandard element. Comment on Poincaré's statement on the impossibility of defining infinite sets having exactly the desired elements: "Le loup qu'on a enfermé dans la bergerie".*[2]

Exercise 4.1.9. *Using* $B(x,y) :\equiv x \in y \wedge \mathrm{fin}(y)$, *show Nelson's Theorem: There exists a finite set* F *containing all standard sets. Can* F *be standard?*

[2] The wolf that was locked in the sheepfold.

Exercise 4.1.10. *Wim Luxemburg in "The Strength of Nonstandard Analysis"* [137] *cites Goethe* [56]*:*

> *Willst du ins Unendliche schreiten? Geh nur im Endlichen nach alle Seiten!* [3]

Establish a relation with the Idealization Axiom.

Exercise 4.1.11. *We let ω always be a positive unlimited integer. Determine the following standardizations.*

1. ${}^{s}(0,1)$.
2. ${}^{s}(0,\omega)$.
3. ${}^{s}(\omega,\infty)$.
4. ${}^{s}\{x\in\mathbb{R}:x^2\simeq 2\}$.
5. ${}^{s}\left\{x\in\mathbb{Q}:x^2\simeq 2\right\}$.
6. ${}^{s}\left\{(x,y)\in\mathbb{R}^2 : x\simeq \frac{u}{\omega}\wedge y\simeq \left(1+\frac{u}{\omega}\right)^{\omega}\wedge u\in\{1,...,\omega\}\right\}$.

Exercise 4.1.12.

1. *Prove that Idealization holds in two dimensions as follows: Let X,Y be standard sets. Let $R\subseteq X\times Y$ be an internal relation such that $R(x,y)$ for all standard $x\in X$ and for all standard $y\in Y$. Then there exist internal sets $I\subseteq X$ and $J\subseteq Y$ such that $R(x,y)$ for all $x\in I$ and for all $y\in J$.*
2. *Deduce the Fehrele Principle (Theorem 4.1.4) from Part 1.*

4.2 The nature of halflines, neutrices and external numbers

External halflines in ordered sets have the lowest possible complexity; i.e. they can only be pregalaxies or prehalos. As a consequence the same is true for neutrices, and *a fortiori* for external numbers.

Definition 4.2.1. *Let S be a totally ordered set. Let $\emptyset\neq A,B\subseteq S$. Then A is said to be a* lower halfline *if $x\in A$ and $y<x$ imply that $y\in A$ and B is said to be an* upper halfline *if $x\in B$ and $x<y$ imply that $y\in B$. An ordered pair (A,B) composed of a lower halfline A and an upper halfline B such that $A\cup B=S$ and $A\cap B=\emptyset$ is called a* cut.

[3] Do you want to go into infinity? Go into the finite, in all directions!

Let S be a totally ordered set. Let T be a subset of S and let a be an element of T. We will write T^a to denote the *convex component* of a within T; i.e. T^a is the maximal set with the following property: if $b \leq a$ and $b \leq d \leq a$ then $d \in T^a$ and if $a \leq c$ and $a \leq d \leq c$ then $d \in T^a$.

Theorem 4.2.1. *Let S be a totally ordered standard set. An external halfline of S is either a galaxy or a halo.*

Proof. Let C be a lower halfline of S and let $a \in C$. By Nelson's Reduction Algorithm there exist standard sets X, Y and an internal family $(I_{xy})_{x \in X, y \in Y}$ such that $C = \bigcap_{\text{st}(x) \in X} \bigcup_{\text{st}(y) \in Y} I_{xy}$. For $x \in X$ we define pregalaxies $G_x := \bigcup_{\text{st}(y) \in Y} I_{xy}$. Let $a \in G_x$. Then G_x^a is also a pregalaxy. Indeed, put $A_y := I_{xy}$ for all $y \in Y$ and let Z be the standard set of all finite subsets of Y. Then using Idealization,

$$
\begin{aligned}
u \in G_x^a &\Leftrightarrow [a, u] \subset G_x \\
&\Leftrightarrow [a, u] \subset \bigcup_{\text{st}(y) \in Y} A_y \\
&\Leftrightarrow \exists^{\text{stfin}} z \subset Y([a, u] \subset B_z) \\
&\Leftrightarrow \exists^{\text{st}} z \in Z(u \in B_z^a) \\
&\Leftrightarrow u \in \bigcup_{\text{st}(z) \in Z} B_z^a.
\end{aligned}
$$

Hence $G_x^a = \bigcup_{\text{st}(z) \in Z} B_z^a$. By the convexity of C we have that $C = \bigcap_{\text{st}(x) \in X} G_x^a$.

There are two possibilities. Either there exists some standard $x \in X$ such that $G_x^a \subseteq C$ or $\forall^{\text{st}} x \in X \exists u \in G_x^a \setminus C$. In the first case we conclude that $C = G_x^a$ and since C is external, it is a galaxy. In the second case, for all standard $x \in X$ we may choose an element $u \in G_x^a \setminus C$. By the Saturation Principle (4.6) there exists an internal function $u : X \to S$ such that $u_x \in G_x^a \setminus C$ for all standard $x \in X$. Then $C = \bigcap_{\text{st}(x) \in X} (-\infty, u_x]$ and we conclude that C is a halo. □

Corollary 4.2.2. *Each neutrix N and each external number α is either a pregalaxy or a prehalo.*

Proof. Let $L = \{x \in \mathbb{R} : \exists y \in N(x \leq y)\}$. By Theorem 4.2.1 the halfline L is a pregalaxy or a prehalo, and by symmetry N is of the same nature as L. Let $\alpha = a + A$ with A a neutrix. Then α is a pregalaxy or a prehalo, according to whether A is a pregalaxy or a prehalo. □

Exercises

Exercise 4.2.1. *Give an example of an external convex set which is neither a galaxy nor a halo.*

Exercise 4.2.2. *Give an example of an external halo H of the form $H = \bigcup_{\mathrm{st}(n)\in\mathbb{N}} A_n$ with the sets A_n non-empty convex and two-by-two disjoint.*

4.3 Generalized Dedekind completeness

Every bounded internal subset S of $\mathbb{R}$ has a supremum. The set S defines a bounded lower halfine L which is either of the form $L = (-\infty, a)$ or of the form $L = (-\infty, a]$, for some $a \in \mathbb{R}$; then a is the supremum of L, and of S. We will see that such representation carries over to external lower halflines, provided that they are external sets in the sense of Axiom 4.1. In fact, as far as definable external halflines are concerned, the completeness property holds for arbitrary ordered standard groups. In this section we will assume that all external sets are definable.

Definition 4.3.1. *Let (A, B) be a cut of a totally ordered commutative standard group S. The set $\Delta := \{d \in S : \exists a \in A, b \in B(d = b - a)\}$ is called the* set of differences *for the cut (A, B). The set $\mathrm{T} := \{z \in S : \forall d \in \Delta(|z| < d)\}$ is called the* thickness of the border *associated with the cut (A, B).*

Lemma 4.3.1. *Let S be a totally ordered commutative standard group. Let G be a galaxy and H be a halo. Let T be the thickness of the border associated with the cut (G, H) of S. Then:*

1. *$G = \bigcup_{g\in G}(g + \mathrm{T})$;*
2. *$H = \bigcup_{h\in H}(h + \mathrm{T})$;*
3. *T is a convex subgroup of S.*

Proof. Part 1 and Part 2 are evident so we will only show Part 3. We have that $\mathrm{T} \neq \emptyset$ for $0 \in \mathrm{T}$. Clearly T is convex. Let $s, t \in \mathrm{T}$ and $g \in G$. Then $g + |s| \in G$, so $g + (|s| + |t|) = (g + |s|) + |t| \in G$. Hence $|s| + |t| \in \mathrm{T}$. Then $s - t \in \mathrm{T}$ because $|s - t| \leq |s| + |t|$. Hence T is a convex subgroup of S. □

Lemma 4.3.2. *Let S be a totally ordered commutative standard group. Let G be a galaxy and H be a halo. Let T be the thickness of the border associated with the cut (G, H) of S. Let X be standard and $f : X \to S$ be an internal function such that $G = \bigcup_{\mathrm{st}(x)\in X}(-\infty, f(x)]$. Then either there exists an internal function $\varphi : X \to S$ such that $\mathrm{T} = \bigcup_{\mathrm{st}(x)\in X}[-\varphi(x), \varphi(x)]$, or there exists an internal function $\psi : X \to S$ such that $\mathrm{T} = \bigcap_{\mathrm{st}(x)\in X}[-\psi(x), \psi(x)]$ and $\psi(x) \in \Delta$ for all standard $x \in X$.*

Proof. Suppose first that G is of the form $(-\infty, a] \cup (a + \mathrm{T})$, for some $a \in G$. Define the internal mapping $\varphi : X \to S$ by

$$\varphi(x) = \begin{cases} f(x) - a, & a \leq f(x) \\ 0, & \text{otherwise.} \end{cases}$$

Then $\mathrm{T} = \bigcup_{\mathrm{st}(x)\in X}[-\varphi(x), \varphi(x)]$. Assume now that there is no $a \in G$ such that $(-\infty, a] \cup (a + \mathrm{T})$. Then for all $u \in G$ there is some $v \in G$ such that $v > u$ and $v - u \notin \mathrm{T}$, i.e. $v - u \in \Delta$. In particular, for all standard $x \in X$ we may choose standard $y \in X$ such that $f(y) - f(x) \in \Delta$. By Standardization we may assume that $y = y(x)$, where the mapping $y : X \to X$ is standard. Let $\psi : X \to S$ be defined by $\psi(x) = f(y(x)) - f(x)$. Then $\psi(x) \in \Delta$, for all standard $x \in X$. Clearly $\mathrm{T} \subseteq \bigcap_{\mathrm{st}(x)\in X}[-\psi(x), \psi(x)]$, for all standard $x \in X$ and for $\psi(x) \in \Delta$. We show that also $\bigcap_{\mathrm{st}(x)\in X}[-\psi(x), \psi(x)] \subseteq \mathrm{T}$. Assume that $|z| \leq \psi(x)$ for all standard $x \in X$. Let $g \in G$ and let $\xi \in X$ be standard and such that $g < f(\xi)$. Then

$$g + |z| \leq f(\xi) + \psi(\xi) = f(\xi) + (f(y(\xi)) - f(\xi)) = f(y(\xi)) \in G.$$

Hence $z \in \mathrm{T}$. So $\bigcap_{\mathrm{st}(x)\in X}[-\psi(x), \psi(x)] \subseteq \mathrm{T}$. We conclude that

$$\bigcap_{\mathrm{st}(x)\in X}[-\psi(x), \psi(x)] = \mathrm{T}.$$

□

Theorem 4.3.3 (Generalized Dedekind completeness). *Let S be a totally ordered standard group. Let (G, H) be a cut of S into a galaxy G and a halo H. Then there exists a unique convex subgroup K of S such that either $G = (-\infty, a] \cup (a + K)$ and K is a galaxy or $G = (-\infty, a] \setminus (a + K)$ and K is a halo.*

Proof. We start by showing the existence. Suppose that there exists $a \in G$ such that $G = (-\infty, a] \cup (a + K)$. Then K is a convex group by Lemma 4.3.1 and a pregalaxy by Lemma 4.3.2. Furthermore, K is external for G is external. Hence K is a galaxy. Now suppose that G is not of the form $(-\infty, a] \cup (a + K)$. By Lemma 4.3.2 there exists an internal function $\psi : X \to S$ such that $K = \bigcap_{\mathrm{st}(x)\in X}[-\psi(x), \psi(x)]$ and $\psi(x) \in \Delta$ for all standard $x \in X$. So K is a prehalo. Furthermore, for each standard $x \in X$ we may choose $h \in H$ such that $h - \psi(x) \in G$. By Saturation we may assume that $h = h(x)$, where $h : X \to S$ is internal. Because the galaxy G must be strictly included in the prehalo $\bigcap_{\mathrm{st}(x)\in X}(-\infty, h(x)]$, there exists some $a \in H$ such that $a < h(x)$, for all standard $x \in X$. We show that $G = (-\infty, a] \setminus (a + K)$. By Part 2 of Lemma 4.3.1 we have that $a + K \subseteq H$, so $G \subseteq (-\infty, a] \setminus (a + K)$. On the other hand, suppose that $u < a + z$, for all $z \in K$. Then $u \leq a - \psi(\xi)$ for some standard $\xi \in X$. Now, $a - \psi(\xi) < h(\xi) - \psi(\xi) \in G$ so $u \in G$, implying that $(-\infty, a] \setminus (a + K) \subseteq G$. Hence $G = (-\infty, a] \setminus (a + K)$. This implies that K is external, and thus is a halo.

We now show uniqueness. It follows from Lemma 4.3.1 that there cannot exist K', a convex subgroup of S such that $K' \subset K$. Let K'' be a convex subgroup of S such that $K \subset K''$. Let d be a positive element of $K'' \setminus K$. Then $d \in \Delta$. Let $g \in G$ and $h \in H$ be such that $g + d = h$. Suppose that G was of the form $(-\infty, a] \cup (a + K'')$. Then $a < g$, so $g - a \in K''$ and also $g - a + d \in K''$, for K'' is a group. But then $a + (g - a + d) = h \in H$, a contradiction. Hence G is not of the form $(-\infty, a] \cup (a + K'')$. Using a similar argument one proves that G cannot be of the form $(-\infty, b] \setminus (b + K'')$ for some $b \in H$. This shows uniqueness.

Finally we show incompatibility. Suppose that G is both of the form $(-\infty, a] \setminus (a + K_1)$ and the form $(-\infty, b] \cup (b + K_2)$; then $K_1 = K_2$ by uniqueness. Hence K is both a galaxy and a halo, in contradiction with Theorem 4.1.4. □

Corollary 4.3.4 (Theorem 4.34 of [130])**.** *Let S be a totally ordered standard group. Let L be an external lower halfline. Then there exists a unique convex subgroup K of S such that either $L = (-\infty, a] \cup (a + K)$ or $L = (-\infty, a] \setminus (a + K)$.*

Proof. By Theorem 4.2.1 every external lower halfline is either a galaxy or a halo. Then the result follows from Theorem 4.3.3. □

The following corollary gives a *normal form* for convex subsets of $\mathbb{R}$, and in particular for neutrices. This may be seen as a generalization of the usual Dedekind completeness of the real numbers system. Given a neutrix N, we let $N^+ := \{x \in N : x \geq 0\}$ and $N^- := \{x \in N : x \leq 0\}$.

Corollary 4.3.5. *Let C be a convex subset of $\mathbb{R}$. Then there exist two unique neutrices K, M and two real numbers a, b such that C takes exactly one of the following forms:*

1. $C = [a, b] \cup (a + M^-) \cup (b + K^+)$;
2. $C = [a, b] \cup (a + M^-) \setminus (b + K^-)$;
3. $C = ([a, b] \setminus (a + M^+)) \cup (b + K^+)$;
4. $C = ([a, b] \setminus (a + M^+)) \setminus (b + K^-)$.

Proof. If C is internal it follows from the Dedekind completeness of $\mathbb{R}$ and the fact that $\mathbb{R}$ itself is a neutrix. If C is external it follows from Corollary 4.3.4. □

For general subsets of $\mathbb{R}$ we may define suprema and infima as follows.

Definition 4.3.2. *Let $S \subseteq \mathbb{R}$ and α be an external number. We say that α is the* weak supremum *of S if $\forall x \in S \exists y \in \alpha (x \leq y)$ and if $\beta \in \mathbb{E}$ has the same property, one has $\alpha \leq \beta$. We write $\alpha := \mathrm{zup}(S)$. We say that α is the* weak infimum *of S if $\forall x \in S \exists y \in \alpha (y \leq x)$ and if $\beta \in \mathbb{E}$ has the same property, one has $\beta < \alpha$ or $\alpha \subseteq \beta$. We write $\alpha := \mathrm{winf}(S)$.*

Definition 4.3.3. *Let $S \subseteq \mathbb{R}$ be non-empty. The* lower halfline L_S defined by S *is given by* $L_S := \{x \in \mathbb{R} : \exists s \in S(x \leq s)\}$ *and the* upper halfline U_S defined by S *is given by* $U_S := \{x \in \mathbb{R} : \exists s \in S(x \geq s)\}$.

Proposition 4.3.6. *Let $S, T \subseteq \mathbb{R}$.*

1. *If S, T are cofinal, then* $\text{zup}(S) = \text{zup}(T)$ *and in particular* $\text{zup}(L_S) = \text{zup}(S)$.

2. *If S, T are coinitial, then* $\text{winf}(S) = \text{winf}(T)$ *and in particular* $\text{winf}(U_S) = \text{winf}(S)$.

Proof. We give the proof of Part 1. The proof of Part 2 is analogous. Let $\beta \in \mathbb{E}$ be such that

$$\forall x \in S \exists y \in \beta(x \leq y). \tag{4.7}$$

Let $t \in T$. Then there exists $s \in S$ such that $t \leq s$. Hence

$$\forall x \in T \exists y \in \beta(x \leq y). \tag{4.8}$$

In the same way one proves that (4.8) implies (4.7). We conclude that $\text{zup}(S) = \text{zup}(T)$, and in particular $\text{zup}(L_S) = \text{zup}(S)$. □

Theorem 4.3.7. *Let $S \in \mathbb{R}$. Then S has a weak infimum and a weak supremum.*

Proof. Let C be the convex closure of S. Then $C = L_S \cap U_S$. Then $\text{winf}(S) = \text{winf}(U_S) = \text{winf}(C)$ and $\text{zup}(S) = \text{zup}(L_S) = \text{zup}(C)$. Let a, b be the real numbers and K, M be the neutrices given by Corollary 4.3.5. Then $a + K = \text{winf}(S)$ and $b + M = \text{zup}(S)$. □

The next propositions are, like Proposition 4.3.6, useful to determine external extrema in practice.

Proposition 4.3.8. *Let f be a bounded internal function, at least defined on some set $C \subseteq \mathbb{R}$. Put $\mu := \text{zup}_{x \in C} f(x)$. Assume $\mu \subseteq L_{\text{Im}(f)}$. Then $(-\infty, \mu] = \bigcup_{c \in C}(-\infty, f(c)]$.*

Proof. Let $\mu = m + M$. Then $f(c) \geq m$ for some $c \in C$. Because $m \in \mu$ is arbitrary one has $(-\infty, \mu] = \bigcup_{c \in C}(-\infty, f(c)]$. □

Proposition 4.3.9. *Let $a, b \in \mathbb{R}$ with $a < b$ and $\gamma \in \mathbb{E}$ such that $a < \gamma < b$. Let $f : [a, b] \to \mathbb{R}$ be a continuous non-decreasing internal function. Then $\sup_{a \leq \gamma} f(x) = \inf_{\gamma < x \leq b} f(x)$.*

Proof. See Exercise 4.3.4. □

Exercises

Exercise 4.3.1. *Write $\{x \in \mathbb{R} : 0 \lnsim x \lesssim y\}$ as an interval bounded by two external numbers.*

Exercise 4.3.2. 1. *Write @ in the form* $C = ([a,b] \setminus (a+M^+)) \cup (b+K^+)$ *with* $a, b \in \mathbb{R}$ *and* M, K *neutrices.*

2. *Show that @ cannot be written in the form* $C = ([a,b] \setminus (a+M)) \cup (b+K)$ *with* $a, b \in \mathbb{R}$ *and* M, K *neutrices.*

3. *Give an external set* S *such that* $\operatorname{zup}(S) < \operatorname{winf}(S)$.

Exercise 4.3.3. *Let* $S \subseteq \mathbb{R}$ *be an external interval.*

1. *Show that* $\operatorname{zup}(S) = \operatorname{winf}(-S)$.

2. *Show that* $\operatorname{zup}(S) - \operatorname{winf}(S) = \operatorname{zup}\{t - s : s, t \in S\}$.

Exercise 4.3.4. *Prove Proposition 4.3.9.*

4.4 Flexible sequences and functions

Flexible functions are definable external functions over the external numbers. The first subsection deals with the form flexible functions can take, with applications to suprema and infima in the sense of the generalized completeness theorem. The second subsection deals with the special case where the domain is $\mathbb{N}$. In the particular case of sequences of real numbers we prove a representation theorem.

4.4.1 Flexible functions

Definition 4.4.1. *Let* $\emptyset \neq X \subseteq \mathbb{R}$ *be an external set. A* flexible function *is a mapping* $F : X \to \mathbb{E}$ *such that its graph* $\Gamma(F) \subseteq X \times \mathbb{R}$ *is definable; i.e. it is of the form*

$$\Gamma(F) = \bigcup_{\operatorname{st}(u)\in U} \bigcap_{\operatorname{st}(v)\in V} I_{uv},$$

where U, V *are standard sets and* $I : U \times V \to X \times \mathcal{P}(\mathbb{R})$ *is an internal set-valued mapping. A flexible function* F *is called* precise *if* $F(x) \in \mathbb{R}$ *for all* x. *A* representative *of* F *is a precise function* f *such that* $f(x) \in F(X)$ *for all* $x \in X$. *The function* $N_F : X \to \mathbb{E}$ *defined by* $N_F(x) = N(F(x))$ *for all* $x \in X$ *is called the* neutrix part *of* F*; we will assume that* $N(F(x)) \neq \mathbb{R}$ *for every* $x \in X$.

A flexible function $F : X \to \mathbb{E}$ *is* internally representable *if*

$$\Gamma(F) = \bigcup \{f : X \to \mathbb{R} : f \text{ is an internal representative of } F\}.$$

Notice that N_F is a flexible function, for $N_F = F - F$ is definable.

Example 4.4.1. 1. Let $F_0 : \mathbb{R} \to \mathbb{E}$ be the mapping defined by $F_0(x) = x + x\oslash$. Then F_0 is a flexible function, for its graph has the representation

$$\Gamma(F_0) = \bigcap_{\mathrm{st}(n)\in\mathbb{N}} \left\{ \{x\} \times \left[x - \frac{1}{n}, x + \frac{1}{n}\right] : x \in \mathbb{R} \right\}.$$

The function F_0 admits the identity function as a representative and its neutrix part is given by the function $N_{F_0} : \mathbb{R} \to \mathbb{E}$ defined by $N_{F_0}(x) = x\oslash$.

2. Let $f : \mathbb{R} \to \mathbb{R}$ be an internal function and ε be a fixed infinitesimal. The functions F_1, F_2 defined by $F_1 = f + \varepsilon \pounds$ and $F_2 = (1 + \oslash)f$ are other examples of flexible functions that are internally representable.

3. It follows from the Fehrele Principle that the flexible function $F_3 : \mathbb{R}^+ \to \mathbb{E}$, defined by $F_3(x) = \oslash$ for limited x and $F_3(x) = 1 + \oslash$ for unlimited x, is not internally representable. The function F_3 can be expressed in the following way:

$$F_3 = \left(\bigcup_{\mathrm{st}(m)\in\mathbb{N}} \bigcap_{\mathrm{st}(n)\in\mathbb{N}} \left\{ [0, m] \times \left[-\frac{1}{n}, \frac{1}{n} \right] \right\} \right) \cup \left(\bigcap_{\mathrm{st}(k)\in\mathbb{N}} \left\{ [k, +\infty) \times \left[1 - \frac{1}{k}, 1 + \frac{1}{k} \right] \right\} \right).$$

One may use Nelson's Reduction Algorithm to express $\Gamma(F_3)$ with only one intersection and one union, which implies that F_3 is a flexible function. Observe that F_3 does not have an internal representative; indeed such a function would be infinitely close to 0 for all limited x, so by the Fehrele Principle up to some unlimited $\omega > 0$, but then it must be infinitely close to 1, a contradiction.

4. Let ε be a positive infinitesimal and let $F_4 : \mathbb{R} \to \mathbb{E}$ be defined by

$$F_4(x) = \begin{cases} e^x + \varepsilon \pounds, & x \in \pounds \\ x^2 + \varepsilon \oslash e^x, & \text{otherwise.} \end{cases}$$

One shows that F_4 is a flexible function which is not internally representable as in the previous example. See Exercise 4.4.2.

If internal representatives of a flexible function have a common property, like being integrable, continuous, S-continuous, or Lipschitz, with some abuse of language we will call the internally representable flexible function F respectively integrable, continuous, S-continuous, Lipschitz, as well. Because we are strongly motivated by approximations and perturbations, we restrict our study to flexible functions F with a (definable) representative f and then $F(x) = f(x) + N_F(x)$ for all $x \in X$.

As in the case of the functions F_1, F_2 and F_3 of Example 4.4.1, many times a flexible function F comes from an internal function f, where, say, a standard number k of complicated explicit terms with narrow range are "neglected", in the sense that such terms are replaced by external numbers containing its range. Then F may be identified with an internal function g of $k+1$ variables, with k variables taking the constant value of an external number, or even an internal function h of $k+1$ variables, with k variables taking the value of a neutrix, which is the neutrix part of such an external number. For instance, let $\varepsilon \simeq 0$ and $f : [0,1] \to \mathbb{R}$ be defined by $f(x) = \frac{e^{-x}}{1+\varepsilon \sin(x)}$. Neglecting infinitesimals we have the simple approximation $f(x) \in F(x)$, with $F : [0,1] \to \mathbb{R}$ defined by $F(x) = (1+\oslash)e^{-x}$. We may define the internal function $g : [0,1] \times \mathbb{R}$ by $g(x,y) = ye^{-x}$. Then $G(x) = g(x, 1+\oslash)$; with $h : [0,1] \times \mathbb{R}$ by $h(x,y) = (1+y)e^{-x}$ we have $G(x) = h(x, \oslash)$.

Definition 4.4.2. *Let $k \in \mathbb{N}$ be standard and $f : \mathbb{R}^k \to \mathbb{R}$ be internal. Let $\alpha_1, ..., \alpha_k$ be external numbers. We define*

$$f(\alpha_1, ..., \alpha_k) := \{f(a_1, ..., a_k) : a_1 \in \alpha_1 \wedge \cdots \wedge a_k \in \alpha_k\}.$$

Definition 4.4.3. *Let $k \in \mathbb{N}$ be standard and $f : \mathbb{R}^{k+1} \to \mathbb{R}$ be internal. Let $\alpha_1, ..., \alpha_k$ be external numbers. We define*

$$\Gamma := \bigcup \{f(x, a_1(x), ..., a_k(x)) : a_1 : \mathbb{R} \to \alpha_1 \wedge \cdots \wedge a_k : \mathbb{R} \to \alpha_k \ \textit{internal}\}.$$

If Γ is the graph of a flexible function F, i.e. $F(x) \in \mathbb{E}$ for all $x \in \mathbb{R}$, we denote F by $x \mapsto F(x) := f(x, \alpha_1, ..., \alpha_k)$.

Let F given by $x \mapsto F(x) := f(x, \alpha_1, ..., \alpha_k)$ be a flexible function in the sense of Definition 4.4.3. If F is internally representable one may as well consider flexible functions of the form $x \mapsto F(x) := f(x, A_1, ..., A_k)$, where $A_1, ..., A_k$ are neutrices. With this convention, the next proposition states that the graph of a flexible function F is the union of the graphs of representatives taking constant values in $A_1, ..., A_k$.

Proposition 4.4.1. *Let $a, b \in \mathbb{R}, a < b$. Let $k \in \mathbb{N}$ be standard, $A_1, ..., A_k > 0$ be bounded neutrices and f be an internal real function of $k+1$ variables, defined at least on a domain $[a,b] \times D$ with $D \supset A_1 \times \cdots \times A_k$ internal. Assume that $F(x) := f(x, A_1, ..., A_k) \in \mathbb{E}$ for all $x \in [a,b]$ is given as in Definition 4.4.3. For $(x, a_1, ..., a_k) \in D$ define $f_{a_1,...,a_k}(x) = f(x, a_1, ..., a_k)$. Then*

$$\Gamma(F) = \bigcup_{a_1 \in A_1, ..., a_k \in A_k} f_{a_1,...,a_k}. \tag{4.9}$$

Proof. Clearly $\bigcup_{a_1 \in A_1,\dots,a_k \in A_k} \Gamma(f, a_1, \dots, a_k) \subseteq \Gamma(F)$. Conversely, let $x \in [a, b]$ and $y \in F(x)$. Then there exist $a_1 \in A_1, \dots, a_k \in A_k$ such that $y = f(a_1, \dots, a_k)$. Hence $\Gamma(F) \subseteq \bigcup_{a_1 \in A_1,\dots,a_k \in A_k} f_{a_1,\dots,a_k}$. Combining, we obtain (4.9). □

It depends on the context whether the use of non-constant representatives of an inner continuous flexible function ϕ is useful or not. If it is, then we may use the notation $\phi(x) := f(x, A_1(x), \dots, A_k(x))$ for $x \in [a, b]$. We will see in Chapters 9 and 10 that solutions of differential equations and recurrent relations with external parameters may be defined similarly as above, and that in many cases constant representatives suffice.

In the remaining part of this subsection we deal with operations on weak suprema and infima for flexible functions.

Proposition 4.4.2. *Let N be a neutrix function, at least defined on some set $C \subseteq \mathbb{R}$. Put $\mu := \operatorname{zup}_{x \in C} N(x)$. Then μ is a neutrix, in fact $\mu = \bigcup_{c \in C} N(c)$.*

Proof. Let $S = \bigcup_{c \in C}(-\infty, N(c)]$. Then $\mu = \operatorname{zup}(S)$, and is of the form $\mu = m + M$ with $m \in \mathbb{R}$ and M a neutrix. We show that $\mu = M$. Note that if $s \in S$, for some $c \in C$ we have $s \in N(c)$ hence $2s \in N(c)$. Now $S = (-\infty, \mu[$ is impossible, for then $\frac{\mu}{2} \in S$, hence $\mu = \frac{2\mu}{2} \in S$. So $\mu \subseteq S$. This means that $2m \in m + M$, i.e. $m \in M$. Hence $\mu = M$. Clearly $\mu = \bigcup_{c \in C} N(c)$. □

Proposition 4.4.3. *Let F be a non-negative flexible function, at least defined on some set $C \subseteq \mathbb{R}$. Let N be a neutrix. Then*

$$\operatorname*{zup}_{c \in C} NF(c) = N \operatorname*{zup}_{c \in C} F(c). \tag{4.10}$$

Proof. Let $\mu = m + M := \operatorname{zup}_{c \in C} F(x)$ and $\nu := \operatorname{zup}_{c \in C} NF(c)$. By Proposition 4.4.2 we have $\nu = \bigcup_{c \in C} NF(c)$. Assume first that $L_F = (-\infty, \mu]$. Then

$$NL_F = N\left(-\infty, \bigcup_{c \in C} F(c)\right] = \left(-\infty, \bigcup_{c \in C} NF(c)\right] = (-\infty, \nu],$$

so (4.10) holds. Secondly, assume that $L_F = (-\infty, \mu[$. Then μ is zeroless. Also $\frac{m}{2} \leq F(c)$ for some $c \in C$. Then $N\mu = N\frac{m}{2} = NF(c) = \nu$, so we derive (4.10) again. □

Proposition 4.4.4. *Let F be a flexible function defined at least on some set $C \subseteq \mathbb{R}$ of the form $F = f + N$ with f internal and N a neutrix function. Put $\lambda := \operatorname{zup}_{c \in C} f(c)$. Assume $\lambda \subseteq L_{\mathrm{Im}(f)}$. Put $\mu := \operatorname{zup}_{c \in C} N(c)$. Then $\operatorname{zup}_{c \in C} F(c) = \lambda + \mu$ and $(-\infty, \lambda + \mu] = \bigcup_{c \in C}(-\infty, F(c)]$.*

Proof. Because F is a flexible function, the representative f is bounded. Then the result follows from Propositions 4.3.8 and 4.4.2. □

Theorem 4.4.5. *Let F be a flexible function, at least defined on some set C, of the form $F = f + N$ with f internal and N a neutrix function. Let $\Phi := \operatorname{zup}_{c \in C} F(c), \sigma = s + S := \operatorname{zup}_{c \in C} f(c)$ and $\nu := \operatorname{zup}_{c \in C} N(c)$. Then $\Phi = \phi + \nu$.*

Proof. Assume first that $\nu \subseteq S$. Then $L_f = L_F$. Indeed, $L_f \subseteq L_F$, because $f(c) \leq F(c)$ for all $c \in C$. On the other hand, let $x \in L_F$. By Proposition 4.4.2 there exists $c \in C$ such that $x \leq f(c) + N(c) \leq f(c) + \nu \leq f(c) + S$. Because L_f is stable under addition by S, we see that $f(c) + S \subseteq L_f$, so $x \in L_f$. Hence $L_f = L_F$. We conclude that $\Phi = \operatorname{zup} L_F = \operatorname{zup} L_f = \sigma = \sigma + \nu$.

Secondly, assume that $S \subset \nu$. Let $d \in \nu$ be such that $S < d$. By Proposition 4.4.2 there exists $c \in C$ be such that $d \leq N(c)$. Let $c \in C$ be such that $\sigma - d < f(c)$. Then $\sigma + \nu = f(c) + \nu$; hence L_F is stable under addition by ν, implying that $L_F = L_{f+\nu} = L_f + \nu$. We conclude that

$$\Phi = \operatorname{zup} L_F = \operatorname{zup}(L_f + \nu) = \operatorname{zup}(L_f) + \nu = \sigma + \nu. \qquad \square$$

Proposition 4.4.6. *Let f, g be two external functions defined at least on some convex set $C \in \mathbb{R}$. Let $\alpha := \sup_{x \in C} g(x)$ and $\beta := \inf_{x \in C} g(x)$.*

1. *Assume $g(x) - f(x) \subseteq N(\alpha)$ for all $x \in C$. Then $\operatorname{zup}_{x \in C} f(x) \subseteq \alpha$.*
2. *Assume $g(x) - f(x) \subseteq N(\beta)$ for all $x \in C$. Then $\operatorname{winf}_{x \in C} f(x) \subseteq \beta$.*

Proof. We prove only Part 1. The proof of Part 2 is analogous. We may assume that $\alpha = A$. Put $\gamma := \operatorname{zup}_{x \in C} f(x)$ and $F := \{y \in \mathbb{R} : \exists x \in C(y \leq f(x))\}$. Suppose that $\gamma \not\leq A$. Let $t \in \mathbb{R}, A < t < \gamma$. If $F = (-\infty, \gamma]$ there exists $x \in C$ such that $t < f(x)$. Because $g(x) - f(x) \subseteq A$, then necessarily $A < g(x)$, a contradiction. If $F = (-\infty, \gamma)$ there exists $x \in C$ such that $-t < g(x)$. Because $g(x) - f(x) \subseteq A$, we must have $-t < f(x)$, a contradiction. Hence $\gamma \leq A$.

Suppose that $\gamma < A$. Let $s \in \mathbb{R}, \gamma < s < A$. Then there exists $x \in C$ such that $s < g(x)$. Because $g(x) - f(x) \subseteq A$, necessarily $\gamma < f(x)$, a contradiction. Combining, we obtain that $\gamma \subseteq A$. $\square$

4.4.2 Flexible sequences

We deal with flexible sequences which are special cases of flexible functions with domain $\mathbb{N}$. We prove a Representation Theorem for precise sequences. General flexible sequences admit an internal partial choice sequence. This is also true for sequences defined on an (external) initial segment, and then the domain of such a choice function enters in the complement of the segment, giving rise to the notion of *internal bridge* of Chapter 5. As we will see, these internal choice sequences are a tool in proving convergence theorems, for they satisfy classical notions of convergence.

Definition 4.4.4. *A* flexible *sequence (u_n) is a flexible function with domain $X = \mathbb{N}$.*

A flexible sequence (u_n) is called *precise*, if $u_n \in \mathbb{R}$ for all $n \in \mathbb{N}$. Being definable its graph $\Gamma(u) := \{(n, y) : n \in \mathbb{N} \wedge y \in u_n\}$ has the form

$$\Gamma(u) = \bigcup_{\operatorname{st}(x) \in X} \bigcap_{\operatorname{st}(y) \in Y} I_{xy}, \tag{4.11}$$

where X and Y are standard sets and $I : X \times Y \to \mathcal{P}(\mathbb{N} \times \mathbb{R})$ is an internal mapping.

Sometimes it is useful to suppose that for fixed $x \in X$, the sets I_{xy} of (4.11) are closed under finite intersections. Then

$$\forall^{\mathrm{stfin}} z \subseteq Y \exists^{\mathrm{st}} \overline{y} \in Y \left(\bigcap_{\mathrm{st}(y) \in z} I_{xy} = I_{x\overline{y}} \right). \tag{4.12}$$

Assume that (u_n) is a precise sequence. In (4.11) put, for all standard $x \in X$, $Q_x = \bigcap_{\mathrm{st}(y) \in Y} I_{xy}$. Then

$$u = \bigcup_{\mathrm{st}(x) \in X} Q_x. \tag{4.13}$$

Since (u_n) is a function, for all standard $x \in X$ the prehalo Q_x is a function on its domain. For every standard $x \in X$, we define the prehalo Δ_x by

$$\Delta_x := \mathrm{dom}(Q_x). \tag{4.14}$$

If u_n is a pregalaxy, at this n the representation (4.11) can be refined. We will use the functional notation also for $I_{xy}(n)$, $Q_x(n)$, etc. to indicate values taken at n.

Proposition 4.4.7. *Let (u_n) be a flexible sequence. Let $n \in \mathbb{N}$. Assume that (4.12) holds and that u_n is a pregalaxy. Then there exists a standard $\overline{y} \in Y$ such that*

$$u_n = \bigcup_{\mathrm{st}(x) \in X} I_{x\overline{y}}(n). \tag{4.15}$$

Proof. Suppose that $Q_x(n)$ is cofinal in u_n for some standard $x \in X$. If its convexification $\overline{Q_x(n)}$ is a halo, we have $\overline{Q_x(n)} = u_n$, in contradiction with the Fehrele Principle. So $\overline{Q_x(n)}$ is internal; hence there exist $s, t \in u_n$ such that $Q_x(n) \subseteq \overline{Q_x(n)} \subseteq [s,t]$. By Idealization there exists standard $\overline{y} \in Y$ such that $I_{x\overline{y}}(n) \subseteq [s,t] \subset u_n$. So $u_n \subseteq \bigcup_{\mathrm{st}(x) \in X} I_{x\overline{y}}(n) \subseteq \bigcup_{\mathrm{st}(x) \in X} \bigcap_{\mathrm{st}(y) \in Y} I_{xy}(n) \subseteq u_n$, which implies (4.15). □

The next lemma expresses that sequences which are either a galaxy or a halo are internal on every internal subdomain, and have an internal extension.

Lemma 4.4.8. *Let E be a non-empty subset of $\mathbb{N}$ and $J \subseteq E$ be internal. Let $u : E \to \mathbb{R}$ be a sequence, which is either a galaxy or a halo. Then $u_{\restriction J}$ is internal, and there exist K such that $E \subseteq K \subseteq \mathbb{N}$ and an internal sequence $v : K \to \mathbb{R}$ such that $v_{\restriction E} = u$.*

Proof. We consider first the case where u is a galaxy. In this case we may suppose that the sets Q_x of formula (4.13) are internal. Then also the sets Δ_x defined by (4.14) are internal. Let $J \subseteq E$ be internal. Now, $\forall k \in J \exists^{\mathrm{st}} x \in$

$X(k \in \Delta_x)$. So by Idealization $\exists^{\text{stfin}} z \subseteq X \forall k \in J \exists^{\text{st}} x \in z(k \in \Delta_x)$. Then $J \subseteq \bigcup_{x \in z} \Delta_x$. For $x \in z$, put $J_x = \Delta_x \cap J$. Then J_x is internal; hence also $u_{\restriction J_x} = Q_{x \restriction J_x}$ is internal. Because $J = \bigcup_{x \in z} J_x$, we have $u_{\restriction J} = \bigcup_{x \in z} u_{\restriction J_x}$, which is internal as it is a standard finite union of internal sets.

To construct an internal extension of u, we note that by the above

$$\forall^{\text{stfin}} z \subseteq X \exists v \forall x \in z$$
$$(v : \mathbb{N} \to \mathbb{R} \text{ is a partial function such that } \forall n \in \Delta_x (v_n = Q_{x \restriction J_x}(n)).$$

By Idealization

$$\exists v \forall^{\text{st}} x \in X$$
$$(v : \mathbb{N} \to \mathbb{R} \text{ is a partial function such that } \forall n \in \Delta_x (v_n = Q_{x \restriction J_x}(n)).$$

Put $K = \text{dom}(v)$. Then v is an internal real-valued function, such that $v_n = u_n$ for all $n \in E$, and $E \subseteq K \subseteq \mathbb{N}$.

We turn now to the case where u is a halo. We construct first an internal extension. We may write $u = \bigcap_{\text{st}(y) \in Y} I_y$, where Y is a standard set and $I : Y \to \mathcal{P}(\mathbb{N} \times \mathbb{R})$ is an internal mapping. Put for every $n \in \mathbb{N}$ and $y \in Y$

$$I_y(n) = \{a \in \mathbb{R} : (n, a) \in I_y\}.$$

Let $n \in \mathbb{N}$ be given, and suppose that $\forall^{\text{stfin}} z \subseteq Y \exists a \in \mathbb{R}(a \in I_y(n) \wedge a \neq u_n)$. By Idealization $\exists a \in \mathbb{R} \forall^{\text{st}} y \in Y(a \in I_y(n) \wedge a \neq u_n)$. Then $(n, a) \in u$, in contradiction with the fact that u is a function on E. We conclude that there exists a standard finite $z \subseteq Y$ such that

$$u_n = \bigcap_{y \in z} I_y(n). \tag{4.16}$$

For arbitrary standard finite $z \subseteq Y$ put

$$w_z := \left\{ \left(n, \bigcap_{y \in z} I_y(n) \right) : n \in \mathbb{N} \wedge \left| \bigcap_{y \in z} I_y(n) \right| = 1 \right\}.$$

Also we define $w := \bigcup \{w_z : z \subseteq Y \text{ standard finite}\}$. Let $n \in E$. By (4.16) for some standard finite $z \subseteq Y$ we have $u_n = \bigcap_{y \in z} I_y(n)$. Then $w_z(n) = u_n$. Let $z' \subseteq Y$ be also standard finite and such that $\left| \bigcap_{y \in z'} I_y(n) \right| = 1$. Now $u_n = \bigcap_{\text{st}(y) \in Y} I_y(n) \subseteq \bigcap_{y \in z'} I_y(n)$, and then $u_n = \bigcap_{y \in z'} I_y(n)$ because u is a function. Hence $w_{z'}(n) = \bigcap_{y \in z'} I_y(n) = u_n = w_z(n)$. Hence w is also a function on its domain, and $u \subseteq w$. Because u is a halo and w a pregalaxy, by the Fehrele Principle there exists an internal function v with $u \subseteq v \subseteq w$. Put $K = \text{dom}(v)$. Then $E \subseteq K \subseteq \mathbb{N}$ and $v_{\restriction E} = u$. To finish the proof, let $J \subseteq E$ be an internal set. Then $u_{\restriction J} = v_{\restriction J}$ is internal. □

We prove a theorem of representation for external sequences of real numbers, using Lemma 4.4.8. The theorem expresses the fact that an external sequence originates from an internal family of internal sequences, for which we consider only sequences with standard indices, each sequence restricted to a prehalic domain. So, a truly external sequence of internal elements must have full complexity (i.e. $\Sigma_2^{\text{st}} = \Pi_2^{\text{st}}$), but with properties coming from internal sequences. The theorem suggests that typically such external sequences are defined by cases. Let us give a simple example.

Example 4.4.2. Let $u : \mathbb{N} \to \mathbb{N}$ be defined by

$$u_n = \begin{cases} 1, & n \in {}^{\sigma}\mathbb{N} \\ 0, & n \in \mathbb{N} \cap \not\infty\,. \end{cases} \tag{4.17}$$

Here formula (4.19) takes the form

$$u = \bigcup_{\text{st}(m) \in \mathbb{N}} \bigcap_{\text{st}(n) \in \mathbb{N}} (I_m \times \{1\} \cup J_n \times \{0\}), \tag{4.18}$$

where $I_m = \{k \in \mathbb{N} : 0 \leq k \leq m\}$ and $J_n = \{p \in \mathbb{N} : p \geq n\}$. On the halo $\mathbb{N} \cap \not\infty$ the external sequence (u_n) is the restriction of, say, the internal sequence which is identically zero. Also, the domains $\{0, ..., m\}$ for $\text{st}(m)$ are internal, and their union ${}^{\sigma}\mathbb{N}$ is a galaxy. On this galaxy the external sequence (u_n) is the restriction of, say, the internal sequence which is identically one. So, it may very well happen that some of the prehalic domains Δ_x are internal, and give way to a union which is a galaxy, say G. By Lemma 4.4.8 the restrictions of the external sequence (u_n) to these internal domains are also internal, so their union is a galaxy, which in turn is the restriction of an internal sequence to the galaxy G.

We recognize in Example 4.4.2 that the representation (4.18) in terms of unions of prehalos reduces indeed to the two cases given in (4.17).

Theorem 4.4.9 (Representation Theorem [45])**.** *Let (u_n) be a precise sequence. If (u_n) is not internal, it can neither be a galaxy, nor a halo. Then there exist a standard set X, an internal family of (internal) sequences $(v_x)_{x \in X}$ and a family of prehalos $(\Delta_x)_{x \in X}$ such that*

$$u = \bigcup_{\text{st}(x) \in X} v_{x \restriction \Delta_x}. \tag{4.19}$$

Proof. By (4.13), to prove the result, we need to show that the prehalos Q_x have internal extensions, and as such are the restrictions to the prehalos Δ_x of these internal extensions.

A strictly external sequence u defined on the internal set $\mathbb{N}$ cannot be a galaxy, for then it would be internal by Lemma 4.4.8. For the same reason it cannot be a halo. Hence it must have full complexity. By (4.13) it has the representation $u = \bigcup_{\text{st}(x)\in X} Q_x$, where for all standard $x \in X$ the prehalo Q_x is a function with domain Δ_x. By Lemma 4.4.8 and the Saturation Principle there exists an internal family of internal sequences $(v_x)_{x\in X}$ such that $v_{x\upharpoonright\Delta_x} = Q_x$ for all standard $x \in X$. This shows (4.19). □

Definition 4.4.5. *Let (u_n) be a flexible sequence. Let $S \subseteq \mathbb{N}$ be cofinal with $\mathbb{N}$. Let $v = \{(n, u_n) : n \in S\}$. Then v is called a* subsequence *of (u_n).*

Definition 4.4.5 coincides with the conventional notion of subsequence in the case where (u_n) is an internal sequence and S is internal. In that case a subsequence is again an internal sequence and we adopt the usual notation (u_{m_n}) to indicate a subsequence of (u_n). If S is external, such a representation is no longer possible.

With some abuse of language we define cofinality and coinitiality also for sequences.

As shown in Proposition 4.4.10 below, a definable sequence of external sets has an internal selection.

Let $S \subset \mathbb{N} \times \mathbb{R}$. We denote by $\pi(S)$ the *projection* of S into $\mathbb{N}$.

Definition 4.4.6. *Let (C, D) be a cut of $\mathbb{N}$ into a non-empty initial segment C and a final segment D. Let (u_n) be an external flexible sequence and v be a subsequence. The sequence v is said to be* cofinal *with u on C if*

$$\forall m \in C \exists n \in C(n \geq m \wedge n \in \text{dom}(v));$$

if D is non-empty, the sequence v is said to be coinitial *with u on D if*

$$\forall \nu \in D \exists \mu \in D(\mu \leq \nu \wedge \mu \in \text{dom}(v)).$$

Proposition 4.4.10. *Let $V = \{(n, V_n)\} \subseteq \mathbb{N} \times \mathcal{P}(\mathbb{R})$ be an external mapping with definable graph, where $\pi(V)$ is cofinal with $\mathbb{N}$. Then there exist an internal set $S \subseteq \mathbb{N}$ cofinal with $\mathbb{N}$ and an internal mapping $v : S \to \mathbb{R}$ such that $v_n \in V_n$ for all $n \in S$.*

Proof. Let X, Y be standard and $I : X \times Y \to \mathcal{P}(\mathbb{N}\times\mathbb{R})$ be an internal mapping such that $V = \bigcup_{\text{st}(x)\in X} \bigcap_{\text{st}(y)\in Y} I_{xy}$. For all $x \in X$, let $H_x = \bigcap_{\text{st}(y)\in Y} I_{xy}$. Because $\pi(V)$ is cofinal with $\mathbb{N}$, there exists $\text{st}(x) \in X$ such that H_x is cofinal with $\mathbb{N}$. It follows that I_{xy} is cofinal with $\mathbb{N}$ for all $\text{st}(y) \in Y$. Hence

$$\forall^{\text{stfin}} z \subseteq Y\, \exists J\, \forall^{\text{st}} y \in z(J \subseteq I_{xy}),$$

where J is cofinal with $\mathbb{N}$. In fact, we can take $J = \bigcap_{\text{st}(y)\in z} I_{xy}$. By Idealization

$$\exists J\, \forall^{\text{st}} y \in Y(J \subseteq I_{xy}),$$

where J is cofinal with $\mathbb{N}$. Put $S = \pi(J)$. Because J is internal, applying the Axiom of Choice, it has an internal selection $v : S \to \mathbb{R}$, i.e. $v_n \in J_n \subseteq V_n$ for all $n \in S$. □

Corollary 4.4.11. *Every precise external sequence* (u_n) *admits an internal subsequence.*

Exercises

Exercise 4.4.1. *Let* $n \in \mathbb{N}$ *be standard and* $\omega \in \mathbb{N}$ *be unlimited. Investigate whether the following functions* F *are flexible functions on their domain. If necessary, restrict their domains in such a way that their values become external numbers.*

1. $F(x) = (1 + \oslash x)^n$.
2. $F(x) = (1 + \oslash x)^\omega$.
3. $F(x) = \log((1 + \oslash)x)$.
4. $F(x) = e^{(1+\oslash)x}$.

Exercise 4.4.2. *Give a representative function and the neutrix part function of the mapping* F_4 *of Part 4 of Example 4.4.1. Show that the function is a flexible function which is not internally representable.*

Exercise 4.4.3. *Show that if* f *is a real continuous internal function such that* $f(\alpha) \in \mathbb{E}$ *for every external number* α*, then* f *must be strictly monotonous and onto.*

Exercise 4.4.4. *For the functions of Exercise 4.4.1 determine weak suprema and infima in* $@$.

4.5 Idempotent neutrices and ideals

We consider two, more advanced, algebraic properties dealing with idempotency and intersections which follow from generalized Dedekind completeness, in fact are equivalent to it.

A neutrix N has the *ring property* if $NN \subseteq N$. Clearly neutrices contained in $\pounds$ are rings. If $N \supset \pounds$ it holds that $NN \supseteq N$, but it is still possible that N is a ring. In that case $NN = N$, meaning that N is idempotent. It follows from generalized Dedekind completeness that every neutrix is the multiple of an idempotent neutrix by a real number; in fact the two statements are equivalent.

If $N \neq 0$ is a ring, due to the absence of zero-divisors it is obviously an integral domain. It certainly has ideals, for example the neutrices $\varepsilon\pounds$ for infinitesimal elements $\varepsilon > 0$ when $N = \pounds$. Moreover, because the ideals are ordered by inclusion N is a valuation domain. Generalized Dedekind completeness proves a maximality property: the intersection of a family of external numbers which is closed under standard finite intersections is non-empty. Again the two statements are equivalent.

4.5.1 Idempotent neutrices

Definition 4.5.1. *A neutrix I is* idempotent *if $II = I$.*

Clearly $\{0\}$ and $\mathbb{R}$ are idempotent, and also $\oslash$ and $\pounds$. Other idempotent neutrices are the microgalaxy m_ε and the microhalo M_ε of Definition 1.4.4, and the neutrices G_ω and H_ω of Example 4.1.1. If a is infinitesimal or unlimited neither $a\oslash$ nor $a\pounds$ are idempotent, but they are obviously a multiple of an idempotent neutrix. With generalized Dedekind completeness, one proves that this property is valid for every (definable) neutrix.

Theorem 4.5.1 ([76]). *Let N be a definable neutrix. Then there exists $\lambda \in \mathbb{R}$ and a unique idempotent neutrix I such that $N = \lambda I$.*

Proof. Let N be a neutrix. Because $\{0\}$ and $\mathbb{R}$ are idempotent, we suppose that $\{0\} \subset N \subset \mathbb{R}$. Put $A = \log(N^+ \setminus \{0\})$. Then A is a lower halfline and $e^A = N^+ \setminus \{0\}$. By the generalized Dedekind completeness theorem there exists $a \in \mathbb{R}$ and a unique external neutrix K such that either $A = (-\infty, a] \cup (a+K)$ or $A = (-\infty, a] \setminus (a+K)$. Let $d > 0$ be limited and $x \in A$. Then $e^x \in N$ hence also $e^{x+d} \in N$ because e^d is limited. Hence $x + d \in A$. We conclude that $K \supseteq \pounds$.

Put $I = \left[-e^K, e^K\right]$ and $J = \left(-e^K, e^K\right)$. Then I is a neutrix because $2I = 2e^K = e^{K+\log(2)} = e^K$. The neutrix I is idempotent because $(e^K)^2 = e^{2K} = e^K$. Let $j \in J, j > 0$. Then $2j \in J$, for $2j < 2e^K = e^{K+\log(2)} = e^K$. Hence J is a neutrix. Also J is idempotent because $j^2 < (e^K)^2 = e^{2K} = e^K$. Put $\lambda = e^a$. Then $N = \lambda I$ if $A = (-\infty, a] \cup (a+K)$ and $N = \lambda J$ if $A = (-\infty, a[\setminus (a+K)$.

The uniqueness of I and J follows from the uniqueness of K. □

As far as external sets are concerned, Theorem 4.5.1 implies the generalized Dedekind completeness theorem.

Theorem 4.5.2 ([134]). *Assume that every external definable neutrix is a multiple of a unique idempotent neutrix I. Then the generalized Dedekind completeness theorem holds for external cuts.*

Proof. Let (A, B) be an external cut. Let $K = \{x \in \mathbb{R} : x + A = A \wedge x + B = B\}$. Then K is an external neutrix. Let $\omega > 0$ be such that $\omega K \supseteq \pounds$. Let $\widehat{A} = \omega A$ and $\widehat{B} = \omega B$. Let $N = \left[-e^{\widehat{A}}, e^{\widehat{A}}\right]$. Then

$$\pounds N = \left[e^{\widehat{A}} \pounds^-, e^{\widehat{A}} \pounds^+\right] = \left[-e^{\widehat{A}}, e^{\widehat{A}}\right].$$

So N is a neutrix. Let I be an idempotent neutrix and $\lambda > 0$ be such that $N = \lambda I$. We distinguish the cases (i) $1 \in I$ and (ii) $1 \notin I$.

(i) One has $\lambda \in N$, so $\log(\lambda) \in \widehat{A}$. Let $\widetilde{I} = \{x \in I : x \geq 1\}$. Because I is idempotent, it holds that $\widetilde{I}^{\pounds^+} = \widetilde{I}$. So $\pounds^+ \log(\widetilde{I}) = \log(\widetilde{I})$. Hence $G := -\log(\widetilde{I}) \cup \log(\widetilde{I})$ is a neutrix. We have $\widehat{A} = (-\infty, \log(\lambda) + G]$; hence $A = (-\infty, \frac{\log(\lambda)}{\omega} + \frac{G}{\omega}]$.

(ii) One has $\lambda > N$, so $\log(\lambda) \in \widehat{B}$. Let $J = \{x \in \mathbb{R} : I < x \leq 1\}$. Let $x \in J$. Then $I = I^2 < x^2 \leq 1$, i.e. $x^2 \in J$. Hence $J \cdot J = J$, which implies that $J^{\pounds^+} = J$. So $\pounds^+ \log(J) = \log(J)$. Hence $H := \log(J) \cup -\log(J)$ is a neutrix. We have $\widehat{B} = [\log(\lambda) + H, +\infty)$; hence $A = (-\infty, \frac{\log(\lambda)}{\omega} + \frac{H}{\omega})$.

The uniqueness of G and H follows from the uniqueness of I. □

Definition 4.5.2. *Let $\alpha = a + A$ be an external number such that $a \geq 0$. If α is a neutrix, by Theorem 4.5.1 there exist a positive real number t and a unique idempotent neutrix I such that $\alpha = tI$. We define the n*-root *of α as follows*

$$\sqrt[n]{\alpha} = \sqrt[n]{t} I. \tag{4.20}$$

If α is zeroless we define its n-th root *in the following way:*

$$\sqrt[n]{\alpha} = \sqrt[n]{a} + \frac{A}{\sqrt[n]{a^{n-1}}}. \tag{4.21}$$

It is not difficult to see that the n-th root of an external number is well-defined (see Exercise 4.5.3). Note that the n-th root was not defined for the case where a is negative.

4.5.2 Ideals and the product of neutrices

Definition 4.5.3. *Let I be an idempotent neutrix. Let $J \subseteq I$ be a neutrix such that $IJ \subseteq J$. Then J is called an* ideal *of I. An ideal $M \subset I$ is* maximal *if for all ideals J such that $M \subseteq J \subseteq I$ it holds that $J = M$ or $J = I$.*

Clearly $\oslash$ is an ideal of $\pounds$, and as such it is maximal. Consider the neutrix $G_\omega := \{x : |x| \leq f(\omega)$, for some standard $f \in F\}$ of Example 4.1.1 for unlimited ω. Then $\mu_{\frac{1}{\omega}}$ is an ideal of G_ω.

An idempotent neutrix with unity has a maximal ideal.

Theorem 4.5.3. *Let I be an idempotent neutrix which contains 1. Put*

$$J = \left\{ \frac{1}{x} : x \in \mathbb{R} \wedge I < |x| \right\} \cup \{0\}.$$

Then

1. *$J < 1$.*

2. *J is an idempotent neutrix.*

3. *J is the maximal ideal of I.*

Proof. If $I = \mathbb{R}$, one has $J = \{0\}$, which is the maximal ideal of $\mathbb{R}$. Assume $I \subset \mathbb{R}$. By construction $J < 1$ and J is convex and symmetric. Let $y \in J, y > 0$. Then $\frac{1}{2y} > I$ because I is a neutrix. Hence $2y \in J$. Hence J is a neutrix. Because $J < 1$ we have $y^2 \in J$. Also $\frac{1}{\sqrt{|y|}} > I$, because I is idempotent. Hence $\sqrt{|y|} \in J$, implying that J is idempotent. To see that J is an ideal of I, let $x \in I$ and $y \in J$. If $|x| < 1$ we have $|xy| < |y|$, hence $xy \in J$. If $|x| > 1$ one has $J < \frac{1}{|m|}$ and because $\sqrt{|y|} \in J$ we have $\frac{1}{|xy|} > \frac{\sqrt{|y|}}{|y|} = \frac{1}{\sqrt{|y|}}$, hence $xy \in J$. Hence J is an ideal of I. Suppose I has an ideal K with $J \subseteq K \subseteq I$. If $1 \in K$, we clearly have $K = I$. If not, suppose there exists positive $x \in K \setminus J, x < 1$. Then $\frac{1}{x} \in I$. Then $\frac{1}{x^2} \in I$. But $\frac{1}{x} = x \cdot \frac{1}{x^2} \in K$, since K is an ideal of I, while $1 < \frac{1}{x}$. So we have a contradiction. As a consequence J is the maximal ideal of I. □

By Theorem 4.5.1 the product of neutrices is determined once the product of idempotent neutrices is known, the latter being equal to one of its factors. In fact, we have the following theorem.

Theorem 4.5.4 (Koudjeti's theorem [75]). *Let $I, J \in \mathbb{R}$ be idempotent neutrices. Then $IJ = I$ or $IJ = J$. More precisely, assume $I \subseteq J$. Then*

1. $IJ = I \Leftrightarrow J \subseteq I : I$
2. $I \subset IJ = J \Leftrightarrow J \supset I : I$

Proof. Observe that $I = II \subseteq IJ \subseteq JJ = J$. Part 1 follows directly from the definition of $I : I$. This also implies that if $I \subset IJ = J$ necessarily $J \supset I : I$. Finally, assume that $J \supset I : I$. Then $1 \in J$ and $IJ \supset I$. Then also $1 \in IJ$, for $I(IJ) = (II)J = IJ \supset I$. Let K be the maximal ideal of J. Then $K < 1$. Now $(IJ)J = I(JJ) = IJ$; hence IJ is an ideal of J. We conclude that $K \subset IJ$. Hence $IJ = J$. □

Corollary 4.5.5. *Let M, N be two neutrices. Then there exist $a, b \in \mathbb{R}$ such that $MN = aN$ or $MN = bM$.*

Proof. By Theorem 4.5.1 there exist idempotent neutrices I, J and $\lambda, \mu \in \mathbb{R}$ such that $M = \lambda I$ and $N = \mu J$. Now $IJ = I$ or $IJ = J$. In the first case $MN = \lambda\mu IJ = \lambda\mu I = \mu M$ and in the second case $MN = \lambda\mu IJ = \lambda\mu J = \lambda N$. □

Dealing with ideals we get a second characterization of the generalized Dedekind completeness, through a non-empty intersection property for families of external numbers satisfying the finite intersection property. This equivalence may be compared with a similar classical equivalence: one uses Dedekind completeness to prove that the intersection of a nested sequence of intervals of real numbers is non-empty, and again the non-empty intersection of a nested sequence of intervals of real numbers may be used to prove Dedekind completeness.

We will adapt the algebraic maximality property of valuation rings.

Definition 4.5.4. *A* valuation domain R *is an integral domain such that the set of its ideals is a chain under inclusion. It is called* maximal *if every family of sets* $\{a_i + L_i : i \in I\}$, *where* $a_i \in R$ *and* L_i *is an ideal of* R *for all elements* i *of some index set* I, *which satisfies the finite intersection property, has non-empty intersection.*

Definition 4.5.5. *An (external)* valuation domain R *is called* S-maximal *if every family of sets* $\{a_i + L_i : i \in I\}$, *where* $a_i \in R$ *and* L_i *is an ideal of* R *for all elements* i *of some index set* I, *which satisfies the standard finite intersection property, has non-empty intersection.*

Proposition 4.5.6. *The set of limited numbers* $\pounds$ *is a valuation domain. A coset* $a + L$ *with* $a \in \pounds$ *and* L *and ideal of* $\pounds$ *is an external number.*

Proof. Firstly $\pounds$ has no zero-divisors other than 0. Secondly a subset of $\pounds$ is an ideal if and only it is a neutrix. Hence a coset $a+L$ with $a \in \pounds$ and L and ideal of $\pounds$ is an external number. Finally, the set of neutrices is ordered by inclusion. □

We show that the generalized Dedekind completeness theorem implies the S-maximality of $\pounds$ and vice-versa.

It is straightforward to check that the non-empty intersection of two external numbers is an external number (see Exercise 3.1.3). By External Induction the non-empty intersection of an S-finite set of external numbers is equal to one of these external numbers.

Let I be a possibly external index set and $F := \{\alpha_i \subseteq \pounds : i \in I\}$ be a family of external numbers with the S-finite intersection property. Clearly $\bigcap F \neq \emptyset$ if $\bigcap F = \alpha_j$ for some $j \in I$. In the remaining case for all $i \in I$ there exists $k \in I, k \neq i$ such that $\alpha_k \subset \alpha_i$. As a consequence $\bigcap F \subset a + \oslash$ for some limited a. So, up to addition with $-a$, we need only to prove the property of non-empty intersection for families F such that all $\alpha_i \subset \oslash$. We will call such families *infinitesimal*.

Let $F := \{\alpha_i \subset \oslash : i \in I\}$ be an infinitesimal family of external numbers. For all $i \in I$ we have $\alpha_i = a_i + N_i$, where $a_i \simeq 0$ and N_i is a neutrix. We define

$$N := \textstyle\bigcap_{i \in I} N_i,$$

$$U := \{y \simeq 0 : \exists i \in I(y > \alpha_i)\},$$

$$D := \{y \simeq 0 : \exists i \in I(y < \alpha_i)\}.$$

Observe that none of the external sets N, U or D is empty.

Theorem 4.5.7 ([134]). *Let* $F := \{\alpha_i \subset \oslash : i \in I\}$ *be an infinitesimal family of external numbers. Then*

1. *$D < U$.*

2. *$\forall \varepsilon > N \exists y \in D \exists z \in U (z - y < \varepsilon)$.*

3. *$N = \{\delta \in \mathbb{R} : D + \delta = D\} = \{x \in \mathbb{R} : U + \delta = U\}$.*

4. *$\exists x \simeq 0 (D = [\oslash, x + N) \wedge U = (x + N, \oslash])$.*

5. *$\exists x \simeq 0 \left(\bigcap F = x + N\right)$.*

6. *£ is an S-maximal valuation domain.*

Proof. 1. Let $y \in D$ and $z \in U$. Let $i \in I$ be such that $y < \alpha_i$ and $j \in I$ be such that $\alpha_j < z$. If $\alpha_i \cap \alpha_j = \alpha_i$ we have $y < \alpha_i < z$ and if $\alpha_i \cap \alpha_j = \alpha_j$ we have $y < \alpha_j < z$. Hence $y < z$.

2. Let $\varepsilon > N$, say $\varepsilon > N_i$, where $i \in I$. Let $a_i \in \alpha_i$. Put $y = a_i - \frac{\varepsilon}{2}$ and $z = a_i + \frac{\varepsilon}{2}$. Then $y < \alpha_i < z$. Hence $y \in D$, $z \in U$ and $z - y = \varepsilon$.

3. Let $y \in D$. Let $i \in I$ be such that $y < \alpha_i$. Then $y + N_i < \alpha_i$. So $y + N < \alpha_i$. Hence $D + N \subseteq D$. As a consequence $N \subseteq \{d \in \mathbb{R} : D + d = D\}$. It follows from Part 2 that $\{d \in \mathbb{R} : D + d = D\} \supseteq N$. We conclude that $N = \{d \in \mathbb{R} : D + d = D\}$. The proof that $N = \{x \in \mathbb{R} : U + d = U\}$ is analogous.

4. It follows from Part 3 and the generalized Dedekind completeness theorem that there exists $x \simeq 0$ such that $D = [\oslash, x + N)$ or $D = [\oslash, x + N]$. In the latter case $x \in D$, which means that for some $i \in I$, one has $x < \alpha_i$. But then $x + N_i < \alpha_i$, which implies that $x + N_i \subseteq D$. Since $N_i \supset N$, we derive that $D \cap U \neq \emptyset$, a contradiction with Part 1. Hence $D = [\oslash, x + N)$. In an analogous way one proves that $U = (z + N, \oslash]$ for some $z \simeq 0$. Suppose $z - x \notin N$. Then $\varepsilon := z - x > N$ by Part 1. Let $y \in D$ be such that $y + \varepsilon \in U$. Then $y + \varepsilon > z + N$, so $y > x + N$, a contradiction. Hence $z - x \in N$. This concludes the proof of Part 4.

5. From Part 4.

6. From Part 5. □

Next we show the converse to Theorem 4.5.7.

Theorem 4.5.8. *The S-maximality of £ implies the generalized Dedekind completeness theorem.*

Proof. Let (G, H) be a cut of $\mathbb{R}$. Let $K = \{d \in \mathbb{R} : G + d = G \wedge H + d = H\}$. Then K is clearly a neutrix. Up to a change of scale we may suppose that $K \subset \oslash$. Define $\mathcal{F} \subseteq \mathbb{E}$ by

$$\mathcal{F} = \{\alpha \in \mathbb{E} : \alpha \cap G \neq \emptyset \wedge \alpha \cap H \neq \emptyset\}.$$

Let $n \in \mathbb{N}$ be standard and $\alpha_1, \ldots, \alpha_n \in \mathcal{F}$. One shows by External Induction that $\alpha_1, \ldots, \alpha_n$ have a common interval, i.e. $\bigcap_{1 \leq i \leq n} \alpha_i \neq \emptyset$. Then $\beta := \bigcap \mathcal{F} \neq \emptyset$ by the S-maximality of £. Let $x \in \bigcap \mathcal{F}$ and $N = \bigcap_{\alpha \in \mathcal{F}} N(\alpha)$. Then N is a neutrix and

$$\beta = \bigcap \{\alpha : \alpha \in \mathcal{F}\} = \bigcap \{x + N(\alpha) : \alpha \in \mathcal{F}\} = x + \bigcap \{N(\alpha) : \alpha \in \mathcal{F}\} = x + N.$$

Hence $\beta \in \mathbb{E}$. Observe that $y < \beta$ implies that $y \in G$ and that $z > \beta$ implies that $z \in H$. We distinguish the cases (i) $\beta \notin \mathcal{F}$ and (ii) $\beta \in \mathcal{F}$.

(i) Then $\beta \subseteq G$ or $\beta \subseteq H$. In the first case $G = (-\infty, \beta]$ and in the second case $H = [\beta, +\infty)$, hence $G = (-\infty, \beta)$. We observe that the case $\beta \in \mathbb{R}$ (i.e. $N = \{0\}$) corresponds to the usual Dedekind completeness of $\mathbb{R}$.

(ii) We prove that there exists $\varepsilon > 0$ such that $N = \varepsilon\pounds$ and $K = \varepsilon\oslash$. If not, there exists a neutrix M such that $K \subset M \subset N$. Because $K \subset M$ there are $\xi \in G$, $\eta \in H$ with $\eta - \xi \in M$. Hence $\xi + M \in \mathcal{F}$ and $\xi + M \subset \xi + N = \beta$, a contradiction. Define $G_{x,\varepsilon} = \frac{-x+G}{\varepsilon}$ and $H_{x,\varepsilon} = \frac{-x+H}{\varepsilon}$. Because $(G_{x,\varepsilon}, G_{x,\varepsilon})$ is a cut of $\mathbb{R}$, by Standardization and Transfer $({}^sG_{x,\varepsilon}, {}^sH_{x,\varepsilon})$ is a standard cut of $\mathbb{R}$. Now $G_{x,\varepsilon} \cap \pounds \neq \emptyset$ and $H_{x,\varepsilon} \cap \pounds \neq \emptyset$, so ${}^sG_{x,\varepsilon}$ and ${}^sH_{x,\varepsilon}$ are non-empty. Let $s = \sup {}^sG_{x,\varepsilon}$. Then s is standard. Also $s \in {}^sG_{x,\varepsilon} \Leftrightarrow {}^sG_{x,\varepsilon} = (-\infty, s] \Leftrightarrow G_{x,\varepsilon} = (-\infty, s + \oslash]$ and $s \in {}^sH_{x,\varepsilon} \Leftrightarrow {}^sH_{x,\varepsilon} = [s, +\infty) \Leftrightarrow H_{x,\varepsilon} = [s + \oslash, +\infty)$. In the first case $G = (-\infty, x + \varepsilon s + \varepsilon\oslash]$ and in the second case $H = [x + \varepsilon s + \varepsilon\oslash, +\infty)$; hence $G = (-\infty, x + \varepsilon s + \varepsilon\oslash)$. □

Exercises

Exercise 4.5.1. *Let $J < 1$ be an idempotent neutrix. Put $I = J : J$. Show that I is an idempotent neutrix and J is the maximal ideal of I.*

Exercise 4.5.2. *Prove Koudjeti's theorem in its original form: Let I, J be idempotent neutrices. Then $IJ = I$ or $IJ = J$. More precisely, if $1 \in J$, let K be the maximal ideal of J. Without restriction of generality we may assume that $I \subseteq J$. Then*

1. $IJ = I \Leftrightarrow (J < 1 \vee (1 \in J \wedge I \subseteq K))$.
2. $IJ = J \Leftrightarrow (1 \in I \vee (1 \in J \wedge K \subseteq I))$.

Exercise 4.5.3. *Show that* (4.20) *and* (4.21) *do not depend on the choice of the representatives and that* $(\sqrt[n]{\alpha})^n = \alpha$.

Exercise 4.5.4. *Let α, β be external numbers. Are the following equalities true for* $\operatorname{st}(n)$*? And for all n? In each case give either a proof or a counterexample.*

1. $\sqrt[n]{\alpha\beta} = \sqrt[n]{\alpha}\sqrt[n]{\beta}$.
2. $\sqrt[n]{\frac{\alpha}{\beta}} = \frac{\sqrt[n]{\alpha}}{\sqrt[n]{\beta}}$.

Exercise 4.5.5. *Consider the microgalaxy m_ε and the microhalo M_ε of Definition 1.4.4, and the neutrices G_ω or μ_ε of Example 4.1.1, with $\varepsilon \simeq 0, \varepsilon > 0$ and $\omega > 0$ unlimited. Investigate the products $m_\varepsilon \cdot M_\eta$ if also $\eta \simeq 0, \eta > 0$, and $G_\omega \cdot H_\nu$ for unlimited $\nu > 0$.*

Exercise 4.5.6. *Let $N \subset \mathbb{R} \times \oslash$ be a convex additive group. Let*

$$L := \{t \in \mathbb{R} : \exists m, n \in N(\|m - n\| = t)\}.$$

We assume that $L \supseteq £$. For $\omega \in L, \omega > 1$ put $F_\omega := \{y : (\omega, y) \in N\}$.

1. *Show that L is a neutrix and the sets F_ω are external numbers.*
2. *Why is $\Lambda := \bigcap_{\omega \in L} F_\omega \neq \emptyset$?*
3. *Show that $N = L(1, \lambda) \oplus W(1, 0)$, where $W := N_\Lambda$ and $\lambda \simeq 0$, and that for $\lambda \neq 0$ have also the orthogonal decomposition $N = L(1, \lambda) \oplus W\left(1, \frac{-1}{\lambda}\right)$ (see [134]).*

Chapter 5

Sequences. Convergence up to a neutrix

5.1 Notions of convergence for flexible sequences 118
5.1.1 Convergence for infinite sequences 118
5.1.2 Convergence with respect to an initial segment 125
5.2 Operations on flexible sequences 129
5.2.1 Boundedness and monotonicity 129
5.2.2 Operations 131
5.3 Cauchy flexible sequences 135

Let N be a neutrix. We study two types of convergence of flexible sequences, *N-convergence* and *strong convergence* introduced in [45]. N-convergence expresses convergence to an external number with tolerance N; i.e. we approach the external number within distance ε for all $\varepsilon > N$. Strong convergence expresses that the sequence enters the external number in (nonstandard) finite time, meaning that from a certain natural number onwards the terms of the sequence are included in the limit set; observe that the sequence still may exhibit fluctuations which are less than the neutrix of the external number. So, in particular the limit is not unique.

We point out that in our setting strong convergence results are only provable for definable sequences. Due to the fact that the convergence properties may happen in finite time, a refinement is also considered, taking into consideration the segment on which the convergence actually happens.

Some applications dealing with the Borel-Ritt Theorem on asymptotic expansions and the problem of matching in singular perturbations will be given in Chapter 9.

As shown in Section 5.1, N-convergence implies strong convergence in a rather universal way. Loosely translated into classical terms, this means that if we approach an $O(\cdot)$, or an $o(\cdot)$-neighbourhood of some object, we enter the neighbourhood. The strong convergence theorems are particularly useful for the study of to what extent the properties of limits of sequences of real numbers are valid or can be adapted to flexible sequences. We do this in Section 5.2.

Finally, in Section 5.3 we consider Cauchy properties.

5.1 Notions of convergence for flexible sequences

In this section we introduce two types of convergence for sequences of external numbers which may be seen as "outer" ($\varepsilon - n_0$ like) and "inner" (entrance into the limit set) forms of convergence. The main results of this section express that outer convergence induces inner convergence except in the case where the sequence is infinite and the limit is precise (Theorem 5.1.11 and Theorem 5.1.12).

Let (u_n) be a standard sequence. A well-known theorem of nonstandard analysis (see Exercise 4.1.2) expresses that if (u_n) converges to a limit a then $u_n \simeq a$ for all unlimited $n \in \mathbb{N}$. Reformulated in the sense of Definitions 5.1.1 and 5.1.2 below this property implies that if (u_n) is $\oslash$-convergent to $a + \oslash$, it converges strongly to $a + \oslash$. The property is of interest, for instance because operations with classical convergence involve epsilontics, and operations with the nonstandard criterion of convergence use simple algebraic properties of infinitesimal numbers. Theorem 5.1.11 extends this property to all external definable sequences and all external numbers. Again the theorem is of interest in relation to behaviour under operations as we shall see in Section 5.2: operations with N-convergence involve epsilontics, and operations with strong convergence may profit from the algebraic properties of external numbers.

5.1.1 Convergence for infinite sequences

Definition 5.1.1. *Let N be a neutrix. We say that a flexible sequence u: $\mathbb{N} \to \mathbb{E}$ N-converges to $\alpha \in \mathbb{E}$ if*

$$\forall \varepsilon > N \exists n_0 \in \mathbb{N} \forall n \in \mathbb{N}(n \geq n_0 \Rightarrow |u_n - \alpha| < \varepsilon). \tag{5.1}$$

We write $u_n \underset{N}{\longrightarrow} \alpha$ or N-$\lim u_n = \alpha$. If $N = N(\alpha)$, we simply say that u converges to α and write $u_n \longrightarrow \alpha$ or $\lim u_n = \alpha$. A flexible sequence which is not N-convergent to any element $\alpha \in \mathbb{E}$ is called N-divergent.

Definition 5.1.2. *Let (u_n) be a flexible sequence and α be an external number. The sequence (u_n) is said to be* strongly convergent to α *if*

$$\exists n_0 \in \mathbb{N} \forall n \in \mathbb{N}(n \geq n_0 \Rightarrow u_n \subseteq \alpha).$$

We write $u_n \rightsquigarrow \alpha$. The external number α is called a strong limit *of (u_n).*

Clearly every flexible sequence is $\mathbb{R}$-convergent. To avoid this trivial case, whenever we refer to N-convergence we always assume that $N \neq \mathbb{R}$. Similarly, every flexible sequence strongly converges to $\mathbb{R}$, so we consider strong limits to be different from $\mathbb{R}$. The other extreme case for N-convergence is when $N = 0$ which corresponds to the usual notion of convergence. In such case we adopt the usual notation and terminology (i.e. we say that (u_n) converges to α and write $u_n \longrightarrow \alpha$ or $\lim u_n = \alpha$).

Remark 5.1.1. 1. In the condition of N-convergence one may assume the elements ε to be *precise*, i.e. to be real numbers (see Exercise 5.1.1). In fact, (5.1) is equivalent to:

$$\forall \varepsilon \in \mathbb{R} \, (\varepsilon > N \Rightarrow \exists n_0 \in \mathbb{N}(n \geq n_0 \Rightarrow |u_n - \alpha| < \varepsilon)) \,. \qquad (5.2)$$

This equivalence often allows for simplifications in proofs and in calculations.

2. Let $\alpha = a+N$ and (u_n) be a flexible sequence. Then (u_n) is N-convergent to α if and only if $(u_n - a)$ is N-convergent to N. So when investigating N-convergence, without loss of generality we may consider α to be a neutrix.

Definition 5.1.1 generalizes the usual $\varepsilon - n_0$ criterion for convergence

$$\forall \varepsilon > 0 \exists n_0 \in \mathbb{N} \forall n \in \mathbb{N}(n \geq n_0 \Rightarrow |u_n - \alpha| < \varepsilon).$$

Here it is asked that for $n \geq n_0$ the terms u_n enter into some interval, which is partly outside the limit set, justifying the term "outer convergence". In the case of strong convergence the terms of the sequence reach the limit set within a finite time, leading to a form of inner convergence indeed.

Let us illustrate Definitions 5.1.1 and 5.1.2 with some examples. These examples are also a source of motivation for some of the results in this section.

Let N be a neutrix. The first example illustrates the fact that N-convergent flexible sequences do not have a unique limit. In Proposition 5.1.4 we show that N-convergence is in fact unable to distinguish between elements whose distance is less than or equal to N. However, N-limits are unique modulo N in the sense that if (u_n) is N-convergent to two (possibly distinct) elements $\alpha, \beta \in \mathbb{E}$ then the absolute value of their difference must be less than N; in particular if $u_n \underset{N}{\longrightarrow} \alpha$, then $u_n \underset{N}{\longrightarrow} a$, for all $a \in \alpha$ (see Proposition 5.1.5).

Example 5.1.1. Consider the flexible sequence defined by $u_n = \frac{1}{n} + \oslash$ and let $\delta \simeq 0$. One has both $u_n \underset{\oslash}{\longrightarrow} \delta$ and $u_n \longrightarrow \oslash$ because $|u_n - \delta| = |u_n - \oslash| = \left|\frac{1}{n} + \oslash\right|$.

The next example illustrates the fact that for every sequence there is a neutrix N such that the sequence is N-convergent. The intuition behind it is the rather obvious fact that if it is possible to approximate a certain quantity with a given imprecision, then it must also be possible to approximate it with a bigger imprecision. This intuition is confirmed by Proposition 5.1.2 expressing that if a flexible sequence is convergent with respect to a given neutrix, then it is convergent with respect to every neutrix which is larger than the original neutrix. Moreover, there exists a smallest neutrix such that the given sequence is convergent with respect to it. For this reason, when discussing N-convergence of a sequence (u_n) we will implicitly be looking for the smallest neutrix N for which (u_n) is N-convergent.

Example 5.1.2. Consider the classically divergent sequence defined by $u_n = (-1)^n$ for all $n \in \mathbb{N}$. Clearly (u_n) is $\pounds$-convergent. We show that (u_n) is $\oslash$-divergent. Since there are no neutrices between $\oslash$ and $\pounds$, it follows that $N = \pounds$ is the smallest neutrix that gives N-convergence for (u_n). Indeed, suppose that $u_n \underset{\oslash}{\longrightarrow} \alpha$ with $\alpha = a + A$ an external number. We have $N(\alpha) = A \leq \oslash$. If $\alpha = 1 + A \subseteq 1 + \oslash$, then taking $\varepsilon_0 = 1$ and choosing $n_0 = 2n + 1 > n$ we have $|u_{n_0} - \alpha| = |-1-1+A| = 2 + A > \varepsilon_0$, a contradiction. If $\alpha \cap (1+\oslash) = \emptyset$, we take $\varepsilon_0 = \frac{|1-a|}{3} > \oslash$. Since $A \subseteq \oslash$, we have $-\varepsilon_0 < A < \varepsilon_0$. For all $n \in \mathbb{N}$, choosing $n_0 = 2n > n$ we obtain that

$$|u_n - \alpha| = |1 - a + A| = |1 - a| + A = 3\varepsilon_0 - \varepsilon_0 = 2\varepsilon_0 > \varepsilon_0,$$

a contradiction. Hence (u_n) is $\oslash$-divergent.

Proposition 5.1.2. *Let N be a neutrix and (u_n) be a flexible sequence that N-converges to some element $\alpha \in \mathbb{E}$. Let M be a neutrix such that $N \subseteq M$. Then (u_n) also M-converges to α. Moreover, there exists a unique neutrix M_0 such that (u_n) is M_0-convergent and $M_0 \subseteq N$ for every neutrix N such that (u_n) is N-convergent.*

Proof. Let $\varepsilon > M$. Then $\varepsilon > N$. Since (u_n) N-converges to α, there exists $n_0 \in \mathbb{N}$ such that $|u_n - \alpha| < \varepsilon$ for all $n \geq n_0$. Hence (u_n) M-converges to α. Without loss of generality, we may assume that (u_n) is N-convergent to a neutrix. Let $L := \{M : M \text{ is a neutrix and } (u_n) \text{ is } M\text{-convergent}\}$ and M_0 be the infimum of L. By Theorem 4.5.1 there exist a positive real number p and a unique idempotent neutrix such that $M_0 = pI$.

Assume first that $M_0 = p\oslash$ for some positive real number p. If $M_0 \notin L$, all elements M of L satisfy $L \supseteq p\pounds$; hence M_0 cannot be the infimum of L, a contradiction. Hence $M_0 \in L$ and so $M_0 \leq N$ for every neutrix N such that (u_n) is N-convergent.

Assume now that M_0 is not a multiple of $\oslash$. Suppose that the sequence is not M_0-convergent. Let $\varepsilon > M_0$. Note that $M_0 \subseteq \varepsilon\oslash$, for $\varepsilon\oslash$ is the biggest neutrix less than ε; hence $M_0 \subset \varepsilon\oslash$, for M_0 is not isomorphic to $\varepsilon\oslash$. For all $n \in \mathbb{N}$ there exists an element $m \geq n$ such that $\varepsilon \leq u_m$. Because $\varepsilon\oslash < \varepsilon$, the sequence (u_n) is not $\varepsilon\oslash$-convergent. So $\varepsilon\oslash \notin L$. It follows from the definition of L that for every neutrix $M \in L$ we have $\varepsilon\oslash \subset M$. It follows that $\varepsilon\oslash \subseteq M_0$, a contradiction. Hence (u_n) is M_0-convergent. □

The following proposition entails, as already suggested by Proposition 5.1.2, that when dealing with N-convergence, it is possible to neglect quantities in the terms which are contained in N.

Proposition 5.1.3. *Let M, N be neutrices such that $M \subseteq N$; let $\alpha \in \mathbb{E}$ and let (u_n) be a flexible sequence. Then the following are equivalent*

1. $(u_n + M) \underset{N}{\longrightarrow} \alpha$.

2. $u_n \underset{N}{\longrightarrow} \alpha$.

Proof. See Exercise 5.1.2. □

As illustrated by Example 5.1.1, N-convergence is unable to distinguish between elements whose distance is less than or equal to the neutrix N. As a consequence, the N-limits of a flexible sequence are not unique. However, N-limits are unique modulo N in the sense that if a flexible sequence (u_n) is N-convergent to two (possibly distinct) elements $\alpha, \beta \in \mathbb{E}$ then the absolute value of their difference must be less than N. Moreover, if (u_n) is N-convergent to some $\alpha \in \mathbb{E}$ then N is the best possible, meaning that $N(\alpha) \subseteq N$ and the largest limit set is $\alpha + N$.

Proposition 5.1.4. *Let* $u : \mathbb{N} \longrightarrow \mathbb{E}$ *be a flexible sequence,* $\alpha, \beta \in \mathbb{E}$ *and* M *be a neutrix such that* $u_n \underset{M}{\longrightarrow} \alpha$ *and* $u_n \underset{M}{\longrightarrow} \beta$. *Then*

1. $N(\alpha) \subseteq M$.
2. $|\alpha - \beta| \leq M$.
3. $u_n \underset{M}{\longrightarrow} \alpha + M$.
4. $u_n \underset{M}{\longrightarrow} \gamma$ *if and only if* $\gamma \subseteq \alpha + M$.

Proof. 1. Suppose that $M \subset N(\alpha)$. Let $\varepsilon \in \mathbb{R}$ be such that $M < \varepsilon \leq N(\alpha)$. Then there exists $n_0 \in \mathbb{N}$ such that for $n \geq n_0$ we have $N(\alpha) \leq N(|u_n - \alpha|) \leq |u_n - \alpha| < \varepsilon \leq N(\alpha)$, which is a contradiction. Hence $N(\alpha) \subseteq M$.

2. Suppose $u_n \underset{M}{\longrightarrow} \alpha$ and $u_n \underset{M}{\longrightarrow} \beta$ with $|\alpha - \beta| > M$. For $\varepsilon_0 = \frac{|\alpha-\beta|}{2} > M$, there exists $n_0 \in \mathbb{N}$ such that $|u_n - \alpha| < \varepsilon_0$ for $n \geq n_0$ and there exists $n_1 \in \mathbb{N}$ such that $|u_n - \beta| < \varepsilon_0$ for $n \geq n_1$. Let $k = \max\{n_0, n_1\}$. Then for $n \geq k$, we have $|\alpha - \beta| \leq |\alpha - u_n + u_n - \beta| \leq |\alpha - u_n| + |u_n - \beta| < 2\varepsilon_0 = |\alpha - \beta|$, a contradiction. Hence $|\alpha - \beta| \leq M$.

3. Let $\varepsilon > M$. Then $\frac{\varepsilon}{2} > \frac{M}{2} = M$. Because $u_n \underset{M}{\longrightarrow} \alpha$, there exists $n_0 \in \mathbb{N}$ such that $|u_n - \alpha| < \frac{\varepsilon}{2}$ for all $n \geq n_0$. It follows that $|u_n - \alpha + M| = |u_n - \alpha| + M < \frac{\varepsilon}{2} + \frac{\varepsilon}{2} = \varepsilon$, for all $n \geq n_0$. Hence $u_n \underset{M}{\longrightarrow} \alpha + M$.

4. If $\gamma \subseteq \alpha + M$ then $|u_n - \gamma| \leq |u_n - \alpha + M| < \varepsilon$ for all $n \geq n_0$, and hence $u_n \underset{M}{\longrightarrow} \gamma$. Conversely, assume that $u_n \underset{M}{\longrightarrow} \gamma$. Then by Part 1 we have $N(\gamma) \subseteq M$. By Part 2 it holds that $|\alpha - \gamma| \subseteq M$. So $\gamma - \alpha \subseteq M$. Then $\gamma \subseteq \gamma + N(\alpha) \subseteq \alpha + M$. □

In Part 2 of Proposition 5.1.4, if $M = 0$, then $\alpha = \beta$; i.e. we have uniqueness of the limit. This was to be expected because as mentioned above this case corresponds to the usual notion of convergence.

Definition 5.1.3. *Let (u_n) be a flexible sequence such that $u_n = a_n + A_n$, for every $n \in \mathbb{N}$ where the elements a_n are precise and $A_n = N(u_n)$ for every $n \in \mathbb{N}$. Then we call the precise sequence (a_n) a* sequence of representatives *of the sequence (u_n).*

Let N be a neutrix. Proposition 5.1.6 shows that a different choice for the sequence of representatives still leads to the same N-limit. First, we show that a flexible sequence (u_n) is N-convergent to α if and only if both (A_n) is N-convergent to the neutrix part of α and a sequence of representatives of (u_n) is N-convergent to an element of α.

Proposition 5.1.5. *Let (u_n) be a flexible sequence such that $u_n = a_n + N(u_n)$ for all $n \in \mathbb{N}$, where (a_n) is a sequence of representatives. Let N be a neutrix and let $\alpha = a + N(\alpha) \in \mathbb{E}$. Then $u_n \underset{N}{\longrightarrow} \alpha$ if and only if $a_n \underset{N}{\longrightarrow} a$ and $N(u_n) \underset{N}{\longrightarrow} N(\alpha)$. In particular if $u_n \underset{N}{\longrightarrow} a$ and $N(\alpha) \leq N$, then $u_n \underset{N}{\longrightarrow} \alpha$.*

Proof. Assume first that $a_n \underset{N}{\longrightarrow} a$ and $N(u_n) \underset{N}{\longrightarrow} N(\alpha)$. Let $\varepsilon > N$. Then there exist $n_1, n_2 \in \mathbb{N}$ such that $|a_n - a| < \frac{\varepsilon}{2}$ for $n \geq n_1$ and $|N(u_n) + N(\alpha)| = |N(u_n) - N(\alpha)| < \frac{\varepsilon}{2}$ for $n \geq n_2$. Let $n_0 = \max\{n_1, n_2\}$. Then for $n \geq n_0$ we have $|u_n - \alpha| = |a_n + N(u_n) - a + N(\alpha)| \leq |a_n - a| + |N(u_n) - N(\alpha)| < \frac{\varepsilon}{2} + \frac{\varepsilon}{2} = \varepsilon$. Hence $u_n \underset{N}{\longrightarrow} \alpha$.

Conversely, assume that $u_n \underset{N}{\longrightarrow} \alpha$. Let $\varepsilon > N$. Then there exist $n_1, n_2 \in \mathbb{N}$ such that $|u_n - \alpha| < \varepsilon$ for $n \geq n_1$ and $|u_n - \alpha| < \frac{\varepsilon}{2}$ for $n \geq n_2$. Then for $n \geq n_1$ we have $|a_n - a| \leq |a_n + N(u_n) - a + N(\alpha)| = |u_n - \alpha| < \varepsilon$ and for $n \geq n_2$ we have $|N(u_n) - N(\alpha)| = |u_n - u_n + \alpha - \alpha| \leq |u_n - \alpha| + |u_n - \alpha| < \frac{\varepsilon}{2} + \frac{\varepsilon}{2} = \varepsilon$. Hence $a_n \underset{N}{\longrightarrow} a$ and $N(u_n) \underset{N}{\longrightarrow} N(\alpha)$. □

Proposition 5.1.6. *Let (u_n) be a flexible sequence. Let $(a_n), (b_n)$ be precise sequences and (A_n) be a sequence of neutrices such that for all $n \in \mathbb{N}$ it holds that $u_n = a_n + A_n$ and $u_n = b_n + A_n$. Let $\alpha = a + A$ be an external number and N be a neutrix such that $u_n \underset{N}{\longrightarrow} \alpha$. Then $a_n \underset{N}{\longrightarrow} a$ if and only if $b_n \underset{N}{\longrightarrow} a$.*

Proof. Suppose that $a_n \underset{N}{\longrightarrow} a$. Let $\varepsilon > N$. Then there exists $n_1 \in \mathbb{N}$ such that for $n \geq n_1$ it holds that $|a_n - a| < \frac{\varepsilon}{2}$. Also, since $u_n \underset{N}{\longrightarrow} \alpha$, by Proposition 5.1.5 it holds that $A_n \underset{N}{\longrightarrow} A$, so there exists $n_2 \in \mathbb{N}$ such that for $n \geq n_2$ it holds that $|A_n - A| < \frac{\varepsilon}{2}$. Let $n_0 = \max\{n_1, n_2\}$. Observe that for each $n \in \mathbb{N}$ we have $b_n - a_n \in A_n$ because a_n and b_n are both representatives of the same external number. Then for $n \geq n_0$ it holds that

$$\begin{aligned}|b_n - a| = |b_n - a_n + a_n - a| \leq |b_n - a_n| + |a_n - a| < A_n + \frac{\varepsilon}{2}\\ \leq |A_n - A| + \frac{\varepsilon}{2} < \frac{\varepsilon}{2} + \frac{\varepsilon}{2} = \varepsilon.\end{aligned}$$

Hence $b_n \underset{N}{\longrightarrow} a$. The other implication is analogous. □

Let N be a neutrix. As the results above point out, most cases of N-convergence to an external number α reduce to the case where $N = N(\alpha)$, so we can talk about convergence instead of N-convergence and write simply $u_n \longrightarrow \alpha$ as indicated in Definition 5.1.1.

We prove the following useful properties relating convergence and divergence of a sequence to corresponding properties of subsequences.

Proposition 5.1.7. *Let N be a neutrix, $\alpha \in \mathbb{E}$ and let (u_n) be a flexible sequence. Then $u_n \underset{N}{\longrightarrow} \alpha$ if and only if $u_{n\restriction_J} \underset{N}{\longrightarrow} \alpha$ for every set J cofinal with $\mathbb{N}$.*

Proof. Assume first that (u_n) is N-convergent to α. Let J be cofinal with $\mathbb{N}$ and let $\varepsilon > N$. Then there exists $n_0 \in \mathbb{N}$ such that for all $n \geq n_0$, we have $|u_n - \alpha| < \varepsilon$. Because J is cofinal with $\mathbb{N}$, there exists $m_0 \in J$ such that $m_0 \geq n_0$. So for all $m \in J$ such that $m \geq m_0 \geq n_0$ we have $|u_m - \alpha| < \varepsilon$. We conclude that $u_{n\restriction_J} \underset{N}{\longrightarrow} \alpha$. The other implication is obvious because (u_n) is a subsequence of itself. □

Proposition 5.1.8. *Every bounded precise sequence admits an internal convergent subsequence.*

Proof. Let (a_n) be a bounded precise sequence. By Corollary 4.4.11 there exists an internal subsequence (a_{n_m}) of (a_n) which is clearly also bounded. So there exists a convergent subsequence (b_n) of (a_{n_m}), which is also a subsequence of (a_n). □

The next lemma expresses that an internal sequence which remains outside a neutrix does not have this neutrix as a limit. It is crucial in proving the main result of this section (Theorem 5.1.11).

Lemma 5.1.9. *Let (a_n) be an internal precise sequence. Let $N \neq 0$ be a neutrix and $n_0 \in \mathbb{N}$ be such that $a_n \notin N$ for all $n \geq n_0$. Then (a_n) is not N-convergent to N.*

Proof. By Proposition 5.1.3 we may assume that $a_n > N$ for all $n \geq n_0$. We consider two cases, (i) the sequence (a_n) is convergent to some $a \in \mathbb{R}$, (ii) the sequence (a_n) is divergent.

(i) We have $a \notin N$. Indeed, suppose that $a \in N$. Let $\varepsilon \in N$ be such that $\varepsilon > 0$. Then there exists $n_1 \in \mathbb{N}$ such that $|a_n - a| < \varepsilon$ for all $n \geq n_1$. So, $|a_n| < |a| + \varepsilon$ for all $n \geq n_2$, where $n_2 = \max\{n_0, n_1\}$. This means that $|a_n| < |a| + \varepsilon \in N$. By our assumption and by the convexity of N it follows that $a_n \in N$ for all $n \geq n_2$, and in particular $a_{n_2} \in N$, which contradicts the assumption. We prove now that $a > N$. Suppose towards a contradiction that $a < N$. Then $a < 0$ and so $|a| = -a > N$. Let $\eta = \frac{|a|}{2}$. Clearly $\frac{|a|}{2} > N$. Then there exists $n_3 \in \mathbb{N}$ such that $|a_n - a| < \eta$ for $n \geq n_3$. Let $n_4 = \max\{n_0, n_3\}$. This implies that $a_n < a + \eta = \frac{a}{2} < 0$ for all $n \geq n_4$, a contradiction. Hence $a > N$. Suppose that $a_n \underset{N}{\longrightarrow} N$. Let $\varepsilon = \frac{a}{2}$. Clearly $\varepsilon > N$. Then there exists $n_5 \in \mathbb{N}$ such that for $n \geq n_5$ it holds that $|a_n - N| < \varepsilon$. This means that $\frac{a}{2} = a - \varepsilon < a_n < a + \varepsilon$ for all $n \geq n_5$. Observe that since $a \notin N$ and $\frac{a}{4} > N$ then $-\frac{a}{4} < N$ and we have that $\frac{a}{4} = \frac{a}{2} - \frac{a}{4} < \frac{a}{2} + N$. It follows that $|a_n - N| = a_n + N > \frac{a}{2} + N > \frac{a}{4} > N$ for all $n \geq n_6$, where $n_6 = \max\{n_0, n_5\}$, which is a contradiction. Hence N is not an N-limit of (a_n).

(ii) Suppose towards a contradiction that $a_n \underset{N}{\longrightarrow} N$. Since (a_n) is an internal real sequence, it follows that (a_n) is bounded. So there exists a subsequence $(a_{m_n}) \subset (a_n)$ such that (a_{m_n}) has a limit $b \in \mathbb{R}$. By (i) the subsequence (a_{m_n}) does not N-converge to N, in contradiction with Proposition 5.1.7. Hence N is not an N-limit of (a_n). □

We show that N-convergence entails strong convergence for sequences of neutrices.

Proposition 5.1.10. *Let (A_n) be a sequence of neutrices and $N \neq 0$ be a neutrix. If $A_n \longrightarrow N$ then $A_n \rightsquigarrow N$.*

Proof. Suppose that for all $n \in \mathbb{N}$ there exists $p_n \geq n$ such that $N \subset A_{p_n}$. Let $D = \{(n, A_n \backslash N) : N \subset A_n\}$. Then D is cofinal with A_n. By Proposition 4.4.10 there exists an internal sequence (b_k) such that $b_k \in A_k \setminus N$ for all $k \in \pi(D)$. Since $A_n \longrightarrow N$, we have $b_k \underset{N}{\longrightarrow} N$, in contradiction with Lemma 5.1.9. Then there exists $n_0 \in \mathbb{N}$ such that $A_n \subseteq N$ for all $n \geq n_0$. Hence $A_n \rightsquigarrow N$. □

We finish this section by showing that convergence implies strong convergence whenever the limit is not precise.

Theorem 5.1.11. *Let (u_n) be a flexible sequence and α be an external number with $N(\alpha) \neq 0$. Then $u_n \longrightarrow \alpha$ if and only if $u_n \rightsquigarrow \alpha$.*

Proof. Without loss of generality we may assume that $\alpha = N(\alpha) = A$. Clearly, if there exists $n_0 \in \mathbb{N}$ such that for all $n \geq n_0$ we have $u_n \subseteq A$, then $u_n \longrightarrow A$.

Conversely, assume that $u_n \longrightarrow A$. For all $n \in N$, put $A_n = N(u_n)$. Then (A_n) is a definable sequence of neutrices, which by Proposition 5.1.5 converges to A. Then $(A_n) \rightsquigarrow A$ by Proposition 5.1.10. Hence there exists $m \in \mathbb{N}$ such that $A_n \subseteq A$ for all $n \geq m$. Suppose u does not converge strongly to A. Then $C = \{n \in \mathbb{N} : u_n \nsubseteq N\}$ is cofinal in $\mathbb{N}$. Put $D = \{n \in C : n \geq m\}$. Then $u_n \cap A = \emptyset$ for all $n \in D$. By Proposition 4.4.10 there exists an internal set $J \subseteq D$ cofinal with $\mathbb{N}$ and an internal sequence $(a_n)_{n \in J}$ such that $a_n \in u_n$ for all $n \in J$. By Lemma 5.1.9 the sequence $(a_n)_{n \in J}$ does not converge to A; hence also $(u_n)_{n \in J}$ does not converge to A, in contradiction with Proposition 5.1.7. Hence u converges strongly to A. □

Example 5.1.3. Consider again the flexible sequence (u_n) defined by $u_n = \frac{1}{n} + \oslash$. Then $u_n \rightsquigarrow \oslash$. In fact $u_n \subseteq \oslash$ for all unlimited $n \in \mathbb{N}$.

5.1.2 Convergence with respect to an initial segment

In this section we turn our attention to sequences which are defined on subsets of $\mathbb{N}$ that are typically not cofinal with $\mathbb{N}$.

Let C be an initial segment of $\mathbb{N}$. If C is internal it is just a discrete interval $\{0,...,c\}$ with $c = \max C$. If C is a galaxy, it may be written in the form $C = \bigcup_{\mathrm{st}(z)\in Z}\{0,...,c_z\}$, where Z is a standard ordered set and $c : Z \to \mathbb{N}$ is increasing, at least on the standard elements of Z. If C is a halo, it may be written in the form $C = \bigcap_{\mathrm{st}(w)\in W}\{0,...,c_w\}$, where W is a standard ordered set and $c : W \to \mathbb{N}$ is decreasing, at least on the standard elements of W.

Definition 5.1.4. *A* local sequence *is a flexible function $u : C \to \mathbb{E}$, where $C \subseteq \mathbb{N}$ is an initial segment.*

If it is clear from the context, we may omit the predicate "local". A local sequence still admits a partial internal choice sequence, but if C is external, by Cauchy's Principle the domain of this sequence must have non-empty intersection with the complement of C; such a sequence will be called an *internal bridge.*

Definition 5.1.5. *Let $C \subseteq \mathbb{N}$ be an initial segment, not cofinal with $\mathbb{N}$, let $D = \mathbb{N} \setminus C$, let $E \subseteq C$ be cofinal with C and $u : E \to \mathbb{E}$ be a local flexible sequence. Let K be an internal set such that $K\cap C$ is cofinal with C and $K \cap D$ is coinitial with D. An internal sequence $b : K \to \mathbb{R}$ such that $\{n \in K \cap C : b_n \in u_n\}$ is cofinal with C is called an* internal bridge *from C to D along u.*

If u is the restriction of an internal sequence to an external initial segment an internal bridge is simply given by this internal sequence. The existence of internal bridges was established in [45, Theorem 2.13] using Saturation.

Taking b as in Definition 5.1.5, we have that $b_{\restriction(K\cap C)}$ is a local sequence cofinal with C and $b_{\restriction(K\cap D)}$ is a local sequence coinitial with D, in line with Definition 4.4.6.

As it turns out, a flexible external sequence (u_n) which converges to $\alpha \in \mathbb{E}$ reaches α in a finite time (see Theorem 5.1.11 below). Put

$$C := \{n \in \mathbb{N} : \exists p > n(u_p \notin \alpha)\}.$$

Then $C \subset \mathbb{N}$ is an initial segment of $\mathbb{N}$. If $C \neq \emptyset$, one may say that (u_n) converges to α already with respect to C: afterwards the terms u_n stay within the limit set α. This property remains true if one truncates the sequence up to an index $\nu \in \mathbb{N} \setminus C$. So one may define convergence also for sequences with finitely many terms. This gives also the liberty to consider convergence for initial parts of a sequence, that afterwards may exhibit different behaviour, a situation that is quite natural in the context of dynamical systems (see Chapter 9).

Below we will extend the notions of convergence of Definitions 5.1.1 and 5.1.2 to convergence with respect to an initial segment of $\mathbb{N}$. Note that by the above we may as well define this notion for local sequences. As shown in Theorem 5.1.12 again convergence implies strong convergence. However in some cases there is a difference with infinite sequences, in that the strong convergence may not happen on the initial segment, but just beyond, along internal bridges.

Definition 5.1.6. *Let $S, C \subseteq \mathbb{N}$ be initial segments of $\mathbb{N}$ with $C \subseteq S$, and let $D = S \setminus C$. Let $u : S \to \mathbb{R}$ be an external local sequence. Let $\alpha = a + A \in \mathbb{E}$. We say that u* converges *to α with respect to C if*

$$\forall \varepsilon > A \exists m \in C \forall n \in S (n \geq m \Rightarrow |u_n - \alpha| < \varepsilon).$$

We say that u strongly converges *to α with respect to C if there exists a non-empty final segment T of S such that $D \subseteq T$ and*

$$\forall n \in T (u_n \in \alpha).$$

In the first case we write $\lim_{n \to D} u_n = \alpha$ and in the second case $\operatorname{Lim}_{n \to D} u_n = \alpha$.

According to Definition 5.1.6, local convergence has to happen on the initial segment C of the domain S of the sequence, but in the remaining part of S the sequence may not deviate. In the case of strong convergence, observe that if $C = S$, we have already $u_n \in \alpha$ on a final segment of C. If $C \subset S$, we have $u_n \in \alpha$ for all $n \in S \setminus C$. So the strong convergence already happens on C, or immediately beyond, and must continue on the whole of S.

The property that (u_n) already strongly converges to α within the initial segment C, or just beyond, depends on the nature of the external sets C and A. If the strong convergence happens just beyond C, either C is a galaxy, while A is a halo, or C is a halo, while A is a galaxy. We illustrate this with an example.

Example 5.1.4. Let $u : \mathbb{N} \to \mathbb{R}$ be defined by $u_n = \frac{1}{n}$. If $C = {}^{\sigma}\mathbb{N}$ and $D = \not\infty \cap \mathbb{N}$, we have $\lim_{n \to D} u_n = \oslash$ and $\operatorname{Lim}_{n \to D} u_n = \oslash$, because $u_n \simeq 0$ for all $n \simeq \infty$. Let $\omega \in \mathbb{N}$ be unlimited. If $C' = \oslash\omega$ and $D' = \mathbb{N} \setminus C'$, we have $\lim_{n \to D'} u_n = \operatorname{Lim}_{n \to D'} u_n = \frac{£}{\omega}$. The limits are not unique, because also $\lim_{n \to D'} u_n = \operatorname{Lim}_{n \to D'} u_n = \oslash$. However $\frac{£}{\omega}$ is minimal with respect to the latter limits, for if $\alpha \in \mathbb{E}$ such that $\alpha \subset \frac{£}{\omega}$ there exists $\nu \in \mathbb{N}$ with $\alpha < \frac{1}{\nu} < \frac{£}{\omega}$, so $u_\nu \notin \alpha$ (see Proposition 5.1.2).

Let v be the restriction of the sequence u to C. Again we have $\lim_{n \to D} u_n = \oslash$. To have strong convergence we have to construct an internal bridge from C to D. This can be u itself or $(u_n)_{n \leq \nu}$, for arbitrary unlimited ν.

In Theorem 5.1.12 we summarize all possibilities of strong convergence of flexible sequences with respect to an initial segment.

Theorem 5.1.12. *Let $S, C \subseteq \mathbb{N}$ be non-empty initial segments of $\mathbb{N}$ with $C \subseteq S$. Let $D = \mathbb{N} \setminus C$ be non-empty and $\alpha = a + A \in \mathbb{E}$. Let $u : S \to \mathbb{R}$ be a local flexible sequence which converges to α on S with respect to C. Then u*

converges strongly to α *with respect to* C *if* $C \subset S$ *or if* $C = S$ *and* C *and* A *are both prehalos or both pregalaxies.*

If C *is a galaxy, while* A *is a halo, or* C *is a halo, while* A *is a galaxy, for every internal bridge* $b : K \to \mathbb{R}$ *from* C *to* D *along* u*, with* $K \subseteq \mathbb{N}$*, there exists* $\nu \in D$ *such that* $b_{\restriction K\cap\{0,\ldots,\nu\}}$ *is strongly convergent with respect to* $K \cap C$*.*

We observe that, contrary to Theorem 5.1.11, convergence of a local sequence to a precise number also implies strong convergence, a property which is obvious for internal finite sequences.

We refer to [45] for a detailed proof of Theorem 5.1.12. Here we just prove the case which deals with internal sequences, for which convergence is only assumed on an external initial segment C of their domain S. It considers the cases where strong convergence possibly does not happen on C, but by conveniently restricting the domain S, they are strongly convergent with respect to this initial segment. Thus extensions by internal bridges are not needed. Theorem 5.1.13 is a sort of Matching Principle, enabling an overall description for the coexistence of different kinds of behaviour for, say, solutions of singular perturbations (see Chapter 9).

Theorem 5.1.13. *Let* (C, D) *be a cut of* $\mathbb{N}$ *into an external initial segment* C *and a final segment* D*. Let* S *be an internal initial segment of* $\mathbb{N}$ *such that* $C \subset S$*. Let* $u : S \to \mathbb{R}$ *be an internal sequence and* $\alpha = a + A \in \mathbb{E}$ *be such that* $u_{\restriction C}$ *converges to* α *with respect to* C*. If* C *is a pregalaxy and* A *is a halo, or if* C *is a halo and* A *is a galaxy, there exists* $\nu \in S \cap D$ *such that the sequence* u *converges strongly to* α *on* $\{0, ..., \nu\}$ *with respect to* C*.*

Proof. Without loss of generality we may assume that $\alpha = A$. Observe that $S \setminus C \neq \emptyset$, for C is external and S is internal. Assume first that C is a galaxy and A is a halo. Then there is a standard set W and an internal function $a : W \to \mathbb{R}^+$ such that $A = \bigcap_{\mathrm{st}(w)\in W}[-a_w, a_w]$, and a standard set Z and an internal function $c : Z \to \mathbb{N}$ such that $C = \bigcup_{\mathrm{st}(z)\in Z}\{0, ..., c_z\}$. Then

$$\forall^{\mathrm{stfin}} I \subseteq W \exists^{\mathrm{st}} z \in Z \forall w \in I \forall n \in \mathbb{N}(c_z \leq n \in C \to |u_n| \leq a_w).$$

Because u is internal and C is external, by Cauchy's Principle we obtain that

$$\forall^{\mathrm{stfin}} I \subseteq W \exists^{\mathrm{st}} z \in Z \exists f \in D \forall w \in I \forall n \in \mathbb{N}(c_z \leq n \leq f \to |u_n| \leq a_w).$$

By Standardization, respectively the Saturation Principle, there exists a standard function $\widetilde{z} : W \to Z$, respectively an internal function $\widetilde{f} : W \to \mathbb{N}$ such that $\widetilde{f}(w) \in D$ for all standard $w \in W$ and

$$\forall^{\mathrm{st}} w \in W \forall n \in \mathbb{N}(c_{\widetilde{z}(w)} \leq n \leq \widetilde{f}(w) \to |u_n| \leq a_w).$$

Put $G = \bigcup_{\mathrm{st}(w)\in W}\{\widetilde{f}(w), ..., \max S\}$. Because G is a pregalaxy included in the halo $S \cap D$, by the Fehrele Principle it is strictly included in $S \cap D$. Let

$\nu \in (S \cap D) \setminus G$. Observe that $\bigcup_{\mathrm{st}(w) \in W} \{0, ..., c_{\widetilde{z}(w)}\} \subseteq C$. Hence $u_n \in A$ at least for all $n \in \{0, ..., \nu\} \cap D$. Then u converges strongly to A on $\{0, ..., \nu\}$ with respect to C.

Secondly, assume that C is a halo and A is a pregalaxy. Let Z be a standard set and $c : Z \to \mathbb{N}$ be an internal function such that $C = \bigcap_{\mathrm{st}(z) \in Z} \{0, ..., c_z\}$; we may assume that c is decreasing at least on ${}^{\sigma}Z$. Suppose that there does not exist $\nu \in S \cap D$ such that u converges strongly to A on $\{0, ..., \nu\}$ with respect to C. Then

$$\forall^{\mathrm{st}} z \in Z \exists n \in D(n < c_z \wedge |u_n| > A).$$

By the Saturation Principle there exists an internal function $\widetilde{n} : Z \to \mathbb{N}$ such that $\widetilde{n}(z) \in D, \widetilde{n}(z) < c_z$ and $\left|u_{\widetilde{n}(z)}\right| > A$ for all standard $z \in Z$. By Idealization if $A = 0$, and if A is a galaxy, by the Fehrele Principle, because $\bigcap_{\mathrm{st}(z) \in Z} \left[0, \left|u_{\widetilde{n}(z)}\right|\right]$ is a prehalo, there exists $\varepsilon \in \mathbb{R}$ such that $A < \varepsilon < \left|u_{\widetilde{n}(z)}\right|$ for all standard $z \in Z$. Because $u_{\restriction C}$ converges to A with respect to C, there exists $\overline{c} \in C$ such that $|u_n| < \varepsilon$ for all $n \in C, n \geq \overline{c}$. Choose standard $\overline{z} \in Z$. Now

$$\{\overline{c}, ..., \widetilde{n}(\overline{z})\} \cap D \subseteq \{n \in \mathbb{N} : \exists m \in \mathbb{N}(\overline{c} \leq m \leq n \wedge |u_m| > \varepsilon)\}.$$

By Cauchy's Principle the inclusion is strict. So there exists $n \in C$ such that $|u_m| > \varepsilon$, for some m with $\overline{c} \leq m \leq n$, a contradiction. Hence u converges strongly to A on some interval $\{0, ..., \nu\}$ with respect to C, with $\nu \in D \cap S$. □

Exercises

Exercise 5.1.1. *Give a detailed proof of the claims of Remark 5.1.1.*

Exercise 5.1.2. *Prove Proposition 5.1.3.*

Exercise 5.1.3. *Show that given a flexible sequence (u_n) and a neutrix N, if $u_n \underset{N}{\longrightarrow} \alpha$ then $|u_n| \underset{N}{\longrightarrow} |\alpha|$. I.e. the absolute value preserves N-convergence.*

Exercise 5.1.4. *Give an example of a neutrix N and a sequence which is N-convergent but not strongly convergent.*

Exercise 5.1.5. *Show that restricting the sequence (u_n) of Example 5.1.4 to $S = \{0, ..., \omega\}$, using a similar argument, the sequence strongly converges to $\oslash$ with respect to ${}^{\sigma}\mathbb{N}$ and to $\frac{£}{\omega}$ with respect to $\oslash\omega$.*

Exercise 5.1.6. *Let $\varepsilon \simeq 0$, $\varepsilon > 0$ and $t_n = n!\varepsilon^n$, for all $n \in \mathbb{N}$. Give a local subsequence (τ_n) and an initial segment C, such that with $D = \mathbb{N} \setminus C$ (i) $\mathrm{Lim}_{n \to D} \tau_n = \oslash$; (ii) $\mathrm{Lim}_{n \to D} \tau_n = M_\varepsilon$; (iii) $\mathrm{Lim}_{n \to D} \tau_n = m_\varepsilon$.*

5.2 Operations on flexible sequences

In this section we show that the usual properties of the convergence of sequences remain valid or may be adapted to the context of flexible sequences. Subsection 5.2.1 deals with boundedness and ordering and Subsection 5.2.2 with the behaviour under operations, where we note that the size of the neutrix N in the N-convergence condition may change. The strong convergence theorems appear to be particularly useful for the proof.

5.2.1 Boundedness and monotonicity

Definition 5.2.1. *We say that a flexible sequence* (u_n) *is* bounded *if there exists* $\alpha \in \mathbb{E}$ *such that* $\alpha \neq \mathbb{R}$ *and* $|u_n| \leq \alpha$*, for all* $n \in \mathbb{N}$.

Clearly the element α in Definition 5.2.1 may be supposed to be precise.

In classical analysis convergent sequences are bounded. This is no longer the case for flexible sequences, as shown by Example 5.2.1. However, it is possible to give an adapted version of this result, proving that an N-convergent flexible sequence is bounded for large enough indices (Theorem 5.2.3). Such sequences will be called *eventually bounded.*

Definition 5.2.2. *A flexible sequence* (u_n) *is said to be* eventually bounded *if there exist* $n_0 \in \mathbb{N}$ *and* $\alpha \neq \mathbb{R}$ *such that* $|u_n| \leq \alpha$ *for all* $n \geq n_0$.

Example 5.2.1. Let (u_n) be the flexible sequence defined by $u_n = \begin{cases} \mathbb{R}, & n \in \pounds \\ 0, & n \notin \pounds. \end{cases}$ This flexible sequence converges to 0 but is not bounded.

The sequence in Example 5.2.1 above is eventually bounded but not bounded, so the two notions of boundedness are not equivalent in general. However the two notions do coincide if the sequence is precise.

Proposition 5.2.1. *A precise sequence* (a_n) *defined on a finite set is bounded.*

Proof. If the sequence a is internal the result is obvious. If a is not internal, by the Representation Theorem for sequences there exist a standard set X, an internal family of (internal) finite sequences $(u_x)_{x \in X}$ and a family of prehalos $(\Delta_x)_{x \in X}$ such that $a = \bigcup_{\mathrm{st}(x) \in X} u_{x \restriction \Delta_x}$. Then $u_{x \restriction \Delta_x}$ has an upper bound for all $x \in X$. By Corollary 4.1.6 there exists M such that $M > u_x$ for all standard $x \in X$. As a consequence the sequence (a_n) is bounded. □

Corollary 5.2.2. *A precise sequence* (a_n) *is bounded if and only if it is eventually bounded.*

Proof. Because the sequence a is eventually bounded there exist $k \in \mathbb{N}$ and $M_0 \in \mathbb{R}$ such that $|a_n| < M_0$, for $n \geq k$. By Proposition 5.2.1 there exists

M_1 such that $a_{\restriction[0,\ldots,k]} < M_1$. Hence a is bounded. The other implication is obvious. □

Theorem 5.2.3. *Every N-convergent sequence is eventually bounded.*

Proof. Let N be a neutrix and (u_n) be a flexible sequence such that $u_n \underset{N}{\longrightarrow} \alpha$. Let $\varepsilon > N$. Then there exists $n_0 \in \mathbb{N}$ such that for $n \geq n_0$ we have $|u_n| - |\alpha| \leq |u_n - \alpha| < \varepsilon$ for all $n \geq n_0$. Hence $|u_n| \leq \varepsilon + |\alpha|$ for all $n \geq n_0$ and we conclude that (u_n) is eventually bounded. □

Let (u_n) be an internal sequence of real numbers and let $n_0 \in \mathbb{N}$. It is well-known that if $u_n \geq 0$ for $n \geq n_0$ and $u_n \longrightarrow a$ for some $a \in \mathbb{R}$ then $a \geq 0$. The following proposition gives an adapted version of this result for flexible sequences.

Proposition 5.2.4. *Let N be a neutrix, and $\alpha = a + A \in \mathbb{E}$. Let (u_n) be a flexible sequence such that $u_n \underset{N}{\longrightarrow} \alpha$. If there exists $n_0 \in \mathbb{N}$ such that $N \leq u_n$ for $n \geq n_0$, then $N \leq \alpha + N$.*

Proof. Observe that $N(\alpha) = A \subseteq N$, so $\alpha + N = a + N$ and $\alpha + N$ is an N-limit of the sequence (u_n). If $\alpha + N$ is a neutrix, then $\alpha + N = N$ and we are done. If not, we may assume that N and $\alpha + N$ are disjoint. Suppose that $\alpha + N = a + N < N$. Let $\varepsilon = -\frac{a}{2} > N$. Then there exists $k_0 \in \mathbb{N}, k_0 \geq n_0$ such that for $n \geq k_0$ it holds that $|u_n - a + N| < \varepsilon$. So $a - \varepsilon < u_n + N < a + \varepsilon$ for all $n \geq k_0$. This means that $\frac{3}{2}a < u_n < \frac{a}{2} < N$ for all $n \geq k_0$, in contradiction with the assumption. Hence $N < \alpha + N$. Combining the two cases we conclude that $N \leq \alpha + N$. □

We show next an adapted version of the classical result expressing that the product of a bounded sequence with an infinitesimal sequence (sequence which converges to zero) is an infinitesimal sequence. This may also be seen as a special case of Theorem 5.2.9 where we deal with the product of two sequences.

Proposition 5.2.5. *Let N be a neutrix and let $(u_n), (v_n)$ be flexible sequences. If $u_n \underset{N}{\longrightarrow} 0$ and (v_n) is eventually bounded by $\alpha \in \mathbb{E}$, then $(u_n v_n) \underset{\alpha N}{\longrightarrow} 0$.*

Proof. Since (v_n) is eventually bounded by $\alpha \in \mathbb{E}$, we have that $\alpha \neq \mathbb{R}$ and there exists $n_1 \in \mathbb{N}$ such that $|v_n| < \alpha$, for all $n \geq n_1$. Without loss of generality we may assume that α is precise. Let $\varepsilon > \alpha N$. Then there exists $n_2 \in \mathbb{N}$ such that $|u_n| < \frac{\varepsilon}{\alpha}$ for $n \geq n_0$. Let $n_0 = \max\{n_1, n_2\}$. Then $|u_n v_n| = |u_n|\,|v_n| < \frac{\varepsilon}{\alpha}\alpha = \varepsilon$ for $n \geq n_0$. Hence $(u_n v_n) \underset{\alpha N}{\longrightarrow} 0$. □

We finish with a version of the squeeze theorem. As in the classical case, this theorem enables one to calculate the N-limit of a sequence (v_n) by comparison with two other flexible sequences whose flexible limits are equal and already known.

Theorem 5.2.6 (Squeeze Theorem). *Let M, N be neutrices and let $(u_n), (v_n)$ and (w_n) be flexible sequences such that $u_n \underset{N}{\longrightarrow} \alpha$, $w_n \underset{M}{\longrightarrow} \alpha$, for some $\alpha \in \mathbb{E}$. If there is $n_0 \in \mathbb{N}$ such that $u_n \leq v_n \leq w_n$ for $n \geq n_0$, then $v_n \underset{N+M}{\longrightarrow} \alpha$. In particular, if $N = M$ then $v_n \underset{N}{\longrightarrow} \alpha$.*

Proof. Assume that there exists $n_0 \in \mathbb{N}$ such that $u_n \leq v_n \leq w_n$ for $n \geq n_0$. Let $\varepsilon \in \mathbb{R}$ be such that $\varepsilon > N + M$. Then $\varepsilon > N$ and $\varepsilon > M$. There exist $n_1, n_2 \in \mathbb{N}$ such that for $n \geq n_1$ we have $|u_n - \alpha| < \varepsilon$ and for $n \geq n_2$ we have $|w_n - \alpha| < \varepsilon$. Let $n_3 = \max\{n_0, n_1, n_2\}$. Then for $n \geq n_3$ we have $-\varepsilon < u_n - \alpha \leq v_n - \alpha \leq w_n - \alpha < \varepsilon$. So $|v_n - \alpha| < \varepsilon$. Hence, by Remark 5.1.1, one concludes that $v_n \underset{N+M}{\longrightarrow} \alpha$. □

5.2.2 Operations

In this section we show that the usual rules of operations of limits of sequences are valid in the context of definable flexible sequences and for strong convergence. The former is in fact a consequence of the latter and of Theorem 5.1.11.

Definition 5.2.3. *A flexible sequence (u_n) is said to be* zeroless *if $u_n \neq N(u_n)$, for all $n \in \mathbb{N}$, i.e. if $0 \notin u_n$, for all $n \in \mathbb{N}$.*

Theorem 5.2.7. *Let $c \in \mathbb{R}$. Let $(u_n), (v_n)$ be flexible sequences such that $u_n \rightsquigarrow \alpha$ and $v_n \rightsquigarrow \beta$ for some $\alpha, \beta \in \mathbb{E}$. Then*

1. *$u_n + v_n \rightsquigarrow \alpha + \beta$.*
2. *$u_n - v_n \rightsquigarrow \alpha - \beta$.*
3. *$cu_n \rightsquigarrow c\alpha$.*
4. *$(u_n v_n) \rightsquigarrow \alpha\beta$.*
5. *If (v_n) and β are zeroless then $\frac{u_n}{v_n} \rightsquigarrow \frac{\alpha}{\beta}$.*

Proof. By the assumptions there exist $n_1, n_2 \in \mathbb{N}$ such that $u_n \subseteq \alpha$ for all $n \geq n_1$, and $v_n \subseteq \beta$ for all $n \geq n_2$. Then $u_n + v_n \subseteq \alpha + \beta$, $u_n - v_n \subseteq \alpha - \beta$, $cu_n \subseteq c\alpha$, $u_n v_n \subseteq \alpha\beta$ and $\frac{u_n}{v_n} \subseteq \frac{\alpha}{\beta}$ for all $n \geq n_0$, where $n_0 = \max\{n_1, n_2\}$. This implies the result. □

We prove next an analogue of Theorem 5.2.7 for N-convergence. We need the following preparatory result.

Lemma 5.2.8. *Let N be a neutrix and let (a_n) be a zeroless precise sequence such that $a_n \underset{N}{\longrightarrow} a$, for some $a \in \mathbb{R}, |a| > N$. Then there exists $n_0 \in \mathbb{N}$ such that for $n \geq n_0$ it holds that $\frac{|a|}{2} < |a_n| < 2\,|a|$.*

Proof. Let $\varepsilon = \frac{|a|}{2}$. Clearly $\frac{|a|}{2} > N$. Then there exists $n_0 \in \mathbb{N}$ such that for all $n \geq n_0$ one has $|a_n - a| < \varepsilon = \frac{|a|}{2}$. It follows that $|a| - \frac{|a|}{2} < |a_n| < |a| + \frac{|a|}{2} = \frac{3}{2}|a|$. Hence $\frac{|a|}{2} < |a_n| < \frac{3}{2}|a| \leq 2|a|$, for all $n \geq n_0$. □

Theorem 5.2.9. *Let N, M be neutrices and let $c \in \mathbb{R}$. Let $(u_n), (v_n)$ be flexible sequences such that $u_n \underset{N}{\longrightarrow} \alpha$ and $v_n \underset{M}{\longrightarrow} \beta$ for some $\alpha, \beta \in \mathbb{E}$. Then*

1. $(u_n + v_n) \underset{N+M}{\longrightarrow} \alpha + \beta$.

2. $(u_n - v_n) \underset{N+M}{\longrightarrow} \alpha - \beta$.

3. $(cu_n) \underset{cN}{\longrightarrow} c\alpha$.

4. $(u_n v_n) \underset{K}{\longrightarrow} \alpha\beta$, *with* $K = \alpha M + \beta N + MN$.

5. *if (u_n) and α are zeroless, then $\left(\frac{1}{u_n}\right)$ is $\frac{N}{a^2}$-convergent to $\frac{1}{\alpha}$.*

In particular, if $M = N$, then $(u_n + v_n) \underset{N}{\longrightarrow} \alpha + \beta$ and $(u_n - v_n) \underset{N}{\longrightarrow} \alpha - \beta$.

Proof. 1. Let $\varepsilon > N + M$. Then $\varepsilon > N$ and $\varepsilon > M$. So there exist $n_1, n_2 \in \mathbb{N}$ such that $|u_n - \alpha| < \frac{\varepsilon}{2}$ for $n \geq n_1$ and $|v_n - \beta| < \frac{\varepsilon}{2}$ for $n \geq n_2$. Let $n_0 = \max\{n_1, n_2\}$. Then for $n \geq n_0$ we have $|(u_n + v_n) - (\alpha + \beta)| \leq |u_n - \alpha| + |v_n - \beta| < \frac{\varepsilon}{2} + \frac{\varepsilon}{2} = \varepsilon$. Hence $(u_n + v_n) \underset{N+M}{\longrightarrow} \alpha + \beta$.

If $N = M$, then $N + M = N$. Hence $(u_n + v_n) \underset{N}{\longrightarrow} \alpha + \beta$.

2. The proof is analogous to the proof of Part 1.

3. If $c = 0$, then $cu_n = 0$ for all $n \in \mathbb{N}$. Hence (cu_n) converges to 0. Assume $c \neq 0$. Let $\varepsilon > cN$. Then $\frac{\varepsilon}{|c|} > N$. So there exists $n_0 \in \mathbb{N}$ such that $|u_n - \alpha| < \frac{\varepsilon}{|c|}$ for all $n \geq n_0$. Because $c \in \mathbb{R}$, distributivity holds. Hence $|cu_n - c\alpha| = |c(u_n - \alpha)| = |c|\,|u_n - \alpha| < |c|\,\frac{\varepsilon}{|c|} = \varepsilon$. We conclude that $(cu_n) \underset{cN}{\longrightarrow} c\alpha$.

4. Let $\alpha = a + A, \beta = b + B$. We will consider three cases: (a) M, N are both non-zero, (b) exactly one of the neutrices M, N is zero and (c) M, N are both zero.

(a) By Proposition 5.1.4 we have that $u_n \underset{N}{\longrightarrow} a + N$ and $v_n \underset{M}{\longrightarrow} b + M$. Then $u_n \longrightarrow a + N$ and $v_n \longrightarrow b + M$. Hence $u_n \rightsquigarrow a + N$ and $v_n \rightsquigarrow b + N$ by Theorem 5.1.11 and the result follows from Theorem 5.2.7.

(b) Without loss of generality we assume that $M = 0$. By Proposition 5.1.3 we may assume that $\alpha = a + N$ and $\beta = b \in \mathbb{R}$. Then $K = bN$. Let $\varepsilon > K$. By Theorem 5.1.11 there exists $n_1 \in \mathbb{N}$ such that $u_n \subseteq \alpha$ for all $n \geq n_1$. Then for all $n \geq n_1$ we have

$$|u_n v_n - \alpha\beta| = |u_n v_n - \alpha b| \leq |\alpha v_n - \alpha b| = |a||v_n - b| + Nv_n + Nb. \quad (5.3)$$

We show that there exists $n_2 \in \mathbb{N}$ such that

$$|a||v_n - b| < \frac{\varepsilon}{3} \quad (5.4)$$

for all $n \geq n_2$. If $a = 0$ the inequality (5.4) is trivial. Otherwise, because $\varepsilon > K$, it holds that $\frac{\varepsilon}{3|a|} > 0$. Since $v_n \longrightarrow b$, there exists $n_2 \in \mathbb{N}$ such that $|a||v_n - b| < \frac{\varepsilon}{3}$ for all $n \geq n_1$.

To estimate the term Nv_n we consider two cases: $b = 0$ and $b \neq 0$.

If $b \neq 0$, there exists $n_3 \in \mathbb{N}$ such that $|v_n| < 2|b|$ for all $n \geq n_3$. It follows that

$$Nv_n \leq N \cdot (2|b|) = K < \frac{\varepsilon}{3}. \tag{5.5}$$

From (5.3)-(5.5) we obtain $|u_n v_n - \alpha\beta| < \varepsilon$, for all $n \geq n_4$, where $n_4 = \max\{n_1, n_2, n_3\}$. Hence $u_n v_n \underset{K}{\longrightarrow} \alpha\beta$.

If $b = 0$, then $K = 0$. Because $N \neq \mathbb{R}$, there exists $\delta > 0$ such that $\frac{\varepsilon}{\delta} > N$. Also, since $v_n \longrightarrow 0$, there exists $n_5 \in \mathbb{N}$ such that $|v_n| < \frac{\delta}{3}$ for all $n \geq n_5$. It follows that

$$v_n N < \frac{\delta}{3} N < \frac{\varepsilon}{3} \tag{5.6}$$

for all $n \geq n_5$. Put $n_0 = \max\{n_1, n_2, n_5\}$. Then, from (5.3), (5.4) and (5.6) we obtain that $|u_n v_n - \alpha\beta| < \varepsilon$ for all $n \geq n_0$. Hence $u_n v_n \underset{K}{\longrightarrow} \alpha\beta$.

(c) We put $A_n = N(u_n)$ and $B_n = N(v_n)$. By Proposition 5.1.5 it holds that $A_n \longrightarrow 0$ and $B_n \longrightarrow 0$.

By Theorem 5.2.3 the sequences $(u_n), (v_n)$ are eventually bounded. Then there exist $K, L \in \mathbb{R}$ and $n_0 \in \mathbb{N}$ such that $|u_n| \leq K, |v_n| \leq L$ for all $n \geq n_0$. Let $\varepsilon > 0$. Since $(u_n), (v_n)$ are convergent there exist $n_1, n_2, n_3 \in \mathbb{N}$ such that $|u_n - a| < \frac{\varepsilon}{3}(|L| + 1)$ for $n \geq n_1$, $|v_n - b| < \frac{\varepsilon}{3}(|a| + 1)$ for $n \geq n_2$ and $|B_n| < \frac{\varepsilon}{3}(|K| + |a| + 1)$ for $n \geq n_3$. Let $n_4 = \max\{n_0, n_1, n_2, n_3\}$. Using Theorem 3.3.1, for $n \geq n_4$ we have

$$\begin{aligned}|u_n v_n - ab| &\leq |v_n u_n - av_n + av_n - ab| = |v_n(u_n - a)| + B_n u_n + B_n a + a(v_n - b)| \\ &\leq |v_n(u_n - a)| + (|u_n| + |a|)B_n + |a||v_n - b| < \frac{\varepsilon}{3} + \frac{\varepsilon}{3} + \frac{\varepsilon}{3} = \varepsilon.\end{aligned}$$

Hence $u_n v_n \longrightarrow ab$.

5. Let $\varepsilon > \frac{N}{a^2}$. Then $\frac{a^2}{b}\varepsilon > N$ for all appreciable b. Because (u_n) is N-convergent to α, it holds that (A_n) is N-convergent to N. Then there exists $n_0 \in \mathbb{N}$ such that for all $n \geq n_0$ we have $|a - u_n| < \frac{a^2}{6}\varepsilon$ and $|A_n - N| < \frac{a^2}{3}\varepsilon$; hence also $|A_n| \leq \frac{a^2}{3}\varepsilon$. By Lemma 5.2.8 there exists $n_1 \in \mathbb{N}$ such that $\frac{|a|}{2} < |a_n| < 2|a|$ for all $n \geq n_1$. Let $k = \max\{n_0, n_1\}$ and $n \geq k$ be arbitrary. There exists $a_n \in \mathbb{R}$ such that $u_n = a_n + A_n$. Then by Part 2 of Theorem 1.4.2

$$\begin{aligned}\left|\frac{1}{u_n} - \frac{1}{\alpha}\right| &= \left|\frac{u_n}{a_n^2} - \frac{\alpha}{a^2}\right| = \frac{|a^2 u_n - a_n^2 \alpha|}{a_n^2 a^2} = \frac{|a^2 a_n - a_n^2 a|}{a_n^2 a^2} + \frac{a^2 A_n + a_n^2 N}{a^2 a_n^2} \\ &\leq \frac{|a - a_n|}{a_n a} + \frac{a^2 A_n + 4|a|^2 N}{\frac{a^4}{4}} \leq \frac{|a - u_n|}{\frac{a^2}{2}} + \frac{A_n + N}{a^2} < \frac{\varepsilon}{3} + \frac{2\varepsilon}{3} = \varepsilon.\end{aligned}$$

Hence $\left(\frac{1}{u_n}\right)$ is $\frac{N}{a^2}$-convergent to $\frac{1}{\alpha}$. □

We end this section with some examples which illustrate how the neutrix of convergence changes under operations.

Example 5.2.2. 1. Let $(u_n), (v_n), (w_n)$ be flexible sequences defined respectively by $u_n = \frac{1}{n} + \oslash$, $v_n = \frac{1}{n^2} + \varepsilon\pounds$ and $w_n = \omega^2 + \omega\pounds$, where ε is a fixed infinitesimal and ω a fixed infinitely large real number. Then $u_n v_n = \frac{1}{n^3} + \frac{\varepsilon\pounds}{n} + \frac{\oslash}{n^2} + \varepsilon\oslash$, which is $(\varepsilon\oslash)$-convergent to 0 and $u_n w_n = \frac{\omega^2}{n} + \frac{\omega\pounds}{n} + \omega^2\oslash$ which is $(\omega^2\oslash)$-convergent to 0.

2. Let ω be unlimited and let (w_n) be the constant sequence from the previous example. We have that $w_n \underset{\omega\pounds}{\longrightarrow} \omega^2$ and $(w_n w_n) \underset{\omega^3\pounds}{\longrightarrow} \omega^4$. Consider the two constant representatives $a = \omega^2$ and $b = \omega^2 + \omega$. Then $ab - a^2 = \omega^3$ which is bigger than the neutrix of convergence $\omega\pounds$. So the sequence is $\omega\pounds$-divergent.

Exercises

Exercise 5.2.1. *Let (u_n) be a flexible sequence and α an external number. Show that if there exists $n_0 \in \mathbb{N}$ such that $u_n = \alpha$ for $n \geq n_0$, then $u_n \longrightarrow \alpha$.*

Exercise 5.2.2. *Show that the conclusion in Proposition 5.2.4 remains true if one replaces $\leq$ by $\geq$.*

Exercise 5.2.3. *Let N be a neutrix and $(u_n), (v_n)$ be two flexible sequences such that $u_n \underset{N}{\longrightarrow} \alpha$, $v_n \underset{N}{\longrightarrow} \beta$, for some $\alpha, \beta \in \mathbb{E}$. Show that if there exists $n_0 \in \mathbb{N}$ such that $u_n \leq v_n$ for all $n \geq n_0$, then $\alpha + N \leq \beta + N$.*

Exercise 5.2.4. *Let $\omega \simeq +\infty$ and let $(u_n), (v_n), (w_n), (x_n), (y_n)$ be the flexible sequences defined respectively by $u_n = \frac{1}{n} + \oslash$, $v_n = 3$, $w_n = \omega$, $x_n = \omega + \pounds\sin(n)$, $y_n = \omega^2 + \omega\pounds$, $z_n = \frac{1}{\omega}$.*

1. *Verify that*
   - *(a)* $u_n \underset{\oslash}{\longrightarrow} 0$.
   - *(b)* $v_n \longrightarrow 3$.
   - *(c)* $w_n \longrightarrow \omega$.
   - *(d)* $x_n \underset{\pounds}{\longrightarrow} \omega$.
   - *(e)* $y_n \underset{\omega\pounds}{\longrightarrow} \omega^2$.
   - *(f)* $z_n \longrightarrow \frac{1}{\omega}$.
2. *Show that*
   - *(a)* $(u_n v_n) \underset{\oslash}{\longrightarrow} 0$.

(b) $(v_n w_n) \longrightarrow 3\omega$.

(c) $(w_n x_n) \underset{\omega \pounds}{\longrightarrow} \omega^2$.

(d) $(x_n y_n) \underset{\omega^2 \pounds}{\longrightarrow} \omega^3$.

(e) $(x_n z_n) \underset{\frac{\pounds}{\omega}}{\longrightarrow} 1$.

(f) $(y_n y_n) \underset{\omega^3 \pounds}{\longrightarrow} \omega^4$.

3. *What is the smallest neutrix N for which the sequences $(u_n + y_n)$ and $(y_n - x_n)$ are N-convergent?*

5.3 Cauchy flexible sequences

In this section we prove that Cauchy's characterization of convergence for sequences, i.e. the fact that a sequence is convergent if and only if it is a Cauchy sequence, still holds for the convergence of flexible sequences.

Intuitively, Cauchy sequences are sequences whose terms become arbitrarily close to each other as the sequence progresses. In other words the difference between terms of the sequence converges to 0. For flexible sequences this means that the terms of the sequence are close to each other up to a neutrix N, i.e. the difference between terms of the sequence N-converges to 0.

Definition 5.3.1. *Let N be a neutrix and (u_n) be a flexible sequence. We say that (u_n) is N-*Cauchy *if and only if*

$$\forall \varepsilon > N \exists k \in \mathbb{N} \forall m \in \mathbb{N} \forall n \in \mathbb{N} (k \leq m \wedge k \leq n \Rightarrow |u_m - u_n| < \varepsilon). \tag{5.7}$$

If $N = 0$ we simply say that (u_n) is a Cauchy sequence.

Definition 5.3.2. *Let N be a neutrix and (u_n) be a flexible sequence. We say that (u_n) is* strongly N-Cauchy *if and only if*

$$\exists k \in \mathbb{N} \forall n \in \mathbb{N} \forall m \in \mathbb{N} (k \leq m \wedge k \leq n \Rightarrow u_n - u_m \subseteq N). \tag{5.8}$$

It is not difficult to show that in (5.7) the element ε can be taken as precise.

Let N be a neutrix and let (u_n) be an N-Cauchy sequence. Clearly, for all $\varepsilon > N$ and for all $k \in \mathbb{N}$ there exists $p \in \mathbb{N}$ such that $|u_{n+k} - u_n| < \varepsilon$ for $n > p$.

Example 5.3.1. Let N be a neutrix. Let (u_n) be the flexible sequence defined by $u_n = s_n + N$, where (s_n) is a Cauchy sequence. We show that (u_n) is

N-Cauchy. Let $\varepsilon > N$. Then there exists $n_0 \in \mathbb{N}$ such that for $m, n > n_0$ it holds that $|s_m - s_n| < \delta$, for $0 < \delta < N$. Then

$$\begin{aligned} |u_m - u_n| &= |s_m + N - (s_n + N)| \\ &= |s_m - s_n + N| \leq |s_m - s_n| + N \leq \delta + N \leq N + N = N < \varepsilon. \end{aligned}$$

Hence (u_n) is an N-Cauchy sequence.

Given a neutrix N, Cauchy's criterion holds for both N-convergence and strong convergence.

Theorem 5.3.1. *Let N be a neutrix and (u_n) be a flexible sequence. Then*

1. *(u_n) is N-convergent if and only if it is N-Cauchy.*
2. *Let $\alpha \in \mathbb{E}$. Then (u_n) strongly converges to α if and only if it is strongly $N(\alpha)$-Cauchy.*

As a consequence we obtain that if N is a non-zero neutrix, the conditions (5.7) and (5.8) are equivalent.

Corollary 5.3.2. *Let (u_n) be a flexible sequence and $N \neq 0$ be a neutrix. Then (u_n) is N-Cauchy if and only if it is strongly N-Cauchy.*

Proof. Using Theorems 5.3.1 and 5.1.11 one derives that a sequence (u_n) is N-Cauchy if and only if it is N-convergent, if and only if it is N-strongly convergent, and if and only if it is strongly N-Cauchy. □

Before we prove Theorem 5.3.1 we study some properties of N-Cauchy sequences and strongly N-Cauchy sequences and present some auxiliary results.

Proposition 5.3.3. *Let N be a neutrix. Every N-Cauchy flexible sequence is eventually bounded.*

Proof. Let (u_n) be an N-Cauchy sequence. Let $\varepsilon > N$. Then there exists $k \in \mathbb{N}$ such that $|u_n - u_m| < \varepsilon$ for all $m, n \geq k$. So for $m, n \geq k$ we have $|u_n| \leq |u_n - u_m| + |u_m| < |u_m| + \varepsilon$. Taking $m = k$ one concludes that (u_n) is eventually bounded. □

We show that the components of a flexible N-Cauchy sequence are also N-Cauchy and the components of a strongly N-Cauchy sequence are strongly N-Cauchy sequences.

Proposition 5.3.4. *Let N be a neutrix. Let (a_n) be a precise sequence and (A_n) be a sequence of neutrices. Let (u_n) be a flexible sequence such that $u_n = a_n + A_n$ for all $n \in \mathbb{N}$. Then*

1. *the sequence (u_n) is N-Cauchy if and only if (a_n) and (A_n) are both N-Cauchy sequences.*

2. *the sequence (u_n) is strongly N-Cauchy if and only (a_n) and (A_n) are both strongly N-Cauchy sequences.*

Proof. 1. Assume that (u_n) is N-Cauchy. Let $\varepsilon > N$. Then there exists $n_0 \in \mathbb{N}$ such that $|a_n + A_n - (a_m + A_m)| < \varepsilon$ for all $m, n \geq n_0$. It follows that $|a_n - a_m| < \varepsilon$ and $|A_n - A_m| < \varepsilon$ for all $m, n \geq n_0$. Hence $(a_n), (A_n)$ are both N-Cauchy.

Conversely, assume that $(a_n), (A_n)$ are both N-Cauchy. Let $\varepsilon > N$. Then there exists $n_0 \in \mathbb{N}$ such that $|a_n - a_m| < \frac{\varepsilon}{2}$ and $|A_n - A_m| < \frac{\varepsilon}{2}$ for all $m, n \geq n_0$. It follows that for all $n, m \geq n_0$ we have $|a_n + A_n - (a_m + A_m)| = |a_n - a_m| + |A_n - A_m| < \frac{\varepsilon}{2} + \frac{\varepsilon}{2} = \varepsilon$. Hence (u_n) is N-Cauchy.

2. The result follows from the fact that $a_n + A_n - a_m + A_m \subseteq N$ if and only if $a_n - a_m \in N$ and $A_n + A_m \subseteq N$. □

Lemma 5.3.5. *Let N be a neutrix and (A_n) be a flexible sequence of neutrices. If (A_n) is N-Cauchy then $A_n \longrightarrow N$.*

Proof. Let $\varepsilon > N$. Then $\frac{\varepsilon}{2} > N$. Because (A_n) is N-Cauchy there exists $n_0 \in \mathbb{N}$ such that $|A_n - A_{n+p}| < \frac{\varepsilon}{2}$ for all $n \geq n_0$ and for all $p > 0$. It follows that for all $n \geq n_0$ and all $p > 0$ we have $|A_n - N| \leq |A_{n+p} + A_n - N| = |A_{n+p} - A_n| + N < \frac{\varepsilon}{2} + \frac{\varepsilon}{2} = \varepsilon$. Hence $A_n \longrightarrow N$. □

Lemma 5.3.6. *Let N be a neutrix and (a_n) be an internal precise sequence. If (a_n) is N-Cauchy then (a_n) is N-convergent.*

Proof. Because (a_n) is precise and N-Cauchy, we have that (a_n) is bounded. Also, since (a_n) is internal, it admits a subsequence (a_{m_n}) convergent to some $a \in \mathbb{R}$. Let $\varepsilon > N$. Then $\frac{\varepsilon}{2} > N$. There exist $n_1, n_2 \in \mathbb{N}$ such that for all $m_n \geq n_1$ we have $|a_{m_n} - a| < \frac{\varepsilon}{2}$ and for all $m, n \geq n_2$ and for all $p \geq 0$ we have $|a_m - a_n| < \frac{\varepsilon}{2}$. Let $n_0 = \max\{n_1, n_2\}$. Then for all $n \geq n_0$ we have

$$|a_n - a| = |a_n - a_{m_n} + a_{m_n} - a| \leq |a_n - a_{m_n}| + |a_{m_n} - a| < \varepsilon.$$

Hence (a_n) is N-convergent. □

Lemma 5.3.7. *Let N be a neutrix and let (u_n) be an N-Cauchy sequence. If (u_n) has an internal N-convergent subsequence, it is N-convergent.*

Proof. Let $A_n = N(u_n)$ for all $n \in \mathbb{N}$. Let (a_{m_n}) be an internal subsequence of (u_n), i.e. $a_{m_n} \in u_{m_n}$ for all $n \in \mathbb{N}$. Assume that $a_{m_n} \underset{N}{\longrightarrow} a_0$ for some $a_0 \in \mathbb{R}$. Let $\varepsilon > N$. Then $\frac{\varepsilon}{2} > N$ and there exists $n_1 \in \mathbb{N}$ such that for $m_n \geq n_1$ we have $|a_{m_n} - a_0| \leq \frac{\varepsilon}{4}$. Also, since (u_n) is N-Cauchy, there exist $n_2, n_3 \in \mathbb{N}$ such that $|u_m - u_n| < \frac{\varepsilon}{2}$ for all $m, n \geq n_2$ and $A_n < \frac{\varepsilon}{4}$ for all $n \geq n_3$. Let $n_4 = \max\{n_1, n_3\}$. Then for all $m_n \geq n_4$ we have $|u_{m_n} - a| = |a_{m_n} + A_{m_n} - a_0| < \frac{\varepsilon}{2}$. Let $n_0 = \max\{n_4, n_2\}$. Then for all $n \geq m_{n_0} \geq n_0$ we have

$$|u_n - a_0| \leq |u_n - u_{m_{n_0}} + u_{m_{n_0}} - a_0| \leq |u_{m_{n_0}} - u_n| + |u_{m_0} - a_0| < \varepsilon.$$

Hence (u_n) N-converges to a_0. □

Proof of Theorem 5.3.1. 1. Assume that $u_n \underset{N}{\longrightarrow} \alpha$, for some $\alpha \in \mathbb{E}$. Then there exists $n_0 \in \mathbb{N}$ such that for $n \geq n_0$ we have $|u_n - \alpha| < \frac{\varepsilon}{2}$. Then for $m, n \geq n_0$ we have

$$\begin{aligned} |u_m - u_n| &\leq |u_m - u_n + N(\alpha)| = |u_m - \alpha - u_n + \alpha| \\ &\leq |u_m - \alpha| + |u_n - \alpha| < \frac{\varepsilon}{2} + \frac{\varepsilon}{2} = \varepsilon. \end{aligned}$$

Hence (u_n) is N-Cauchy.

Conversely, assume that (u_n) is N-Cauchy. By Proposition 4.4.10, there is an internal subsequence (a_{n_m}) of (u_n) which is also N-Cauchy. Then (a_{n_m}) is N-convergent by Lemma 5.3.6. By Lemma 5.3.7 we conclude that (u_n) is N-convergent.

2. Assume that (u_n) is strongly convergent to some $\alpha = a + A \in \mathbb{E}$. Then there exists $n_0 \in \mathbb{N}$ such that $u_n \subseteq \alpha$ for all $n \geq n_0$. It follows that $u_n - a \subseteq A$ for all $n \geq n_0$. So for all $n, m \geq n_0$ we have $u_n - u_m = u_n - a + a - u_m \subseteq A + A = A$. Hence (u_n) is strongly A-Cauchy.

Conversely, assume that N is a neutrix and (u_n) is strongly N-Cauchy. If $N = 0$, there exists $n_0 \in \mathbb{N}$ such that $u_n = a$ for all $n \geq n_0$. So $u_n \rightsquigarrow a$. If $N \neq 0$, the sequence (u_n) is N-Cauchy. By Part 1 of Theorem 5.3.1 the sequence (u_n) is N-convergent and by Theorem 5.1.11 it is strongly convergent. □

We finish with a result concerning the N-Cauchy property for initial segments which is useful to prove a nonstandard version of the Borel-Ritt theorem (see Chapter 9). We refer to [45] for a proof.

Theorem 5.3.8. *Let $S, C \subseteq \mathbb{N}$ be initial segments of $\mathbb{N}$ such that $C \subseteq S$. Let N be a neutrix and $u : S \to \mathbb{E}$ be an N-Cauchy sequence on S with respect to C. Then there exists $a \in \mathbb{R}$ such that (u_n) converges to $a + N$ on S with respect to C.*

Exercises

Exercise 5.3.1. *Show, using the definition, that the sequences defined in Exercise 5.2.4 are N-Cauchy, indicating in each case what is the corresponding neutrix N.*

Exercise 5.3.2. *Give an example of a sequence which is N-Cauchy but not strongly N-Cauchy.*

Chapter 6

Functions of external numbers

6.1 Limits of flexible functions 140
6.1.1 Relation with convergence for sequences; strong convergence 146
6.2 Flexible continuity 149
6.2.1 Outer continuity 150
6.2.2 Inner continuity 153
6.3 $M \times N$-derivation of flexible functions 156
6.4 Weak extrema and monotonicity 160

By Definition 4.4.1 flexible functions are definable functions of real arguments whose values are external numbers. As in Chapter 4 we restrict our study to flexible functions $F : X \subseteq \mathbb{R} \to \mathbb{E}$ with a (definable) representative f so that $F(x) = f(x) + N_F(x)$ for all $x \in X$, where $N_F(x)$ is the neutrix part of $F(x)$. We study these functions with respect to notions of limit and continuity. Much of the material originates from the Ph.D. thesis of Nam Van Tran [146].

We start with some topological notions. Let $\alpha \in \mathbb{E}$ such that $N(\alpha) \subset \mathbb{R}$. Let $r \in \mathbb{R}$, $r > N(\alpha)$. We call

$$B(\alpha, r) := \{x \in \mathbb{R} : |x - \alpha| < r\}$$

the *open ball* centered in α of radius $r > N(\alpha)$. We say that V is a *neighbourhood* of α if there exists an open ball $B(\alpha, r)$ with $r > N(\alpha)$ such that $B(\alpha, r) \subseteq V$. For example, the open interval $]-1, 1[$ is a neighbourhood of the external point $\alpha = \frac{1}{2} + \oslash$.

Definition 6.0.1. *Let $x_0 \in \mathbb{R}$ and N be a neutrix. If $x \in \mathbb{R}$ is such that $|x - x_0| \in N$ we say that x is N*-close *to x_0.*

Definition 6.0.2. *Let $\emptyset \neq X \subseteq \mathbb{R}$, $x_0 \in \mathbb{R}$ and let N be a neutrix. We call x_0 a N*-limit point *of X if $\forall \delta > N (B(x_0, \delta) \cap (X \setminus (x_0 + N)) \neq \emptyset)$.*

For example, the numbers 0 and $-1 + \varepsilon$, with $\varepsilon \simeq 0$, are $\oslash$-close to $(-1, 1)$. Both points are $\oslash$-limit points of $(-1, 1)$, and also of $(-1 + \oslash, 1)$. For $\varepsilon \simeq 0$ the point $-1 + \varepsilon$ is not $\oslash$-close to $(-1 + \oslash, 1)$. The case where $N = 0$ in Definition 6.0.2 corresponds to the usual notion of limit point.

6.1 Limits of flexible functions

We extend the notion of limit to flexible functions. In the same spirit as what was done for sequences in Chapter 5 (see also [45]) these limits depend on neutrices, say M, N, and so we call them *$M \times N$-limits*. Operations and properties of the $M \times N$-limits are investigated.

Definition 6.1.1. *Let M, N be neutrices, $\emptyset \neq X \subseteq \mathbb{R}$ and $F : X \to \mathbb{E}$ be a flexible function. Let $x_0 \in \mathbb{R}$ be a M-limit point of X. We say that an external number α is a $M \times N$-*limit *of F at x_0, and write ${}^{N}_{M}\lim\limits_{x \to x_0} F(x) = \alpha$ if*

$$\forall \varepsilon > N \exists \delta > M \forall x \in X (0 < |x - x_0| < \delta \Rightarrow |F(x) - \alpha| < \varepsilon). \tag{6.1}$$

If $M = N \neq 0$ we write simply $M\text{-}\lim\limits_{x \to x_0} F(x)$. The case where $M = N = 0$ corresponds to the usual notion of limit, so we will use the conventional notation $\lim\limits_{x \to x_0} F(x)$.

The notions of Definition 6.1.1 may be adapted in an obvious way to one-sided limits.

Example 6.1.1. Let $F : \mathbb{R} \to \mathbb{E}$ be the flexible function defined by $F(x) = x + x\oslash$. Then $\oslash\text{-}\lim\limits_{x \to 1} F(x) = 1 + \oslash$. To see this, let $\varepsilon > \oslash$ and put $\delta = \frac{\varepsilon}{2}$. Clearly $\delta > \oslash$. Then for all $x \in \mathbb{R}$ such that $|x - 1| < \delta$ one has $|F(x) - (1 + \oslash)| = |x - 1| + \oslash < \frac{\varepsilon}{2} + \frac{\varepsilon}{2} = \varepsilon$. Similarly, one shows that $\oslash\text{-}\lim\limits_{x \to 0} F(x) = \oslash$. Also $\lim\limits_{x \to 0} F(x) = 0$.

When considering $M \times N$-limits we are implicitly working with the largest possible neutrix M and the smallest possible neutrix N. The justification for this is given by the next proposition.

Proposition 6.1.1. *Let M_1, N_1, M_2, N_2 be neutrices such that $M_2 \subseteq M_1$ and $N_1 \subseteq N_2$. Let α be an external number, $F : X \subseteq \mathbb{R} \to \mathbb{E}$ be a flexible function and $x_0 \in \mathbb{R}^n$ be a M_1-limit point of X. If ${}^{N_1}_{M_1}\lim\limits_{x \to x_0} F(x) = \alpha$, then ${}^{N_2}_{M_2}\lim\limits_{x \to x_0} F(x) = \alpha$.*

Proof. Let $\varepsilon > N_2$. Then $\varepsilon > N_1$, so there exists $\delta > M_1$ such that $|F(x) - \alpha| < \varepsilon$ for all $x \in X$ such that $0 < |x - x_0| < \delta$. Because $M_2 \subseteq M_1$, it holds that $\delta > M_2$, and we conclude that ${}^{N_2}_{M_2}\lim\limits_{x \to x_0} F(x) = \alpha$. □

Many properties of limits of sequences carry over to $M \times N$-limits of functions, with similar proofs, which are often stated as exercises. The convergence with respect to an initial segment of Definition 5.1.6 is reflected by the fact that the N-limit should be attained when entering the complement of M.

Clearly, in (6.1), the element δ can be taken as precise. By Exercise 6.1.2 the element ε can also be taken as precise.

Proposition 6.1.2. *If* ${}_{M}^{N}\lim\limits_{x \to x_0} F(x) = \alpha$, *then* $N(\alpha) \subseteq N$.

Proof. See Exercise 6.1.1. □

The $M \times N$-limits are not unique. Similarly to flexible sequences (see Proposition 5.1.4), if α is an $M \times N$-limit of F at x_0 and $\beta \subseteq \alpha + N$, then β is also an $M \times N$-limit of F at x_0 (see Exercise 6.1.4), and $M \times N$-limits are unique up to the neutrix N.

Proposition 6.1.3. *Let* M, N *be neutrices and let* α *be an external number. If* α *is an* $M \times N$*-limit of* F *at* $x_0 \in \mathbb{R}$, *then* $\alpha + N$ *is also an* $M \times N$*-limit of* F *at* x_0. *Additionally, every external number* β *such that* $\beta - \alpha \subseteq N$ *is an* $M \times N$*-limit of* F *at* x_0.

Proof. Let $\varepsilon > N$. Then $\frac{\varepsilon}{2} > N$ and there exists $\delta > M$ such that for all $x \in X$, if $0 < |x - x_0| < \delta$, then $|F(x) - \alpha| < \frac{\varepsilon}{2}$. So $|F(x) - (\alpha + N)| = |F(x) - \alpha| + N < \frac{\varepsilon}{2} + \frac{\varepsilon}{2} = \varepsilon$. Hence $\alpha + N$ is an $M \times N$-limit of F at x_0.

Since $N(\alpha) \subseteq N$, by Proposition 6.1.2, if $\beta - \alpha \subseteq N$ then $\beta \subseteq \alpha + N$, and arguing as above we obtain that β is an $M \times N$-limit of F at x_0. □

Proposition 6.1.4. *Let* M, N *be neutrices,* α, β *be external numbers and* F *be a flexible function such that* ${}_{M}^{N}\lim\limits_{x \to x_0} F(x) = \alpha$ *and* ${}_{M}^{N}\lim\limits_{x \to x_0} F(x) = \beta$, *for some* $x_0 \in \mathbb{R}$. *Then* $|\alpha - \beta| \leq N$.

Proof. Suppose that $|\alpha - \beta| > N$. Let $\varepsilon = \frac{|\alpha - \beta|}{2}$. Clearly $\varepsilon > N$. Then there exist $\delta_0, \delta_1 > M$ such that for all x it holds that if $|x - \gamma| < \delta_0$ then $|F(x) - \alpha| < \varepsilon$ and if $|x - \gamma| < \delta_1$ then $|F(x) - \beta| < \varepsilon$. Let $\delta := \min\{\delta_0, \delta_1\}$. If $|x - \gamma| < \delta$, then

$$|\alpha - \beta| \leq |\alpha - f(x) + f(x) - \beta| \leq |\alpha - f(x)| + |f(x) - \beta| < 2\varepsilon = |\alpha - \beta|,$$

a contradiction. Hence $|\alpha - \beta| \leq N$. □

Corollary 6.1.5. *Let* M, N *be neutrices,* $\alpha, \beta \in \mathbb{E}$ *be such that* $N(\alpha) = N(\beta) = N$. *If* F *is a flexible function such that* ${}_{M}^{N}\lim\limits_{x \to x_0} F(x) = \alpha$ *and* ${}_{M}^{N}\lim\limits_{x \to x_0} F(x) = \beta$, *for some* $x_0 \in \mathbb{R}$, *then* $\alpha = \beta$.

For flexible functions the notion of flexible limit can be equivalently expressed in terms of neighbourhoods.

Proposition 6.1.6. *Let* N *be a neutrix,* $F : X \subseteq \mathbb{R} \to \mathbb{E}$ *be a flexible function and* $x_0 \in \mathbb{R}$ *be a limit point of* X. *Let* $\alpha = a + A \in \mathbb{E}$. *Then* $N\text{-}\lim\limits_{x \to x_0} F(x) = \alpha$ *if and only if for each neighbourhood* V *of* α *one has*

$$\exists \delta > N \forall x \in X \setminus \{x_0\} (N < |x - x_0| < \delta \Rightarrow F(x) \subset V).$$

Proof. Assume that $N\text{-}\lim_{x \to x_0} F(x) = \alpha$ and V is a neighbourhood of α. Then there exists $\varepsilon > A$ such that $(a - \varepsilon, a + \varepsilon) \subseteq V$. For such ε, there exists $\delta > N$ such that whenever $x \in X$, if $0 < |x - x_0| < \delta$ then $|F(x) - \alpha| < \varepsilon$, that is $a - \varepsilon < F(x) + N(\alpha) < a + \varepsilon$. Hence $F(x) \subseteq F(x) + N(\alpha) \subset (a - \varepsilon, a + \varepsilon) \subseteq V$.

Conversely, let $\varepsilon > A$. Then $(a - \varepsilon, a + \varepsilon)$ is a neighbourhood of α. So there exists $\delta > N$ such that for all $x \in X$, if $0 < |x - x_0| < \delta$ one has $F(x) \subset (a - \varepsilon, a + \varepsilon)$. It follows that $F(x) - A \subset (a - \varepsilon, a + \varepsilon)$ for all $x \in X, 0 < |x - x_0| < \delta$. So $|F(x) - \alpha| < \varepsilon$ for all $x \in X, 0 < |x - x_0| < \delta$. Hence $N\text{-}\lim_{x \to x_0} F(x) = \alpha$. □

Next theorem shows that a flexible function converges if and only if both its neutrix part and its representatives converge.

Theorem 6.1.7. *Let M, N be neutrices and $F : X \subseteq \mathbb{R} \to \mathbb{E}$ be a flexible function. Let f be a representative of F. Let $\alpha = a + A$ be an external number and x_0 be an M-limit point of X. Then ${}_M^N\lim_{x \to x_0} F(x) = \alpha$ if and only if ${}_M^N\lim_{x \to x_0} f(x) = a$ and ${}_M^N\lim_{x \to x_0} N_F(x) = A$.*

Proof. Assume first that ${}_M^N\lim_{x \to x_0} F(x) = \alpha$. Let $\varepsilon > N$. Then there exists $\delta > M$ such that for all $x \in X$, if $0 < |x - x_0| < \delta$ then $|F(x) - \alpha| = |f(x) - a + N_F(x) + A| < \varepsilon$. It follows that $|f(x) - a| < \varepsilon$ and $N_F(x) + A \leq N_F(x) + N < \varepsilon$, by Proposition 6.1.2. Hence ${}_M^N\lim_{x \to x_0} f(x) = a$ and ${}_M^N\lim_{x \to x_0} N_F(x) = A$.

Assume now that ${}_M^N\lim_{x \to x_0} f(x) = a$ and ${}_M^N\lim_{x \to x_0} N_F(x) = A$. Let $\varepsilon > N$. Then $\frac{\varepsilon}{2} > N$ and therefore there exist $\delta_1, \delta_2 > M$ such that whenever $x \in X$, if $0 < |x - x_0| < \delta_1$ one has $|f(x) - a| < \frac{\varepsilon}{2}$, and if $0 < |x - x_0| < \delta_2$ then $N_F(x) + A < \frac{\varepsilon}{2}$. Put $\delta := \min\{\delta_1, \delta_2\}$. Then for all $x \in X$ such that $0 < |x - x_0| < \delta$ it holds that $|F(x) - \alpha| = |f(x) - a| + N_F(x) + A < \frac{\varepsilon}{2} + \frac{\varepsilon}{2} = \varepsilon$. Hence ${}_M^N\lim_{x \to x_0} F(x) = \alpha$. □

The next theorem considers the behaviour of $M \times N$-limits of flexible functions under operations. As shown in the next section, the relation between convergence and strong convergence is not as obvious as in the case of sequences. So we avoid the algebraic advantages of strong convergence, and present the conditions and arguments given in [146] for convergence. The corollaries following the theorem consider special cases which enable to lessen the rather cumbersome conditions for convergence of the product and division.

Theorem 6.1.8. *Let M, N_1, N_2 be neutrices and α, β be external numbers. Let $F, G : X \subseteq \mathbb{R} \to \mathbb{E}$ be flexible functions such that ${}_M^{N_1}\lim_{x \to x_0} F(x) = \alpha$ and ${}_M^{N_2}\lim_{x \to x_0} G(x) = \beta$, where x_0 is a M-limit point of X. Then*

1. *${}_M^N\lim_{x \to x_0} (F + G)(x) = \alpha + \beta$, where $N = N_1 + N_2$.*

2. ${}^{N}_{M}\lim_{x\to x_0}(F-G)(x)=\alpha-\beta$, *where* $N=N_1+N_2$.

3. *Let* $k\in\mathbb{R}$. *Then* ${}^{kN_1}_{M}\lim_{x\to x_0}(kF)(x)=k\alpha$.

4. ${}^{N_1}_{M}\lim_{x\to x_0}|F(x)|=|\alpha|$.

5. ${}^{K}_{M}\lim_{x\to x_0}(FG)(x)=\alpha\beta$, *where* $K=N_1+N_2+N_1^2+N_2^2+\alpha N_2+\beta N_1$.

6. *Assume that* α *is a zeroless external number and that there exists* $\delta>M$ *such that* $F(x)$ *is zeroless for all* $x\in X$ *such that* $0<|x-x_0|<\delta$. *Then* ${}^{K}_{M}\lim_{x\to x_0}\frac{1}{F(x)}=\frac{1}{\alpha}$, *where* $K=\frac{N_1}{\alpha^2}$.

7. *If* $G(x)$ *is zeroless in an* M*-neighbourhood of* x_0 *and* β *is zeroless then* ${}^{K}_{M}\lim_{x\to x_0}\left(\frac{F}{G}\right)(x)=\frac{\alpha}{\beta}$, *where* $K=N_1+\frac{N_2}{\beta^2}+N_1^2+\left(\frac{N_2}{\beta^2}\right)^2+\alpha\frac{N_2}{\beta^2}+\beta N_1$.

Proof. 1. Let $\varepsilon>N$. Then $\varepsilon>N_1$ and $\varepsilon>N_2$ and there exist $\delta_1,\delta_2>M$ such that for all $x\in X$, if $0<|x-x_0|<\delta_1$, then $|F(x)-\alpha|<\frac{\varepsilon}{2}$ and if $0<|x-x_0|<\delta_2$, then $|G(x)-\beta|<\frac{\varepsilon}{2}$. Let $\delta:=\min\{\delta_1,\delta_2\}$. Clearly $\delta>M$ and for all $x\in X$, if $0<|x-x_0|<\delta$, then

$$|F(x)+G(x)-(\alpha+\beta)|\le|F(x)-\alpha|+|G(x)-\beta|<\frac{\varepsilon}{2}+\frac{\varepsilon}{2}=\varepsilon.$$

Hence ${}^{N}_{M}\lim_{x\to x_0}(F+G)(x)=\alpha+\beta$.

2. The proof is similar to the proof of Part 1.

3. If $k=0$, the result is trivial, so we assume that $k\neq 0$. Let $\varepsilon>kN_1$. Then $\frac{\varepsilon}{|k|}>N_1$. So there exists $\delta>M$ such that for all $x\in X$, if $0<|x-x_0|<\delta$, then $|F(x)-\alpha|<\frac{\varepsilon}{|k|}$. Hence

$$|kF(x)-k\alpha|=|k||F(x)-\alpha|<|k|\frac{\varepsilon}{|k|}=\varepsilon,$$

and we conclude that ${}^{(kN_1)}_{M}\lim_{x\to x_0}(kF)(x)=k\alpha$.

4. The result is a consequence of the fact that $||F(x)|-|\alpha||\le|F(x)-\alpha|$.

5. For all $x\in X$ we write $F(x)=f(x)+N_F(x)$, $G(x)=g(x)+N_G(x)$, $\alpha=a+A$ and $\beta=b+B$. Then

$$\begin{aligned}|F(x)G(x)-\alpha\beta|&=|f(x)G(x)-\alpha\beta+N_F(x)G(x)|\\&\le|f(x)G(x)-f(x)\beta+f(x)\beta-\alpha\beta+N_F(x)G(x)|\end{aligned}$$
$$\le|f(x)||G(x)-\beta|+|b||f(x)-\alpha|+Bf(x)+B\alpha+N_F(x)g(x)+N_F(x)N_G(x).$$

Let $\varepsilon>K$. Note that $\sqrt{\varepsilon}>N_1$ and $\sqrt{\varepsilon}>N_2$. Firstly we show that

$$\exists\delta_1>M\forall x\in X\left(0<|x-x_0|<\delta_1\Rightarrow|f(x)||G(x)-\beta|<\frac{\varepsilon}{6}\right).$$

If α is not zeroless we may take $a = 0$. Then there exist $\eta_1, \eta_2 > M$ such that for all $x \in X$, if $0 < |x - x_0| < \eta_1$, then $|f(x)| < \sqrt{\varepsilon}$ and if $0 < |x - x_0| < \eta_2$, then $|G(x) - \beta| < \frac{\sqrt{\varepsilon}}{6}$. Let $\delta_1 := \min\{\eta_1, \eta_2\}$. Then $\delta_1 > M$ and for all $x \in X$ such that $0 < |x - x_0| < \delta_1$ it holds that $|f(x)||G(x) - \beta| < \frac{\varepsilon}{6}$.

If α is zeroless, then $|a| > A$. Then there exists $\eta_3 > M$ such that for $x \in X$, if $0 < |x - x_0| < \eta_3$, then $|f(x) - a| < |a|$, which implies that $|f(x)| < 2|a|$. Also, $\varepsilon > |a|N_2$, so $\frac{\varepsilon}{12|a|} > B$. Then there exists $\eta_4 > M$ such that for $x \in X$, if $0 < |x - x_0| < \eta_4$, then $|G(x) - \beta| < \frac{\varepsilon}{12|a|}$. Put $\delta_1 := \min\{\eta_3, \eta_4\}$. Then for all $x \in X$, if $0 < |x - x_0| < \delta_1$, then $|f(x)||G(x) - \beta| \le 2|a||G(x) - \beta| < \frac{\varepsilon}{6}$.

Secondly, we prove that

$$\exists \delta_2 > M \forall x \in X \left(0 < |x - x_0| < \delta_2 \Rightarrow |b||f(x) - \alpha| < \frac{\varepsilon}{6}\right).$$

We assume that β is zeroless because otherwise we may choose $b = 0$ and the result follows trivially. Then $|b| > B$. Since $\varepsilon > K$ we have that $\varepsilon > \beta N_1 = bN_1$ and therefore $\frac{\varepsilon}{6|b|} > N_1$. Then there exists $\delta_2 > M$ such that for all $x \in X$, if $0 < |x - x_0| < \delta_2$, then $|f(x) - \alpha| < \frac{\varepsilon}{6|b|}$ and the result also follows.

Thirdly, we show that

$$\exists \delta_3 > M \forall x \in X \left(0 < |x - x_0| < \delta_3 \Rightarrow f(x)B < \frac{\varepsilon}{6}\right).$$

If α is not zeroless, we saw that $|f(x)| < \sqrt{\varepsilon}$ for all $x \in X$ with $0 < |x - x_0| < \eta_1$. By Proposition 6.1.2 we conclude that for all $x \in X$, if $0 < |x - x_0| < \eta_1$, then $f(x)B = |f(x)|B \le |f(x)|N_2 < \sqrt{\varepsilon}\frac{\sqrt{\varepsilon}}{6} = \frac{\varepsilon}{6}$.

If α is zeroless, taking η_3 as above, using Proposition 6.1.2 we see that for all $x \in X$, if $0 < |x - x_0| < \eta_3$, then $|f(x)|B \le |f(x)|N_2 < 2|a|N_2 < \frac{\varepsilon}{6}$. So the result follows with $\delta_3 := \min\{\eta_1, \eta_3\}$.

Fourthly, we show that

$$\exists \delta_4 > M \forall x \in X \left(0 < |x - x_0| < \delta_4 \Rightarrow |g(x)|N_F(x) < \frac{\varepsilon}{6}\right).$$

If β is not zeroless, again we may choose $b = 0$, and then ${}^{N_2}_{M}\lim\limits_{x \to x_0} g(x) = 0$. Then there exist $\eta_5, \eta_6 > M$ such that for all $x \in X$, if $0 < |x - x_0| < \eta_5$, then $|g(x)| < \sqrt{\frac{\varepsilon}{6}}$ and if $0 < |x - x_0| < \eta_6$, then $N_F(x) \le \sqrt{\frac{\varepsilon}{6}}$. Let $\delta_4 := \min\{\eta_5, \eta_6\}$. Then for all $x \in X$, if $0 < |x - x_0| < \delta_6$, then $|g(x)|N_F(x) < \frac{\varepsilon}{6}$.

If β is zeroless we let $\varepsilon = |b| > N_2$. Then there exists $\eta_7 > M$ such that for all $x \in X$, if $0 < |x - x_0| < \eta_7$, then $|g(x) - b| < |b|$, which implies that $|g(x)| < 2|b|$. Furthermore, $\varepsilon > bN_1$, so $N_1 < \frac{\varepsilon}{|b|}$. It follows that $N_1 < \frac{c}{12|b|}$. Then there exists $\eta_8 > M$ such that for all $x \in X$, if $0 < |x - x_0| < \eta_6$, then $|N_F(x)| < \frac{\varepsilon}{12|b|}$. Put $\delta_4 := \min\{\eta_7, \eta_8\}$. Then for all $x \in X$ if $0 < |x - x_0| < \delta_4$, then $|g(x)|N_F(x) \le 2|b|N_F(x) < \frac{\varepsilon}{6}$.

Finally, we show that

$$\exists \delta_5 > M \forall x \in X \left(0 < |x - x_0| < \delta_5 \Rightarrow N_F(x)N_G(x) < \frac{\varepsilon}{6}\right).$$

As we saw above, for all $x \in X$, if $0 < |x - x_0| < \eta_6$, then $N_F(x) < \sqrt{\frac{\varepsilon}{6}}$. Arguing in a similar way we see that there exists η_9 such that for all $x \in X$, if $0 < |x - x_0| < \eta_9$, then $B(x) < \sqrt{\frac{\varepsilon}{6}}$. Put $\delta_5 := \min\{\eta_6, \eta_9\}$. Then for all $x \in X$, if $0 < |x - x_0| < \delta_5$, then $A(x)B(x) < \frac{\varepsilon}{6}$.

Clearly, $N_2\alpha < \frac{\varepsilon}{6}$. Combining, let $\delta := \min\{\delta_1, \delta_2, \delta_3, \delta_4, \delta_5\}$. Then for all $x \in X$, if $0 < |x - x_0| < \delta$, then $|(FG)(x) - \alpha\beta| < \varepsilon$ and we conclude that ${}^K_M\lim_{x \to x_0} (FG)(x) = \alpha\beta$.

6. Let $\varepsilon > K$. So, in particular $\varepsilon > \frac{N}{a^2}$. Then there exists $\eta > M$ such that $\eta \leq \delta$ and for all $x \in X$ such that $0 < |x - x_0| < \eta$ it holds that $|f(x) - a| < a^2\frac{\varepsilon}{4}$. Then

$$\begin{aligned}
\left|\frac{1}{F(x)} - \frac{1}{\alpha}\right| &= \left|\frac{F(x)}{f^2(x)} - \frac{\alpha}{a^2}\right| \\
&= \left|\frac{a^2F(x) - f^2(x)\alpha}{f^2(x)a^2}\right| \\
&\leq \left|\frac{a - f(x)}{f(x)a}\right| + \frac{a^2N_F(x) - f^2(x)A}{f^2(x)a^2} \\
&\leq \left|\frac{a - f(x)}{\frac{a^2}{2}}\right| + \frac{a^2N_F(x) - 4a^2A}{\frac{a^4}{4}} \\
&\leq \left|\frac{a^2\frac{\varepsilon}{4}}{a^2\frac{\varepsilon}{2}}\right| + \frac{N_F(x) + A}{a^2} \\
&< \frac{\varepsilon}{2} + \frac{\varepsilon}{2} = \varepsilon.
\end{aligned}$$

Hence ${}^K_M\lim_{x \to x_0} \frac{1}{F(x)} = \frac{1}{\alpha}$.

7. By Parts 5 and 6. □

Corollary 6.1.9. *If in Part 5 of Theorem 6.1.8 one assumes additionally that $N_1, N_2 \subseteq \pounds$ then $K = N_1 + N_2 + \alpha N_2 + \beta N_1$. If furthermore neither α is an absorber of N_2, nor β is an absorber of N_1, then $K = \alpha N_2 + \beta N_1$.*

Proof. If $N_1, N_2 \subseteq \pounds$ then $N_1^2 \subseteq N_1$ and $N_2^2 \subseteq N_2$, so we may neglect the squared terms N_1^2, N_2^2. If α is not an absorber of N_2 and β is not an absorber of N_1 both $N_1 \subseteq \beta N_1$ and $N_2 \subseteq \alpha N_2$, so we may also neglect the terms N_1, N_2. □

Corollary 6.1.10. *If in Part 7 of Theorem 6.1.8 one assumes additionally that $N_1, N_2 \subseteq \pounds$ then $K = N_1 + \frac{N_2}{\beta^2} + \alpha\frac{N_2}{\beta^2} + \beta N_1$. If furthermore neither α is an absorber of N_2, nor β is an absorber of N_1, then $K = \alpha\frac{N_2}{\beta^2} + \beta N_1$.*

Proof. Similar to the proof of Corollary 6.1.9. □

We end with a version of the Squeeze Theorem for $M \times N$-limits of flexible functions.

Theorem 6.1.11 (Squeeze Theorem). *Let M, N be neutrices and $F, G, H : X \subseteq \mathbb{R}^n \to \mathbb{E}$ be flexible functions such that $F(x) \leq G(x) \leq H(x)$ for all $x \in V \subseteq X$, where V is an M-neighbourhood of x_0. If ${}_M^N\lim\limits_{x\to x_0} F(x) = {}_M^N\lim\limits_{x\to x_0} H(x) = \alpha$ for some external number α, then ${}_M^N\lim\limits_{x\to x_0} G(x) = \alpha$.*

Proof. Let $\varepsilon > N$. Then there exists $\delta > M$ such that for all $x \in V$, if $0 < |x - x_0| < \delta$, then $|F(x) - \alpha| < \varepsilon$ and $|H(x) - \alpha| < \varepsilon$. Moreover, for all $x \in V$ we have that $F(x) - \alpha \leq G(x) - \alpha \leq H(x) - \alpha$. Then for all $x \in V$ it holds that $|G(x) - \alpha| \leq \max\{|F(x) - \alpha|, |H(x) - \alpha|\} < \varepsilon$. Hence ${}_M^N\lim\limits_{x\to x_0} G(x) = \alpha$. □

6.1.1 Relation with convergence for sequences; strong convergence

We start by adapting the definition of convergence with respect to an initial segment and strong convergence of sequences to functions, and then relate convergence of sequences and functions, which are assumed to be continuous.

Definition 6.1.2. *Let (C, D) be a cut of $\mathbb{R}$. Let S be an initial segment of $\mathbb{R}$ such that $C \subseteq S$. Let $F : S \to \mathbb{R}$ be a flexible function and $\alpha = a + A \in \mathbb{E}$. We say that F* converges *to α on S with respect to C if*

$$\forall \varepsilon > A \exists c \in C \forall y \in S (y \geq c \Rightarrow |F(y) - \alpha| < \varepsilon).$$

We may write $\lim_{x\to D} F(x) = \alpha$. If $C = S$, or $S = \mathbb{R}$, we simply speak of convergence with respect to C.

The following theorems deal with the relation between $M \times N$-limits of flexible functions and limits of flexible sequences, which are M-limits for the variable and N-limits for the values. As in the case of real functions, convergence of the $\varepsilon - \delta$-type quite directly implies sequential convergence, but the converse is less obvious, and needs a form of the Axiom of Choice. To this end, in the reciprocal case we restrict ourselves to flexible functions with internal continuous representatives.

Theorem 6.1.12. *Let M, N be neutrices, $F : X \subseteq \mathbb{R} \to \mathbb{E}$ be a flexible function and x_0 be a M-limit point of X. If ${}_M^N\lim\limits_{x\to x_0} F(x) = \alpha$ for some external number α, then N-$\lim F(x_n) = \alpha$ for every flexible sequence (x_n) with values in X which M-converges to x_0.*

Proof. Let $\varepsilon > N$. Then there exists $\delta > M$ such that for all $x \in X$ with $0 < |x - x_0| < \delta$ we have $|F(x) - \alpha| < \varepsilon$. Also, there exists $n_0 \in \mathbb{N}$ such that for all $n \geq n_0$ it holds that $0 < |x_n - x_0| < \delta$. Then $|F(x_n) - \alpha| < \varepsilon$, for all $n \geq n_0$. Hence N-$\lim F(x_n) = \alpha$. □

Theorem 6.1.13. *Let M, N be neutrices, $\overline{x} \in \mathbb{R}$, $\beta \in \mathbb{E}$ and V be an internal neighbourhood of $\overline{x} + M$. Let F be a flexible function defined on a set X with $V \setminus (\overline{x} + M) \subseteq X \subseteq \mathbb{R}$, such that there exists an internal continuous real function f such that $f_{\restriction X}$ is a representative of F. If for every internal sequence (x_n) for which $C := \{n \in N : x_n \notin \overline{x} + M\}$ is a non-empty initial segment and*

$$M\text{-}\lim_{n\to\infty} x_n = \overline{x} \text{ w.r.t. } C \Rightarrow N\text{-}\lim_{n\to\infty} F(x_n) = \beta \text{ w.r.t. } C, \tag{6.2}$$

then ${}^{N}_{M}\lim_{x\to\overline{x}} F(x) = \beta$.

Proof. Let $\beta = b + B$. Suppose that ${}^{N}_{M}\lim_{x\to\overline{x}} F(x) \neq \beta$. Then, working with the representative f, there exists $\varepsilon \in \mathbb{R}$, $\varepsilon > N$, such that

$$\forall \delta > M \exists x \in X(|x - \overline{x}| \leq \delta \wedge |f(x) - b| \geq \varepsilon). \tag{6.3}$$

If $M = \{0\}$, we may obtain by the usual choice argument an internal sequence (x_n) converging to $\overline{x}$, while all terms x_n are different from $\overline{x}$ and $|f(x_n) - b| \geq \varepsilon$ for all $n \in N$. Then $F(x_n)$ cannot N-converge to β, a contradiction.

If M is external, by Cauchy's Principle condition (6.3) holds up to some positive $\overline{\delta} \in M$. Let (δ_n) be an internal sequence such that $\Delta := \{n \in N : \delta_n \notin M\}$ is a non-empty initial segment and $M\text{-}\lim_{n\to\infty} \delta_n = 0$. We define an internal sequence (x_n) such that $M\text{-}\lim_{n\to\infty} x_n = \overline{x}$ with respect to $\{n \in N : x_n \notin \overline{x} + M\}$. As long as $|\delta_n - \overline{x}| \geq \overline{\delta}$, we define $x_n \in \mathbb{R}$ to be such that $|\overline{x} - x_n| \leq \delta_n$, and $|f(x_n) - b| \geq \varepsilon$, where we assume that $|\overline{x} - x_n|$ is maximal; if there are two such values, we take the least of the two. Then $x_n \notin \overline{x} + M$ as long as $\delta_n > M$ by (6.2). Staying inside $\overline{x} + M$, we may complete the thus obtained internal sequence into an internal sequence (x_n) defined for all $n \in \mathbb{N}$. Then $(f(x_n))$; hence also $(F(x_n))$ does not N-converge to β with respect to Δ, a contradiction. Hence $N\text{-}\lim_{x\to x_0} F(x) = \beta$. $\square$

Definition 6.1.3. *Let (C, D) be a cut of $\mathbb{R}$. Let S be an initial segment of $\mathbb{R}$ such that $C \subseteq S$. Let $F : S \to \mathbb{R}$ be a flexible function and $\alpha = a + A \in \mathbb{E}$. We say that F* strongly converges *to α on S with respect to C if there exists a non-empty final segment T of S such that $D \subseteq T$ and $\forall y \in S(y \in T \Rightarrow F(y) \subseteq \alpha)$. We may write $\mathrm{Lim}_{x\to D} f(x) = \alpha$. If $C = S$, or $S = \mathbb{R}$, we simply speak of strong convergence with respect to C.*

External initial segments C are stable under the neutrix, say K, consisting of all $t \in \mathbb{R}$ such that $c + t \in C \Leftrightarrow c \in C$ for all $c \in C$. Then $\frac{C}{t}$ is stable under $\frac{K}{t} \supseteq \mathbb{N}$. For convergence of sequences with respect to an initial segment this represents the minimal thickness for the initial segment to be external. In the case of internal continuous functions, the next lemma relates convergence and strong convergence with respect to such external initial segments.

Proposition 6.1.14. *Let (C, D) be a cut of $\mathbb{R}^+$ into an external initial segment C and a final segment D. Let S be an (internal) initial segment of $\mathbb{R}$ such that $C \subset S$. Let $f : S \to \mathbb{R}$ be an internal continuous function and $\alpha = a + A \in \mathbb{E}$. For every $t \in \mathbb{R}^+$, define $S_t := \{n \in \mathbb{N} : nt \in S\}$ and $u_t : S_t \to \mathbb{R}$ by $u_t(n) = f(tn)$. Then*

1. *f converges to α on S with respect to C if and only if u_t converges to α on S_t with respect to $\frac{C}{t}$ for all $t \in \mathbb{R}^+$ such that $c + t \in C \Leftrightarrow c \in C$ for all $c \in \mathbb{R}$.*

2. *f converges strongly to α on S with respect to C if and only if u_t converges to α on S_t with respect to $\frac{C}{t}$ for all $t \in \mathbb{R}^+$ such that $c + t \in C \Leftrightarrow c \in C$ for all $c \in \mathbb{R}$.*

Proof. Without loss of generality we may assume that $\alpha = A$.

1. Assume first that f converges to α on S with respect to C. Let $\varepsilon \in \mathbb{R}$, $\varepsilon > A$. Then there is $x \in C$ such that $|f(y)| < \varepsilon$ for all $y \in S$, $y \geq x$. Let $t \in \mathbb{R}^+$ be such that $c + t \in C \Leftrightarrow c \in C$ for all $c \in \mathbb{R}$. Let $m \in \mathbb{N}$ be minimal such that $mt \geq x$. Then $|u_t(n)| < \varepsilon$ for all $n \in S_t$, $n \geq m$. Hence u_t converges to A on S_t with respect to $\frac{C}{t}$.

As for the converse, let $\varepsilon \in \mathbb{R}$, $\varepsilon > A$. Let $s \in S$ be such that $C < s$. By uniform continuity of f on the compact interval $[0, s]$ there exists $t > 0$ such that $c + t \in C \Leftrightarrow c \in C$ for all $c \in \mathbb{R}$ and $|f(v) - f(w)| < \frac{\varepsilon}{2}$ for all $v, w \in [0, s]$ with $|v - w| < t$. Because u_t converges to A on S_t with respect to $\frac{C}{t}$, there exists $m \in \frac{C}{t}$ such that $|u_t(n)| < \frac{\varepsilon}{2}$ for all $n \in S_t$, $n \geq m$. Let $x \in [0, s]$ be such that $x > mt$ and $n \in \frac{C}{t}$ be minimal such that $x \leq nt$. Then $|f(x)| \leq |f(x) - f(nt)| + |u_t(n)| < \frac{\varepsilon}{2} + \frac{\varepsilon}{2} = \varepsilon$. Hence f converges to A on $[0, s]$ with respect to C. Because s is arbitrary f converges to A on S with respect to C.

2. The proof is similar to the proof of Part 1. □

In Chapter 9 we deal with the particular case of convergence of internal and flexible functions with respect to the set where they are still outside the limit set, and relate to the well-known matching problem for different behaviour of a function on different, contingent sets.

Exercises

Exercise 6.1.1. *Prove Proposition 6.1.2.*

Exercise 6.1.2. *Show that* (6.1) *is equivalent to each of the following conditions.*

1. *For each real neighbourhood V of α, there exists $\delta > M$ such that for all $x \in X$, if $0 < |x - x_0| < \delta$ then $F(x) \subseteq V$.*

2. $\forall \varepsilon \in \mathbb{R}(\varepsilon > N \Rightarrow (\exists \delta > M \forall x \in X \setminus \{x_0\}(|x - x_0| < \delta \Rightarrow |F(x) - \alpha| < \varepsilon)))$.

Exercise 6.1.3. *Let M, N be neutrices, α, β be external numbers and let F be a flexible function such that $F(x) \le \beta + N$ for all $x \in X$. Show that if* ${}^{M}_{N}\lim_{x \to x_0} F(x) = \alpha$, *then* $\alpha \le \beta + N$.

Exercise 6.1.4. *Let M, N be neutrices and $F : X \subseteq \mathbb{R}^n \to \mathbb{E}$ be a flexible function. Assume that $\alpha \in \mathbb{E}$ is an $M \times N$-limit of F at some $x_0 \in \mathbb{R}^n$.*

1. *Show that $\alpha + N$ is also a $M \times N$-limit of F at x_0.*
2. *Show that, if β is an external number such that $\beta - \alpha \subseteq N$ then β is also an $M \times N$-limit of F at x_0.*

Exercise 6.1.5. *Let $\omega \in \mathbb{R}$ be positive unlimited. Determine the following limits.*

1. $\lim_{x \to \omega + \pounds} \frac{1}{x}$.
2. $\lim_{x \to \omega + \oslash} \frac{1}{x^2}$.
3. $\lim_{x \to \omega + \pounds} \log(x)$.
4. $\lim_{x \to \omega + \frac{\oslash}{\omega}} e^x$.
5. $\lim_{x \to \omega + \oslash} x^3$.
6. $\lim_{x \to \omega + \pounds} \frac{3x_2 + \pounds x + \sqrt{\omega}}{x^2 + \omega x + \pounds}$.

Exercise 6.1.6. *Let $\omega \simeq +\infty$ and let $F : \mathbb{R} \to \mathbb{E}$ be a flexible function defined by $F(x) = e^x + \omega\pounds$. Show that* ${}^{\omega\pounds}_{\oslash}\lim_{x \to 0} F(x) = \omega\pounds$ *but F^2 is not $\oslash \times \omega\pounds$-convergent to $\omega\pounds$.*

6.2 Flexible continuity

A real function is S-continuous if infinitesimal changes in the argument lead to infinitesimal changes in the value. S-continuous functions may be continuous, like all standard continuous functions [29, 34], but need not be, for infinitesimal jumps are allowed. Combining continuity and discontinuity into a single notion S-continuity serves as an intermediate in discretisations and transitions from the discrete to the continuous.

Sometimes S-continuity is called "continuity for a short-sighted person" [34, p.15]. Not all short-sighted people are equally short-sighted, so continuity turns out to be a matter of degree. With this perspective one could generalize to jumps within an arbitrary neutrix N, within a distance contained in some neutrix M, and then we speak of $M\times N$-inner continuity. We also speak of $M\times N$-outer continuity which is defined in an $\varepsilon-\delta$ way. In this sense S-continuity of internal functions corresponds to $\oslash\times\oslash$-continuity (both inner and outer). There is not always equivalence between $M\times N$-inner and outer continuity. Clearly every function is $\mathbb{R}$-continuous so we assume that the neutrices N considered below are always such that $N\neq\mathbb{R}$.

We will deal only with continuity on both sides. The theory is easily adapted to one-sided continuity.

6.2.1 Outer continuity

Definition 6.2.1. *Let $F:X\subseteq\mathbb{R}\to\mathbb{E}$ be a flexible function and M,N be neutrices. Let $x_0\in X$ be an M-limit point of X. We say that F is $M\times N$-(outer)* continuous *at x_0 if and only if ${}^N_M\lim\limits_{x\to x_0}F(x)=F(x_0)$, otherwise said if*

$$\forall\varepsilon>N\exists\delta>M\forall x(|x-x_0|<\delta\Rightarrow|F(x)-F(x_0)|<\varepsilon).$$

*In particular, if $N=N(F(x_0))$, we say that F is M-*continuous *at x_0, and if furthermore $M=0$ we say that F is* continuous *at x_0. We say that F is* continuous *(resp. $M\times N$-*continuous*, M-*continuous*) on X if it is continuous (resp. $M\times N$-continuous, M-continuous) at every point on X.*

By Proposition 6.1.1, if a flexible function F is $M\times N$-continuous, then it is also $M'\times N'$-continuous for all neutrices M',N' such that $M'\subseteq M$ and $N\subseteq N'$.

Example 6.2.1. Let ε be infinitesimal. Let f,g,h,j be functions defined respectively by

1. $f(x)=\arctan\left(\frac{x}{\varepsilon}\right)$.

2. $g(x)=\begin{cases}-\varepsilon, & x<0\\ \varepsilon, & x\geq 0.\end{cases}$

3. $h(x)=\begin{cases}-\varepsilon^2, & x<0\\ \varepsilon^2, & x\geq 0.\end{cases}$

4. $j(x)=\begin{cases}-1, & x<0\\ 1, & x\geq 0.\end{cases}$

5. $k(x)=\begin{cases}0, & x\in\oslash\\ 1, & x\notin\oslash.\end{cases}$

The function f is a well-known example of a continuous function which is not S-continuous. In our terminology, f is neither $\oslash$-continuous nor $\oslash\times\oslash$-

continuous. It is not difficult to see that both g and h are not outer continuous at $x_0 = 0$ but are both $\oslash$-continuous. The function h is in fact $\varepsilon^2 \pounds$-continuous while g is not $\varepsilon^2 \pounds$-continuous. The function j is clearly not continuous nor $\oslash$-continuous at $x_0 = 0$. However j is $\pounds$-continuous at that point. The function k is not $\oslash$-continuous at 0 though it is both continuous and S-continuous at 0. Observe that $f(x) \in f(0) + \oslash$ for all $x \in \oslash$.

Proposition 6.2.1. *Let $M > 0$, N be neutrices and $x_0 \in \mathbb{R}$. If a flexible function $F : \mathbb{R} \to \mathbb{E}$ is $M \times N$-continuous at x_0, then $F(x_0 + M) \subseteq N_{F(x_0)+N}$.*

Proof. Suppose that $F(x_0 + M) \nsubseteq N_{F(x_0)+N}$. Then there exists $x \in x_0 + M$ and $\varepsilon \in \mathbb{R}$, $\varepsilon > N$ such that $F(x) \supset [-\varepsilon, \varepsilon]$, in contradiction with the $M \times N$-continuity at x_0. Hence $F(x_0 + M) \subseteq N_{F(x_0)+N}$. □

Theorem 6.2.2. *Let K, M, N be neutrices. Let $I \subseteq \mathbb{R}$, $J \subseteq \mathbb{R}$. Let $f : I \to J$ be a real function and $G : J \to \mathbb{E}$ and $H : I \to \mathbb{E}$ be flexible functions such that for all $x \in I$ it holds that $H(x) = G(f(x))$. If f is $K \times M$-continuous at x_0 and G is $M \times N$-continuous at $f(x_0)$, then H is $K \times N$-continuous at x_0.*

Proof. Let $\varepsilon > N$. By the $M \times N$-continuity of G at $f(x_0)$, there exists $\eta > M$ such that, for all $y \in J$, if $|y - f(x_0)| < \eta$ then $|G(y) - G(f(a))| < \varepsilon$. Since f is $K \times M$-continuous at x_0, there exists $\delta > M$ such that, for all $x \in I$, if $|x - x_0| < \delta$ then $|f(x) - f(x_0)| < \eta$. Hence, for all $x \in I$, if $|x - x_0| < \delta$ then $|H(x) - H(x_0)| = |G(f(x)) - G(f(x_0))| < \varepsilon$. We conclude that H is $K \times N$-continuous at x_0. □

Important consequences of ordinary continuity are the Intermediate Value Theorem, and the Extreme Value Theorem. We are able to adapt these properties to $M \times N$-continuous flexible functions on a compact interval, say $[a, b]$, provided that M is the maximal neutrix less than $b - a$, i.e. $M = \oslash(b - a)$.

Note that neither the intermediate value property nor the maximum property holds for $M \times N$-continuity with respect to neutrices $M \subset \oslash(b - a)$. Indeed, define $f : [-1, 1] \to \mathbb{R}$ by $f(x) = \begin{cases} x - 1, & x \lesssim 0 \\ x + 1, & x \gtrsim 0 \end{cases}$, and $g : [0, 1] \to \mathbb{R}$ by $g(x) = \begin{cases} \frac{1}{x}, & x \gtrsim 0 \\ 0, & x \simeq 0 \end{cases}$. Both f and g are $M \times \oslash$-continuous for $M \subset \oslash$, but not $\oslash \times \oslash$- continuous. Obviously $f(x) \notin \oslash$ for any $x \in [-1, 1]$, while $g(x)$ assumes arbitrarily large limited values, without attaining a maximum, which should be $\pounds$.

Theorem 6.2.3 (Weak Intermediate Value Theorem). *Let $a, b \in \mathbb{R}$, $a < b$. Let $N \subset \mathbb{R}$ be a neutrix, $M := \oslash(b - a)$ and $F : [a, b] \to \mathbb{E}$ be an $M \times N$-continuous flexible function such that $F(a) < N < F(b)$. Then there exists $c \in [a, b]$ with $F(c) \subseteq N$.*

Proof. Define $L := \{x \in [a, b] : \forall y \leq x(f(y) < N)\}$. Let $\gamma = c + C$ be such that $\gamma = \text{zup}(L)$, then $C \subseteq M$.

Suppose first that $f(c) < N$. By $M \times N$-continuity we have $f(\gamma) \subseteq f(c) + N < N$. There exists $\varepsilon \in \mathbb{R}$ with $0 < \varepsilon$ and $f(c) + \varepsilon < N$. Again by $M \times N$-continuity there exists $\delta \in \mathbb{R}, \delta > C$ such that $F(c+t) \leq F(c) + \varepsilon$ for all t with $\gamma < t \leq \delta$. So $\gamma \neq \text{zup}(L)$, a contradiction.

Second, suppose that $f(c) > N$. By $M \times N$-continuity we have $f(\gamma) \subseteq f(c) + N > N$, and $N < \varepsilon < f(c) + N$ for some $\varepsilon \in \mathbb{R}$ with $0 < \varepsilon$. Again by $M \times N$-continuity there exists $\eta \in \mathbb{R}$, $\eta > 0$ such that $|F(c+t) - F(c)| \leq \varepsilon$ for all t with $|t| < \eta$. So $F(c - \eta) > N$, contradicting the fact that $c - \eta \in L$.

We conclude that $F(c) \cap N \neq \emptyset$. If $F(c) \supset N$, the flexible function F would not be $M \times N$-continuous at c; hence $F(c) \subseteq N$. □

Theorem 6.2.4 (Weak Maximum Value Theorem). *Let $a, b \in \mathbb{R}$, $a < b$. Let $N \subset \mathbb{R}$ be a neutrix, $M := \oslash(b-a)$ and $F : [a,b] \to \mathbb{E}$ be a $M \times N$-continuous flexible function. Then there exists $c \in [a,b]$ with $F(x) \leq F(c) + N$ for all $x \in [a,b]$.*

Proof. Up to a linear transformation we may assume that $a = 0$ and $b = 1$, then $M = \oslash$. Let $L := \bigcup_{x \in [0,1]} (-\infty, F(x)]$ and $\sigma = s + S$ be such that $\sigma = \text{zup}(L)$.

Assume first that $L = (-\infty, \sigma]$. Then there exists $c \in [0,1]$ such that $s \leq F(c)$, and then $s + S \leq F(c) + N$ by $M \times N$-continuity. Hence $S \subseteq N$ and $F(x) \leq \sigma \leq F(c) + N$ for all $x \in [0,1]$.

Second, we assume that $L = (-\infty, \sigma)$. Then $F(x) < \sigma$ for all $x \in [0,1]$. If $S \subset N$, for some $c \in [0,1]$ we have $\sigma \subset F(c) + N$; hence $F(x) \leq \sigma \leq F(c) + N$ for all $x \in [0,1]$. We will see that the remaining case, $N \subseteq S$, leads to a contradiction. Put $I_0 := [0,1]$. If $\text{zup}\left\{F(x) : x \in \left[0, \frac{1}{2}\right]\right\} = \sigma$ we put $I_1 := \left[0, \frac{1}{2}\right]$; if not, it holds that $\text{zup}\left\{F(x) : x \in \left[\frac{1}{2}, 1\right]\right\} = \sigma$ and we put $I_1 := \left[\frac{1}{2}, 1\right]$. Using External Induction we thus find a sequence of nested intervals $(I_n)_{\text{st}(n) \in \mathbb{N}}$ of length $\frac{1}{2^{n+1}}$ with $\text{zup}\{F(x) : x \in I_n\} = \sigma$, which by compactness has an intersection of the form $c + \oslash$ with $\text{st}(c) \in [0,1]$. Now $F(c) < \sigma$; hence also $F(c) + N < \sigma$ because $N \subseteq S$. Then there exists $\varepsilon \in \mathbb{R}$ such that $N < \varepsilon$ and $F(c) + \varepsilon < \sigma$. By $M \times N$-continuity there exists a positive appreciable $\delta \in \mathbb{R}$ such that $F(x) < F(c) + \varepsilon$ for all $x \in [c - \delta, c + \delta]$. For some standard $n \in \mathbb{N}$ we have $I_n \subseteq [c - \delta, c + \delta]$; hence $\text{zup}_{x \in [c-\delta, c+\delta]} F(x) = \sigma$, which implies that $F(c) + \varepsilon \leq F(x)$ for some $x \in [c - \delta, c + \delta]$, a contradiction. So this case cannot happen indeed, which proves the theorem. □

We prove a strong convergence theorem for $M \times N$-convergence in the presence of $M \times N$-continuity at the limit set.

Theorem 6.2.5. *Let $M > 0$, N be neutrices, $x \in \mathbb{R}$, $\beta = b + N \in \mathbb{E}$ and V be an internal neighbourhood of $\overline{x} + M$. Let F be an $M \times N$-continuous flexible function defined on an internal set X with $V \setminus (\overline{x} + M) \subseteq X \subseteq \mathbb{R}$, for which there exists an internal continuous real function f such that $f_{\restriction X}$ is a representative of F. If for every internal sequence (x_n) such that $C := \{n \in \mathbb{N} : x_n \notin \overline{x} + M\}$ is a non-empty initial segment and $M\text{-}\lim_{n \to \infty} x_n = \overline{x}$*

with respect to C one has $\lim_{n\to\infty} F(x_n) = \beta$ with respect to C, then $F(x) \subseteq \beta$ for all $x \in \overline{x} + M \cap X$.

Proof. For internal sequences (x_n) such that C is a non-empty initial segment and M-$\lim_{n\to\infty} x_n = \overline{x}$ with respect to C using Theorem 5.1.11 we obtain that they strongly converge to $\overline{x} + M$; hence $x_\nu \in \overline{x} + M$ for some $\nu \in \mathbb{N}$. Then $x_\nu \notin C$, so $F(x_\nu) \subseteq \beta$; otherwise $F(x_n)$ would not converge to β with respect to C. Then $F(x) \subseteq \beta$ for all $x \in \overline{x} + M \cap X$ by $M \times N$-continuity of F. □

Let M, N be neutrices. We end this section with a version of the Cauchy criterion for $M \times N$-limits of flexible functions which allows us to deduce whether a given flexible function is $M \times N$-convergent without the necessity of calculating the $M \times N$-limit.

Theorem 6.2.6 (Cauchy criterion). *Let M, N be neutrices, $F : X \subseteq \mathbb{R} \to \mathbb{E}$ be a flexible function and x_0 be an M-limit point of X. Then F is $M \times N$-convergent at x_0 if and only if*

$$\forall \varepsilon > N \exists \delta > M \forall x, x' \in X (0 < |x - x_0| < \delta \wedge 0 < |x' - x_0| < \delta \Rightarrow |F(x) - F(x')| < \varepsilon). \tag{6.4}$$

Proof. Assume (6.1). Let $\varepsilon > N$. Then there exists $\delta > M$ such that for all $x, x' \in X$, if $0 < |x - x_0| < \delta$ and $0 < |x' - x_0| < \delta$, then $|F(x) - \alpha| < \frac{\varepsilon}{2}$ and $|F(x') - \alpha| < \frac{\varepsilon}{2}$. Hence $|F(x) - F(x')| \leq |F(x) - \alpha| + |F(x') - \alpha| < \frac{\varepsilon}{2} + \frac{\varepsilon}{2} = \varepsilon$, so (6.4) holds.

Assume now (6.4). Let $(x_n) \subset X$ be an internal sequence converging to x_0. Then the flexible sequence $F(x_n)$ is N-Cauchy. By Theorem 6.2.5 it holds that $F(x_n) \underset{N}{\to} \alpha$, for some $\alpha \in \mathbb{E}$. Let $\varepsilon > N$. Then there exists $\delta > M$ such that for all $x, x' \in X$, if $0 < |x - x_0| < \delta$ and $0 < |x' - x_0| < \delta$, then $|F(x) - F(x')| < \frac{\varepsilon}{2}$. Also, since $\lim x_n = x_0$, there exists $n_1 \in \mathbb{N}$ such that for all $n \geq n_1$ it holds that $0 < |x_n - x_0| < \delta$. Then, there exists $n_2 \in \mathbb{N}$ such that for all $n \geq n_2$ it holds that $|F(x) - \alpha| < \frac{\varepsilon}{2}$. Let $y \in X$ be such that $0 < |y - x_0| < \delta$ and $n_0 := \max\{n_1, n_2\}$. Then $|F(y) - \alpha| \leq |F(y) - F(x_{n_0})| + |F(x_{n_0}) - \alpha| \leq \frac{\varepsilon}{2} + \frac{\varepsilon}{2} = \varepsilon$. Hence (6.1) holds. □

6.2.2 Inner continuity

The graph of a flexible function is not always the union of the graphs of its internal representatives, but if it is, we can take advantage from that fact. We consider here the case where the graph is the union of the graphs of internal continuous functions and study the relation with $M \times N$-continuity.

Definition 6.2.2. *Let $a, b \in \mathbb{R}$, $a < b$. Let $\phi : [a, b] \to \mathbb{E}$ be a flexible function. We say that ϕ is* inner continuous *if there exists an internal family of continuous functions C defined on $[a, b]$ such that $\Gamma(\phi)$ is the union of continuous representatives, i.e. if $\Gamma(\phi) = \bigcup \{f \in C : f \subseteq \Gamma(\phi)\}$.*

Example 6.2.2. 1. The function $\phi : \mathbb{R} \to \mathbb{E}$ defined by $\phi(x) = \oslash x$ is inner continuous for, with $C := \{f_a : \forall x \in \mathbb{R}(f_a(x) = ax)\}$ the set of linear functions, we have $\Gamma(\phi) = \bigcup_{a \in \oslash} f_a$.

2. Let $\psi : \mathbb{R}^+ \to \mathbb{E}$ be defined by $\psi(x) = \begin{cases} \oslash, & x = 0 \\ \pounds, & x \neq 0. \end{cases}$

We define functions $f_{a,b}$ by $f_{0,b} := 0$ and, for $a \neq 0$, $f_{a,b}(x) := \begin{cases} ax, & x \leq b \\ \frac{b}{a}, & x > b. \end{cases}$ Observe that for $a \neq 0$ the functions $f_{a,b}$ are linear functions truncated at $\frac{b}{a}$. We have $\Gamma(\psi) = \bigcup_{a,b \in \pounds} f_{a,b}$. Hence ψ is inner continuous.

Lemma 6.2.7. *Let $\phi : [a, b] \to \mathbb{E}$ be an inner continuous flexible function; let f be a continuous representative of ϕ and $\nu = \phi - f$. Then ν is an inner continuous neutrix function.*

Proof. Directly from Definition 6.2.2. □

An inner continuous flexible function is $M \times N$-continuous for appropriate neutrices M and N. Indeed, let $\phi : [a, b] \to \mathbb{E}$ be an inner continuous bounded flexible function with a continuous representative f and bounded neutrix part $\nu = \phi - f$, and $x_0 \in (a, b)$. Given a neutrix $M > 0$ such that $x_0 + M$ is contained in $[a, b]$, the neutrix function ν is $M \times K$-continuous for the neutrix $K := \bigcup_{x \in M} \nu(x)$. Let $L := \pounds\, \mathrm{zup}_{x,y \in M} |(f(x) - f(y)|$ and $N = K + L$. Both K and L are bounded, hence external. Then ϕ is $M \times N$-continuous with N external.

The next proposition expresses that internal continuous functions depending on external parameters are inner continuous.

Proposition 6.2.8. *Let $a, b \in \mathbb{R}$, $a < b$. Let $k \in \mathbb{N}$ be standard, $\alpha_1, ..., \alpha_k \in \mathbb{E}$ have neutrices different from $\mathbb{R}$ and f be an internal continuous real function of $k + 1$ variables, defined at least for $[a, b] \times \alpha_1 \times \cdots \times \alpha_k$. Assume that $\phi(x) := f(x, \alpha_1, ..., \alpha_k) \in \mathbb{E}$ for all $x \in [a, b]$ is a flexible function. Then $\phi : [a, b] \to \mathbb{E}$ is inner continuous.*

Proof. Put $a_0 := a, b_0 := b$. There exist $a_1, ..., a_k, b_1, ..., b_k \in \mathbb{R}$ with $a_i < b_i$ such that $\alpha \subseteq [a_i, b_i]$ for all i with $1 \leq i \leq k$ and such that f is defined on the compact set $\prod_{i=0}^{k} [a_i, b_i]$. Then f is continuous on $\prod_{i=0}^{k} [a_i, b_i]$. Put

$$C := \{c : \exists x_1 \in [a_1, b_1] \cdots \exists x_k \in [a_k, b_k] \forall x \in [a, b](c(x) = f(x, x_1, ..., x_k))\}.$$

Then C is an internal family of continuous functions such that $\Gamma(\phi) \subseteq C$. Let $(x, y) \in \Gamma(\phi)$. Then there exist $x_1 \in \alpha_1, ..., x_k \in \alpha_k$ such that $f(x, x_1, ..., x_k) = y$. Define $g : [a, b] \to \mathbb{R}$ by $g(x) = f(x, x_1, ..., x_k)$. Then $g \in C$ and $g \subseteq \Gamma(\phi)$. Hence $\Gamma(\phi) = \bigcup \{c \in C : c \subseteq \Gamma(\phi)\}$. It follows that ϕ is inner continuous. □

The following proposition on constant representative functions is a corollary to Proposition 4.4.1.

Proposition 6.2.9. *Let $a, b \in \mathbb{R}$, $a < b$. Let $k \in \mathbb{N}$ be standard, $A_1, ..., A_k > 0$ be neutrices different from $\mathbb{R}$ and f be an internal continuous real function of $k+1$ variables, defined at least on a domain $[a,b] \times D$ with $D \supset A_1 \times \cdots \times A_k$ internal. Assume that $\phi(x) := f(x, A_1, ..., A_k) \in \mathbb{E}$ for all $x \in [a,b]$. For $(x, a_1, ..., a_k) \in D$ define $f_{a_1,...,a_k}(x) = f(x, a_1, ..., a_k)$. Then the functions $f_{a_1,...,a_k}$ are continuous and*

$$\Gamma(\phi) = \bigcup_{a_1 \in A_1, ..., a_k \in A_k} f_{a_1,...,a_k}. \tag{6.5}$$

Proof. Formula (6.5) follows from (4.9). Being sections of a continuous function, the functions $f_{a_1,...,a_k}$ are clearly continuous. □

Exercises

Exercise 6.2.1. *Prove the claims of Example 6.2.1.*

Exercise 6.2.2. *Let $F : \mathbb{R} \to \mathbb{E}$ be the flexible function defined by*

$$F(x) = \begin{cases} \oslash, & x \in \oslash \\ 0, & x \notin \oslash. \end{cases}$$

Show that F is $\oslash$-continuous at 0.

Exercise 6.2.3. *Show that the following are inner continuous.*

1. *$F : \mathbb{R}^+ \to \mathbb{E}$ be defined by* $F(x) = \begin{cases} \oslash, & x \in \oslash \\ £, & x \notin \oslash. \end{cases}$

2. *$G : \mathbb{R}^+ \to \mathbb{E}$ be defined by* $G(x) = \begin{cases} 0, & x = 0 \\ \frac{\oslash}{x}, & x \neq 0. \end{cases}$

Exercise 6.2.4. *Show that the flexible function $F : \mathbb{R} \to \mathbb{E}$ defined by $F(x) = \sin(x) + £\cos(x)$ is inner continuous, and not locally constant.*

Exercise 6.2.5. *Let $F : (0,1] \to \mathbb{E}$ be the flexible function defined by*

$$F(x) = \begin{cases} £, & x \in \left[\frac{1+\oslash}{2n}, \frac{1+\oslash}{2n-1}\right] \\ \oslash, & x \in \left(\frac{1+\oslash}{2n+1}, \frac{1+\oslash}{2n}\right), \end{cases}$$

where $n \in \mathbb{N}, n \geq 1$.

1. *Show that F is $\oslash \times £$-continuous.*

2. *Write F in the form $F = G + H$ in such a way that G is $\frac{\oslash}{n} \times \oslash$-continuous and H is $\frac{\oslash}{n} \times £$-continuous.*

3. *Show that F is inner continuous.*

4. *Show that F is locally constant.*

5. *Extend F to a function $\overline{F} : [0,1] \to \mathbb{E}$ in two manners such that $\overline{F}$ is inner continuous, and in such a way that $\overline{F}$ is not inner continuous. Show that $\overline{F}$ never can be locally constant.*

Exercise 6.2.6. *Show that the flexible function F defined by $F(x) = x^2 + \oslash x$ is inner continuous at $x_0 = 0$ but not $\oslash$-continuous at $x_0 = 0$*

6.3 $M \times N$-derivation of flexible functions

It is not possible to use the classical techniques to build a notion of derivative for flexible functions. In fact, when x approaches x_0 the neutrix part of the expression $\frac{F(x)-F(x_0)}{x-x_0}$ in general approaches $\mathbb{R}$. For instance, if F has constant neutrix part $\oslash$, at any point $x_0 \in \mathbb{R}$ it holds that $\lim_{x \to x_0} \frac{\oslash}{x-x_0} = \mathbb{R}$. So we must keep sufficient distance to x_0, and we will ask that $\frac{1}{x-x_0}$ does not become an absorber of $N_F(x_0)$, in order to have $\frac{N_F(x_0)}{x-x_0} \subseteq N_F(x_0)$. To this end we will introduce a so-called *outer limit*, indicating that a point x_0 is approximated by points which are outside $x_0 + M$ for some neutrix M.

Definition 6.3.1. *Let M, N be neutrices, $F : X \subseteq \mathbb{R} \to \mathbb{E}$ be a flexible function, $x_0 \in \mathbb{R}$ be a limit point of X and $\alpha = a + A$ be an external number. We say that α is an $M \times N$-*outer limit* of F at x_0, and write $N\text{-}\lim_{x \to x_0 + M} F(x) = \alpha$, if and only if*

$$\forall \varepsilon > N \exists \delta > M \forall x \in X (M < |x - x_0| < \delta \Rightarrow |F(x) - \alpha| < \varepsilon).$$

We may also say that $F(x)$ is outer *$M \times N$-convergent to α when x approaches x_0. If $M = 0$ we write simply $N\text{-}\lim_{x \to x_0} F(x)$; if $N = 0$ we write $\lim_{x \to x_0 + M} F(x)$ and if $M = N = 0$ we write $\lim_{x \to x_0} F(x)$. The latter case corresponds to the usual notion of limit which justifies the more conventional notation.*

The difference between an $M \times N$-limit and an $M \times N$-outer limit is that the former considers values of a given function at points $x \in \mathbb{R}$ such that $|x - x_0| \in M$ while the latter does not. As a consequence it is easy to see that if α is an $M \times N$-limit of F at x_0, then it is also an $M \times N$-outer limit of F at x_0. The converse need not be true, in particular if the neutrix part of F for some points of $x_0 + M$ is relatively big, which may happen, e.g. if F is a difference quotient. As for stability under operations, notice that Theorem 6.1.8 still holds, for all limits were determined "from the outside".

Definition 6.3.2. *Let M, N be neutrices and $F : X \subseteq \mathbb{R} \to \mathbb{E}$ be a flexible function. Let $x_0 \in X$ be an M-limit point of X. We say that F is $M \times N$-*differentiable *at x_0 with $M \times N$-*derivative $\frac{d_N F}{d_M x}(x_0)$ *if*

$$\frac{d_N F}{d_M x}(x_0) := N\text{-}\lim_{x \to x_0 + M} \frac{F(x) - F(x_0)}{x - x_0}$$

*exists as an outer limit. If F is $M \times N$-differentiable at x_0 then $\frac{d_N F}{d_M x}(x_0)$ is called an $M \times N$-*derivative *of F at x_0. If $N = N_F(x_0)$ we write simply $\frac{dF}{d_M x}(x_0)$ and speak of an M-*derivative *of F at x_0.*

Remark 6.3.1. Note that if $N \neq 0$ then an $M \times N$-derivative of a flexible function F at a point $x_0 \in \mathbb{R}$ is not unique. In fact, just as the common $M \times N$-limit, an outer limit is unique up to the neutrix N. So if $\alpha, \beta \in \mathbb{E}$ are $M \times N$-derivatives of F then $\alpha - \beta \subseteq N$ and if $\alpha \in \mathbb{E}$ is a $\oslash \times N$-derivative of F at x_0, then $\alpha + N$ is also a $\oslash \times N$-derivative of F at x_0 (see Exercise 6.3.4).

Example 6.3.1. Let $F : \mathbb{R} \to \mathbb{E}$ be defined by $F(x) = \oslash x$, and $x_0 \in \mathbb{R}$. Let $a > 0$ be appreciable. Because $-ax < F(x) < ax$ for all $x \in \mathbb{R}$, it is natural to choose M and N in a way that $-a < \frac{d_N F}{d_M x}(x_0) < a$, and a being arbitrary, we wish to have

$$\frac{d_N F}{d_M x}(x_0) = \oslash. \tag{6.6}$$

We will see that this obtained for $x_0 \neq 0$, $M := \oslash x_0$ and $N := \frac{F(x_0)}{x_0} = \oslash$. Indeed,

$$\begin{aligned}
\frac{d_N F}{d_M x}(x_0) &= N\text{-}\lim_{x \to x_0 + M} \frac{F(x) - F(x_0)}{x - x_0} \\
&= N\text{-}\lim_{x \to x_0 + M} \frac{\oslash x - \oslash x_0}{x - x_0} \\
&= N\text{-}\lim_{x \to x_0 + M} \frac{\oslash (x - x_0) + \oslash x_0}{x - x_0} \\
&= N\text{-}\lim_{x \to x_0 + M} \oslash + \frac{\oslash x_0}{x - x_0} \\
&= N\text{-}\lim_{x \to x_0 + M} \oslash + N\text{-}\lim_{x \to x_0 + M} \frac{\oslash x_0}{x - x_0}
\end{aligned}$$

The first part exists for $N \supseteq \oslash$ and then is equal to $\oslash$. Then the second limit is equal to $\oslash$ for $M = \oslash x_0$, for we may write

$$N\text{-}\lim_{x \to x_0 + \oslash x_0} \frac{\oslash x_0}{@ x_0} = \oslash.$$

Combining, we obtain (6.6), where we take $N = \oslash$ minimal.

Observe that at $x_0 = 0$ the $0 \times \oslash$–outer limit $\frac{d_\oslash F}{d_{\oslash x_0} x}(x_0)$ does not exist.

Example 6.3.2. Let $F:\mathbb{R}\to\mathbb{E}$ be defined by $F(x)=x^2+\oslash$. At $x_0\in\mathbb{R}$ one has (see Exercise 6.3.5)

$$\frac{d_\oslash F}{d_\oslash x}(x_0)=\oslash\text{-}\lim_{x\to x_0+\oslash}\frac{x^2+\oslash-x_0^2+\oslash}{x-x_0}=2x_0+\oslash.$$

The next proposition gives conditions for an M-differentiable function to be M-continuous at some point x_0. Note that we need to add information on F inside x_0+M.

Proposition 6.3.2. *Let $X\subseteq\mathbb{R}$, $X\neq\emptyset$. Let $F:X\to\mathbb{E}$ be a flexible function. Let M be a neutrix and x_0 be an M-interior point of X. Assume that $M,N:=N_F(x_0)\subseteq £$, that $N_F(x)\subseteq F(x_0)$ for all $x\in x_0+M$ and that F is $M\times N$-differentiable at x_0+M, with $M\times N$-derivative $D:=\frac{d_N F}{d_M x}(x_0)$. Then F is M-continuous at x_0.*

Proof. By definition $N\text{-}\lim\frac{F(x)-F(x_0)}{x-x_0}=D$. Also $M\text{-}\lim(x-x_0)=M$. Hence, applying the product rule for limits and noting that $N+DM+M\supseteq N$,

$$\begin{aligned}N\text{-}\lim_{x\to x_0+M}\frac{F(x)-F(x_0)}{x-x_0}&\cdot M\text{-}\lim_{x\to x_0+M}(x-x_0)\\&=(N+DM+M)\times M\text{-}\lim_{x\to x_0+M}F(x)-F(x_0)\\&=N\times M\text{-}\lim_{x\to x_0+M}F(x)-F(x_0)\\&=N=N_F(x_0).\end{aligned}$$

So $N\text{-}\lim_{x\to x_0}F(x)=F(x_0)$. Because $N_F(x)\subseteq F(x_0)$ for all $x\in x_0+M$ we derive that F is M-continuous at x_0. □

An interesting class of functions is those who at microscopic level exhibit rapid oscillations, so for instance, are not S-differentiable, but at a bigger scale they are smoother; a simple example is given by $x\mapsto\frac{\sin(\omega x)}{\omega^2}$ for unlimited $\omega\in\mathbb{R}$. This has been observed by various authors (see [128, 61, 27]), in terms of generalized derivatives which typically may be recognized as $\oslash\times 0$-outer limits. So $M\times N$-derivatives are a still further generalization.

Exercises

Exercise 6.3.1. *At every point $x\in\mathbb{R}$, give external neutrices M,N such that the following flexible functions F are $M\times N$-differentiable.*

1. $F(x)=x^2+x+\oslash$.
2. $F(x)=x^2+\oslash x+1$.
3. $F(x)=\oslash x^2+x+1$.

Exercise 6.3.2. *Consider the flexible function $F:\mathbb{R}\to\mathbb{E}$ defined by*

$$F(x) = |x| + \oslash. \tag{6.7}$$

1. *Show that F is $\oslash \times \oslash$-differentiable at all non-infinitesimal x, and determine the $\oslash \times \oslash$-derivative.*

2. *Show that F is not $\oslash \times \oslash$-differentiable at infinitesimal x.*

3. *Let $\delta \simeq 0$, $\delta > 0$. Show that ϕ is $\delta\oslash\times\oslash$-differentiable at infinitesimal x, and determine the $\delta\oslash\times\oslash$-derivative.*

4. *For $\delta \simeq 0$, $\delta > 0$, determine the $\delta\oslash\times\oslash$-derivative at non-infinitesimal x.*

Exercise 6.3.3. *Verify that Theorem 6.1.8 on behaviour under operations for $M \times N$-limits also holds for $M \times N$-outer limits.*

Exercise 6.3.4. *Prove the claims of Remark 6.3.1.*

Exercise 6.3.5. *Prove the equalities in Example 6.3.2.*

Exercise 6.3.6. 1. *Adapt Definition 6.2.2 to define* inner differentiability.

2. *Show that the functions F given in Exercise 6.3.1 and 6.3.2 are inner differentiable.*

3. *Assume that with respect to the function F of (6.7) we restrict ourselves to the set, say C, of convex representatives f, and define the* inner derivative *at a point $x \in \mathbb{R}$ as $F'(x) := \{f'(x) : f \in C\}$. Show that $F'(x) = \begin{cases} 1+\oslash, & x \gtrsim 0 \\ -1+\oslash, & x \lesssim 0 \end{cases}$. Show also that $\phi'(x)$ is not defined as an external number for $x \simeq 0$.*

4. *Show that if we add the condition that the functions of C are symmetric, we obtain $\phi'(x) = \begin{cases} 1+\oslash, & x > 0 \\ 0, & x = 0 \\ -1+\oslash, & x < 0. \end{cases}$*

Exercise 6.3.7. *Let $\omega \in \mathbb{N}$ be unlimited. Indicate neutrices K, L such that the function $x \mapsto \frac{\sin(\omega x)}{\omega^2}$ is not $K \times L$-differentiable, and bigger neutrices M, N such that it becomes $M \times N$-differentiable.*

6.4 Weak extrema and monotonicity

The derivative of a function acts as an indicator for local minima and maxima, and for being locally increasing or decreasing. The introduction of neutrices is susceptible to weaken these relationships; for instance many flexible functions hide local fluctuations and are locally constant. It is to be expected that $M \times N$-derivatives give information on the slope of the function when taking steps larger than M and movements larger than N. Below we will weaken the notions of extrema to weak extrema up to a neutrix, prove a necessary condition for the existence of weak extrema, and characterize monotonicity over big enough intervals.

Definition 6.4.1. *Let N, M be neutrices and $\emptyset \neq X \subseteq \mathbb{R}$. Let $F : X \to \mathbb{E}$ be a flexible function. An M-interior point x_0 of X is called an* N-minimizer *of the minimization problem* $\min_{x \in X} F(x)$ *if for all $x \in X$ one has $F(x) - F(x_0) \geq N$. Then $F(x_0)$ is called a* near minimum *with respect to N, or an* N-minimum. *A near minimum $F(x_0)$ is* M-local *if it is an N-minimum on some M-neighbourhood of x_0. (Local) near maxima are defined analogously.*

The next theorem generalizes the well-known result stating that at minima the derivative is always 0. An analogous theorem holds for M-local N-maximizers.

Theorem 6.4.1. *Let N, M be neutrices such that M contains all absorbers of N, and $\emptyset \neq X \subseteq \mathbb{R}$. Let $F : X \to \mathbb{E}$ be a flexible function. Assume that x_0 is an M-local N-minimizer of F on X, x_0 is an M-interior point of X and F is $M \times N$-differentiable at x_0. Then*

$$\frac{d^N F}{d^M x}(x_0) = N. \tag{6.8}$$

Proof. Let $\alpha = a + N$ and assume that $\frac{d^N F}{d^M x}(x_0) = \alpha$. Then for all $\varepsilon > N$ there exists $\delta_0 > M$ such that for all $x \in X$, $M < |x - x_0| < \delta_0$ one has $\left| \frac{F(x)-F(x_0)}{x-x_0} - a + N \right| < \varepsilon$. This implies that

$$a + N - \varepsilon < \frac{F(x) - F(x_0)}{x - x_0} + N, \tag{6.9}$$

for all $M < x_0 - x < \delta_0$ and

$$\frac{F(x) - F(x_0)}{x - x_0} + N < \varepsilon + a + N, \tag{6.10}$$

for all $M < x - x_0 < \delta_0$. On the other hand, the point x_0 is an M-local N-minimizer of F on X, so there exists $\delta_1 > M$ such that

$$F(x) - F(x_0) \geq N \tag{6.11}$$

for all $x \in X$, $|x-x_0| \leq \delta_1$. Put $\delta := \min\{\delta_0, \delta_1\}$. If $|x| > M$ is not an absorber of N, (6.11) implies that $\frac{F(x)-F(x_0)}{x-x_0} \leq \frac{N}{x-x_0} \leq N$ whenever $M < x_0 - x < \delta$. Hence

$$\frac{F(x) - F(x_0)}{x - x_0} + N \leq N + N = N, \tag{6.12}$$

for all $x \in X, M < x_0 - x < \delta$. From (6.9) and (6.12) we derive that

$$a + N - \varepsilon < N. \tag{6.13}$$

Similarly, it follows from (6.11) that $\frac{F(x)-F(x_0)}{x-x_0} \geq \frac{N}{x-x_0} \geq N$ for all $M < x - x_0 < \delta_0$. Using (6.10) and analogous arguments we obtain that

$$a + N + \varepsilon > N. \tag{6.14}$$

Because $\varepsilon > N$ is arbitrary, formulas (6.13) and (6.14) imply that $a \in N$. Indeed, if $a < N$, we choose $\varepsilon = -\frac{a}{2} > N$ and then $a + \varepsilon + N = \frac{a}{2} + N < N$, in contradiction with (6.14). If $a > N$, taking $\varepsilon = \frac{a}{2} > N$, we get $a + N - \varepsilon = \frac{a}{2} + N > N$, in contradiction with (6.13). We conclude that (6.8) holds. □

Many flexible functions are locally constant at each point. For example, consider the flexible function defined by $F(x) = x + \oslash$ for $x \in \mathbb{R}$. Let $x_0 \in \mathbb{R}$. Then for all $x \in x_0 + \oslash$ one has $F(x) = F(x_0)$. This means that we should consider monotonicity of flexible functions with a certain order of magnitude for the steps of the variable.

Definition 6.4.2. *Let $x_0 \in \mathbb{R}$, M be a neutrix and let F be a flexible function defined on an M-neighbourhood of x_0. We say that F is*

1. M-increasing *at x_0 if*

$$\exists d > M \forall x \forall y (x_0 - d \leq x < x_0 + M < y \leq x_0 + d \Rightarrow F(x) < F(x_0) < F(y)).$$

2. M-decreasing *at x_0 if*

$$\exists d > M \forall x \forall y (x_0 - d \leq x < x_0 + M < y \leq x_0 + d \Rightarrow F(x) > F(x_0) > F(y)).$$

Example 6.4.1. The flexible function defined by $F(x) = x + \oslash$ is $\oslash$-increasing on $\mathbb{R}$. This follows from the fact that for all $x, y \in \mathbb{R}$, $y - x > \oslash$, one has $F(x) = x + \oslash < x + y - x + \oslash = F(y)$.

We end with two results which relate $M \times N$ derivatives to M-monotonicity defined above.

Proposition 6.4.2. *Let M, N be neutrices such that M contains all absorbers of N, and let F be a flexible function defined on an M-neighbourhood of $x_0 \in \mathbb{R}$. Assume that $F(x)$ is $M \times N$-differentiable at x_0 and $\frac{d_N F}{d_M x}(x_0) > N$. Then there exists $d > M$ such that*

1. $\forall x \in X(M < x_0 - x \leq d \Rightarrow F(x) - F(x_0) < N)$.

2. $\forall x \in X(M < x - x_0 \leq d \Rightarrow F(x) - F(x_0) > N)$.

In particular, if $N_F(x_0) \subseteq N$, the function F is M-increasing at x_0.

Proof. Because $\frac{d_N F}{d_M x}(x_0) > N$, we may take a representative $\varepsilon \in \frac{1}{2}\frac{d_N F}{d_M x}(x_0) > N$. By the definition of $M \times N$-derivative,

$$\exists d > M \forall x \left(M < |x - x_0| \leq d \Rightarrow \left| \frac{F(x) - F(x_0)}{x - x_0} - \frac{d_N F}{d_M x}(x_0) \right| < \varepsilon \right).$$

Then

$$\forall x \left(M < |x - x_0| \leq d \Rightarrow \frac{d_N F}{d_M x}(x_0) - \varepsilon < \frac{F(x) - F(x_0)}{x - x_0} \right).$$

Note that $\frac{d_N F}{d_M x}(x_0) - \varepsilon = \frac{1}{2}\frac{d_N F}{d_M x}(x_0) > N$. Also, if $x - x_0 \notin M$, then $x - x_0$ is not an absorber of N. Hence $F(x) - F(x_0) > N$ for all $M < x - x_0 \leq d$. If in particular $N_F(x) \subseteq N$, then $F(x) > F(x_0)$. Moreover, for all $x \in X$ if $M < x_0 - x \leq d$, then $F(x) - F(x_0) < N$, and if in addition $N_F(x) \subseteq N$, then $F(x) < F(x_0)$. This completes the proof. □

Proposition 6.4.3. *Let M, N be neutrices such that M contains all absorbers of N, and let F be a flexible function defined on an M-neighbourhood of $x_0 \in \mathbb{R}$. Assume that $F(x)$ is $M \times N$-differentiable at x_0 and $\frac{d_N F}{d_M x}(x_0) < N$. Then there exists $d > M$ such that*

1. $\forall x \in X(M < x_0 - x \leq d \Rightarrow F(x) - F(x_0) > N)$.

2. $\forall x \in X(M < x - x_0 \leq d \Rightarrow F(x) - F(x_0) < N)$.

In particular, if $N_F(x_0) \subseteq N$, the function F is M-decreasing at x_0.

Proof. See Exercise 6.4.2. □

Exercises

Exercise 6.4.1. *Investigate the following flexible functions $F : \mathbb{R} \to \mathbb{E}$ for local $\oslash$-minimizers and $\oslash$-monotonicity:*

1. $F(x) = x^2 + x + \oslash$.

2. $F(x) = x^2 + \oslash x + 1$.

3. $F(x) = x^2 + \oslash x + \oslash$.

4. $F(x) = \oslash x^2$.

5. $F(x) = \oslash x^2 + 4x^3$.

Exercise 6.4.2. *Prove Proposition 6.4.3.*

Chapter 7

Integration of functions of external numbers

7.1 Integrals of internal functions on external intervals 163
7.2 Integrals of flexible functions 167
7.3 Elementary properties of integrals 172
7.4 Special integrals and applications 178
7.4.1 Mass and tail of probabilities and integrals 178
7.4.2 On local averaging 180
7.4.3 The concentration lemma and the Laplace method 182

This chapter aims to develop tools to evaluate asymptotic integrals in terms of external numbers. We start with integrals of internal functions over external intervals. These functions are supposed to be integrable in some classical sense, in particular Riemann-integrable or Lebesgue-integrable. We consider also integration for classes of flexible functions, which have internal integrable representatives, first over internal intervals, and then also over external intervals. Some of these external intervals come quite naturally. Typically, given an integral, we can determine the domain where the essential contribution to its value, i.e. the contribution up to infinitesimals, takes place. Usually this is an external interval, called "mass" in Chapter 2. Its complement may then be a model of what informally is called "tail".

Earlier approaches to this subject along these lines are contained in [76, 50]. Here we intend to systematize, and to develop the theory further.

7.1 Integrals of internal functions on external intervals

We define integrals of internal Lebesgue-integrable functions over an external interval I of $\mathbb{R}$ in terms of suprema and infima of integrals over internal intervals. We investigate the case where these extrema can be stated as values of a primitive.

Definition 7.1.1. *Let $a, b \in \mathbb{R}$ with $a < b$ and $f : [a, b] \to \mathbb{R}$ be an internal Lebesgue-integrable function. Let I be an external interval included in $[a, b]$*

and suppose that the function f does not change sign on I. Then we say that f is integrable *on I, and we define the integral of f on I by:*

1. $\int_I f(x)dx := \sup\left\{\int_s^t f(x)dx : s,t \in I \wedge s < t\right\}$, *if $f \geq 0$ on I.*

2. $\int_I f(x)dx := \inf\left\{\int_s^t f(x)dx : s,t \in I \wedge s < t\right\}$, *if $f \leq 0$ on I.*

Clearly $\int_I f(x)dx \geq 0$ if $f \geq 0$ on I. If $f \leq 0$ on I, it is still possible that $\int_I f(x)dx \geq 0$ without being zero, for it can be a neutrix, as seen by the integral $\int_\oslash -1dx = \oslash$. Such problems can be avoided if we consider $-f$ instead of f. So below we focus on non-negative functions.

Definition 7.1.2. *Let $a, b \in \mathbb{R}$ with $a < b$ and $f : [a,b] \to \mathbb{R}$ be a non-negative internal Lebesgue-integrable function. Let $\gamma \subset [a,b]$ be an external number. We write*

$$\int_a^\gamma f(x)dx := \int_{[a,\gamma]} f(x)dx; \qquad \int_a^{\underline{\gamma}} f(x)dx := \int_{[a,\gamma)} f(x)dx;$$
$$\int_\gamma^b f(x)dx := \int_{[\gamma,b]} f(x)dx; \qquad \int_{\overline{\gamma}}^b f(x)dx := \int_{(\gamma,b]} f(x)dx.$$

Example 7.1.1. Let $\omega > 0$ be unlimited.

1. By straightforward verification one sees that $\int_0^{\underline{\omega+\pounds}} x^2dx = \int_0^{\omega+\pounds} x^2dx = \frac{1}{3}\omega^3 + \pounds\omega^2$.

2. $\int_0^{\underline{\omega+\pounds}} e^x dx$ is strictly less than $\int_0^{\omega+\pounds} e^x dx$, for $\int_0^{\omega+\pounds} e^x dx = \pounds e^\omega$, while $\int_0^{\underline{\omega+\pounds}} e^x dx = \oslash e^\omega$.

3. $\int_\pounds e^{-\frac{x^2}{2}} dx = \sqrt{2\pi} + \oslash$.

The next theorem states adapted forms of the *Chasles Relation*, about splitting the interval of integration in two.

Theorem 7.1.1. *Let $a, b \in \mathbb{R}$ with $a < b$. Let f be a non-negative internal real function, Lebesgue-integrable on $[a,b]$. Let γ be such that $a < \gamma < b$. Then*

$$\int_a^b f(x)dx \in \int_a^\gamma f(x)dx + \int_{\overline{\gamma}}^b f(x)dx \leq \int_a^\gamma f(x)dx + \int_\gamma^b f(x)dx. \qquad (7.1)$$

Proof. If $\gamma \in \mathbb{R}$, the theorem reduces to the classical Chasles Relation; in fact we have equalities. So we may assume that γ is external. Consider the set $A := \left\{\int_a^t f(x)dx : t < \gamma\right\}$. Let $\sigma := \int_a^\gamma f(x)dx = \sup A$, with neutrix part N, say. Then $\int_a^\gamma f(x)dx = \int_a^c f(x)dx + N$ for some $c \in [a,\gamma]$, or $\int_{\overline{\gamma}}^b f(x)dx = \int_d^b f(x)dx + N$ for some $d \in (\gamma, b]$. We prove only the first case. The proof of the second case is similar.

We show that

$$\int_{\overline{\gamma}}^{b} f(x)dx = \int_{c}^{b} f(x)dx + N. \tag{7.2}$$

Indeed, by non-negativity, for every $t > \gamma$ we have $\int_t^b f(x)dx \le \int_c^b f(x)dx \le \int_c^b f(x)dx + N$. Hence $\int_{\overline{\gamma}}^b f(x)dx \le \int_c^b f(x)dx + N$. Conversely, let $p \in \mathbb{R}$, $p > N$. Then $\int_a^r f(x)dx < \int_a^c f(x)dx + p$ for all $r \le \gamma$. By Cauchy's Principle there exists $s \in \mathbb{R}$ with $\gamma < s \le b$ such that still $\int_a^s f(x)dx < \int_a^c f(x)dx + p$. Then $\int_s^b f(x)dx > \int_c^b f(x)dx - p$. Hence $\int_{\overline{\gamma}}^b f(x)dx \ge \int_c^b f(x)dx + N$. This implies (7.2). Hence

$$\int_a^b f(x)dx \in \int_a^{\gamma} f(x)dx + \int_{\overline{\gamma}}^b f(x)dx.$$

Additionally,

$$\begin{aligned}\int_a^{\gamma} f(x)dx + \int_{\overline{\gamma}}^b f(x)dx &= \int_a^b f(x)dx + N \\ &= \int_a^c f(x)dx + N + \int_c^b f(x)dx \\ &\le \int_a^{\gamma} f(x)dx + \int_{\gamma}^b f(x)dx.\end{aligned}$$

Combining, we obtain (7.1). □

The next proposition relates the integral of an internal function over an external interval to values of a primitive.

Proposition 7.1.2. *Let $a, b \in \mathbb{R}$ with $a < b$ and $f : [a, b] \to \mathbb{R}$ be an internal non-negative Lebesgue-integrable function. Let F be a primitive of f and let I be an external interval such that $I \subseteq [a, b]$. Then*

$$\int_I f(x)dx = \sup_{t \in I} F(t) - \inf_{s \in I} F(s).$$

Proof. The function F is an internal function which is non-decreasing on the interval $[a, b]$. Let $c \in I$; we may assume that I has elements both less and larger than c. Then

$$\begin{aligned}
\int_I f(x)dx &= \sup\left\{\left\{\int_s^t f(x)dx : s,t \in I \wedge s<t\right\}\right. \\
&= \sup\left(\left\{\left\{\int_s^c f(x)dx : s\in I \wedge s<c\right\}+\left\{\left\{\int_c^t f(x)dx : t\in I \wedge c<t\right\}\right)\right. \\
&= \sup\left(\{F(t)-F(c) : t\in I \wedge c<t\}+\{F(c)-F(s) : s\in I \wedge s<c\}\right) \\
&= \sup_{t\in I} F(t) - F(c) + F(c) - \inf_{s\in I} F(s) \\
&= \sup_{t\in I} F(t) - \inf_{s\in I} F(s).
\end{aligned}$$

□

Assume the external interval I has lower bound $\alpha \in \mathbb{E}$ and upper bound $\beta \in \mathbb{E}$, with $\alpha < \beta$. We may or may not have that $\inf_{s\in I} F(s) = F(\alpha)$ or $\sup_{t\in I} F(t) = F(\beta)$, as shown by the next examples.

Example 7.1.2. 1. Let $n \in \mathbb{N}$ be standard. Then

$$\int_\alpha^\beta x^n dx = \sup_{t\leq\beta} \frac{t^{n+1}}{n+1} - \inf_{s\geq\alpha} \frac{t^{n+1}}{n+1} = \frac{\beta^{n+1}-\alpha^{n+1}}{n+1}.$$

It is straightforward to verify that

$$\int_{\overline{\alpha}}^\beta x^n dx = \int_\alpha^{\overline{\beta}} x^n dx = \int_{\overline{\alpha}}^{\overline{\beta}} x^n dx.$$

2. Let $n \in \mathbb{N}, n \geq 2$ be standard. Since the lower bound of the interval of integration is internal we have

$$\begin{aligned}
\int_1^{\pounds} \frac{1}{x^n}dx &= \sup_{t\leq\pounds} \frac{1}{-n+1}\left(\frac{1}{t^{n-1}}-1\right) = \frac{1}{n-1}\left(1-\inf_{t\leq\pounds}\frac{1}{t^{n-1}}\right) \\
&= \frac{1}{n-1}\left(1-\oslash^{n-1}\right) = \frac{1}{n-1}+\oslash.
\end{aligned}$$

3. Let $\omega > 0$ be unlimited. Then applying Proposition 4.3.9

$$\begin{aligned}
\int_0^{\omega+\pounds} x^{\omega-1}dx &= \sup_{t<\omega+\pounds} \frac{1}{\omega}t^\omega = \inf_{t\in\omega+\pounds} \frac{1}{\omega}t^\omega = \inf_{u\in\pounds} \frac{1}{\omega}(\omega+u)^\omega \\
&= \inf_{u\in\pounds} \omega^{\omega-1}\left(1+\frac{u}{\omega}\right)^\omega \\
&= \omega^{\omega-1} \inf_{u\in\pounds} e^u = \oslash\omega^{\omega-1}.
\end{aligned}$$

Exercises

Exercise 7.1.1. *Let* $\varepsilon \simeq 0$, $\varepsilon > 0$ *and* $\omega \simeq \infty$, $\omega \in \mathbb{N}$. *Determine*

1. $\int_{-1+\oslash}^{1+\oslash} \frac{1}{1+x^2} dx$.
2. $\int_0^{\pounds} \frac{e^{-x}}{1+\varepsilon x} dx$.
3. $\int_{\oslash} \sin\left(\frac{x}{\varepsilon}\right) dx$.
4. $\int_{\oslash}^{\frac{1}{\oslash}} \log(x) dx$.
5. $\int_{\oslash}^{\oslash\omega} \log(x) dx$.

7.2 Integrals of flexible functions

We consider integrals of flexible functions, first defined on internal intervals, and then on external intervals. In line with the previous section, we focus on non-negative functions. In order to avoid obvious problems with the common notation of primitives we denote flexible functions by Greek letters.

Definition 7.2.1. *Let* $a, b \in \mathbb{R}, a < b$. *Let* $\phi : [a, b] \to \mathbb{E}$ *be a non-negative flexible function. We define*

$$\underline{\int_a^b} \phi(x)dx := \sup \left\{ \int_a^b f(x)dx : f \leq \phi \wedge f \text{ internal and integrable} \right\}$$

and

$$\overline{\int_a^b} \phi(x)dx := \inf \left\{ \int_a^b f(x)dx : f \geq \phi \wedge f \text{ internal and integrable} \right\}.$$

We say that ϕ *is* outer-integrable *over* $[a, b]$ *if* $\underline{\int_a^b} \phi(x)dx = \overline{\int_a^b} \phi(x)dx$, *and we write*

$$ {}_a\!\int^b \phi(x)dx := \underline{\int_a^b} \phi(x)dx = \overline{\int_a^b} \phi(x)dx.$$

Definition 7.2.2. *Let* $a, b \in \mathbb{R}$, $a < b$. *Let* $\phi : [a, b] \to \mathbb{E}$ *be a non-negative flexible function. Let* F *be the family of internal (Lebesgue-)integrable functions over* $[a, b]$. *We say that* ϕ *is* inner-integrable *if it is internally representable and* $\sup I_\phi = \inf I_\phi$, *with* $I_\phi := \left\{ \int_a^b f(x)dx : f \in F \wedge f \subseteq \Gamma(\phi) \right\}$; *then* $I_\phi \in \mathbb{E}$ *and we define* $\int_a^b \phi(x)dx := I_\phi$.

So if $\phi : [a,b] \to \mathbb{E}$ is an inner-integrable flexible function, we have

$$\int_a^b \phi(x)dx = \left\{ \int_a^b f(x)dx : f \subseteq \Gamma(\phi) \wedge f \text{ internal and integrable} \right\}.$$

Theorem 7.2.1. *Let $\phi : [a,b] \to \mathbb{E}$ be a non-negative flexible function. If ϕ is inner-integrable then it is also outer-integrable and*

$$\overline{\int_a^b} \phi(x)dx = \int_a^b \phi(x)dx. \tag{7.3}$$

Proof. Let $\mathcal{F}$ be the set of internal and integrable functions $f : [a,b] \to \mathbb{R}$. Let $I := \left\{ \int_a^b f(x)dx : f \in \mathcal{F} \wedge f \leq \phi \right\}$ and $J := \left\{ \int_a^b f(x)dx : f \in \mathcal{F} \subseteq \Gamma(\phi) \right\}$. We prove first that

$$\sup I = \sup J. \tag{7.4}$$

Firstly, let $f \subseteq \Gamma(\phi)$ be integrable. Then $f \in \mathcal{F}$ and $f \leq \phi$. Hence $\sup J \leq \sup I$. Secondly, let $f \leq \phi$ be integrable and $g \in J$. Put $h := \max(f,g)$. Then for all $x \in [a,b]$ one has $h(x) \in \phi(x)$, for if $h(x) = g(x)$ certainly $h(x) \in \phi(x)$, and if $h(x) \neq g(x)$ we cannot have $f(x) < \phi(x)$, because then $f(x) < g(x)$ and $h(x) = g(x)$, a contradiction; hence $h(x) = f(x)$ also implies that $h(x) \in \phi(x)$. Hence $f \leq h \in J$. Now $\int_a^b f(x)dx \leq \int_a^b h(x)dx$. Then $\sup I \leq \sup J$ and hence (7.4) holds. From this we conclude that

$$\int_a^b \phi(x)dx = \sup \left\{ \int_a^b f(x)dx : f \in \mathcal{F} \wedge f \subseteq \Gamma(\phi) \wedge f \text{ integrable} \right\}.$$

Analogously one shows that

$$\int_a^b \phi(x)dx = \inf \left\{ \int_a^b f(x)dx : f \in \mathcal{F} \wedge f \subseteq \Gamma(\phi) \wedge f \text{ integrable} \right\}.$$

Hence ϕ is outer-integrable, and therefore (7.3) holds. □

Discontinuous flexible functions with jumps at external numbers cannot be inner-integrable, because they are not internally representable. This is illustrated by the following example, which happens to be outer-integrable.

Example 7.2.1. Let $f : [0,2] \to \mathbb{R}$ be defined by $f(x) = \begin{cases} 1, & x \leq 1 + \oslash \\ 2, & x > 1 + \oslash \end{cases}$.

Due to the discontinuity at $1+\oslash$ the function f is not internally representable, hence is not inner-integrable.

To see that f is outer-integrable, let $f_n, g_n : [0,2] \to \mathbb{R}$ for standard $n \in \mathbb{N}$ be defined by

$$f_n(x) = \begin{cases} 1, & x \leq 1 + \frac{1}{n} \\ 2, & x > 1 + \frac{1}{n}, \end{cases}$$

$$g_n(x) = \begin{cases} 1, & x \leq 1 - \frac{1}{n} \\ 2, & x > 1 - \frac{1}{n}. \end{cases}$$

Then $\int_0^2 f_n(x)dx = 3 - \frac{1}{n}$ and $\int_0^2 g_n(x)dx = 3 + \frac{1}{n}$ and one sees that both

$$\underline{\int_a^b} f(x)dx := \sup_{\mathrm{st}(n)\in\mathbb{N}} \int_a^b f_n(x)dx = 3 + \oslash.$$

$$\overline{\int_a^b} f(x)dx := \inf_{\mathrm{st}(n)\in\mathbb{N}} \int_a^b g_n(x)dx = 3 + \oslash.$$

Hence f is outer-integrable with outer-integral ${}_a^b\!\int \phi(x)dx = 3 + \oslash$.

A flexible function is *integrable* over an internal interval if it is both inner-integrable and outer-integrable. Because the former implies the latter by Theorem 7.2.1, we will call inner-integrable functions simply integrable.

We will show that the integration of integrable functions reduces to the integration of an (internal) representative and the integration of a neutrix function, taking the sum of both. As we will see, integrals of neutrix functions, even over external intervals, have simpler properties than integrals of general flexible functions: the integral is a neutrix (Proposition 7.2.2), in case of external intervals obtained by unions (Proposition 7.2.5), and the exact Chasles Relation holds (Theorem 7.3.1).

Proposition 7.2.2. *Let $a, b \in \mathbb{R}$, $a < b$. Let $\nu : [a, b] \to \mathbb{E}$ be an integrable neutrix function. Then $\int_a^b \nu(x)dx$ is a neutrix.*

Proof. Let $y \in I_\nu$. Then there exists an internal integrable function $f \subseteq \Gamma(\nu)$ such that $\int_a^b f(x)dx = y$. Let $c \in \mathbb{R}$ be limited. Then also $cf \subseteq \Gamma(\nu)$ and $\int_a^b cf(x)dx = cy$. Hence I_ν is stable under multiplication by limited numbers, which implies that I_ν is a neutrix. □

Theorem 7.2.3. *Let $a, b \in \mathbb{R}$, $a < b$. Let $\phi : [a, b] \to \mathbb{E}$ be a non-negative integrable flexible function. Let f be an integrable representative of ϕ. Define $\nu : [a, b] \to \mathbb{E}$ by $\nu(x) = \phi(x) - f(x)$. Then ν is a neutrix function which is integrable over $[a, b]$ and*

$$\int_a^b \phi(x)dx = \int_a^b f(x)dx + \int_a^b \nu(x)dx. \tag{7.5}$$

Proof. Let $y = \int_a^b f(x)dx$. Then $\int_a^b \phi(x)dx = y + Y$, with Y a neutrix. Then ν is an neutrix integrable function with integral equal to Y because

$$\begin{aligned}\int_a^b \nu(x)dx &= \left\{\int_a^b g(x) - f(x)dx : g \in \Gamma(\phi) \wedge g \text{ internal integrable}\right\}\\ &= \left\{\int_a^b g(x)dx : g \in \Gamma(\phi) \wedge g \text{ internal integrable}\right\} - \int_a^b f(x)dx\\ &= y + Y - y = Y.\end{aligned}$$

Then (7.5) is a consequence. □

The next theorem expresses that one may deduce integrability of a flexible function from the integrability of the associated neutrix function and the existence of an integrable representative.

Theorem 7.2.4. *Let $a, b \in \mathbb{R}$, $a < b$. Let $\phi : [a, b] \to \mathbb{E}$ be a non-negative flexible function. Assume that $\phi = f + \nu$ where f is an integrable representative and ν is an integrable neutrix function. Then ϕ is integrable over $[a, b]$ and*

$$\int_a^b \phi(x)dx = \int_a^b f(x)dx + \int_a^b \nu(x)dx. \tag{7.6}$$

Proof. Because $\Gamma(\phi) = \{f\} \cup \Gamma(\nu)$, the function ϕ is integrable. Then (7.6) follows from the fact that $\phi = f + \nu$. □

The following example shows the decomposition of a flexible function into an integrable internal function and an integrable neutrix function.

Example 7.2.2. One has

$$\begin{aligned}\int_0^1 e^{-u+\oslash}du &= \int_0^1 e^{-u}e^{\oslash}du = \int_0^1 e^{-u}(1+\oslash)du \\ &= \int_0^1 e^{-u}du + \int_0^1 \oslash du = 1 - \frac{1}{e} + \oslash.\end{aligned} \tag{7.7}$$

Integrals of flexible functions defined on external intervals are defined in analogy with Definition 7.1.1.

Definition 7.2.3. *Let $a, b \in \mathbb{R}$ with $a < b$ and $\phi : [a, b] \to \mathbb{E}$ be an integrable flexible function. Let I be an external interval included in $[a, b]$ and suppose that ϕ does not change sign on I. Then we say that ϕ is* integrable on I, *and we define the integral of ϕ on I by:*

1. $\int_I \phi(x)dx := \text{zup}\left\{\int_s^t \phi(x)dx : s, t \in I \wedge s < t\right\}$, *if $\phi \geq 0$ on I.*

2. $\int_I \phi(x)dx := \text{winf}\left\{\int_s^t \phi(x)dx : s, t \in I \wedge s < t\right\}$, *if $\phi \leq 0$ on I.*

As in the case of internal intervals we study only non-negative functions.

We show that, like in the case of internal intervals, the integral of a flexible function over an external interval is equal to the sum of the integral of a representative and the integral of a neutrix function. The latter is the union of integrals over internal subintervals.

Proposition 7.2.5. *Let $a, b \in \mathbb{R}$ with $a < b$ and $\nu : [a, b] \to \mathbb{E}$ be an integrable neutrix function. Let I be an external interval included in $[a, b]$. Then $\int_I \nu(x)dx$ is a neutrix and*

$$\int_I \nu(x)dx = \bigcup_{\substack{s<t \\ s,t\in I}} \int_s^t \nu(x)dx.$$

Proof. The proposition is a consequence of Propositions 7.2.2 and 4.4.2. □

Theorem 7.2.6. *Let $a, b \in \mathbb{R}$ with $a < b$ and $\phi : [a, b] \to \mathbb{E}$ be an integrable non-negative flexible function. Let I be an external interval included in $[a, b]$. Let f be an integrable representative of ϕ. Define $\nu : [a, b] \to \mathbb{E}$ by $\nu(x) = \phi(x) - f(x)$. Then ν is a neutrix function which is integrable over I and*

$$\int_I \phi(x)dx = \int_I f(x)dx + \int_I \nu(x)dx.$$

Proof. Because ϕ and f are integrable over I, also ν is integrable. Applying Theorem 4.4.5 we find

$$\begin{aligned}\int_I \phi(x)dx &= \operatorname{zup}\left\{\int_s^t \phi(x)dx : s, t \in I \wedge s < t \wedge s, t \in \mathbb{R}\right\} \\ &= \operatorname{zup}\left\{\int_s^t f(x) + \nu(x)dx : s, t \in I \wedge s < t \wedge s, t \in \mathbb{R}\right\} \\ &= \operatorname{zup}\left\{\int_s^t f(x)dx : s, t \in I \wedge s < t \wedge s, t \in \mathbb{R}\right\} + \\ &\qquad \operatorname{zup}\left\{\int_s^t \nu(x)dx : s, t \in I \wedge s < t \wedge s, t \in \mathbb{R}\right\} \\ &= \int_I f(x)dx + \int_I \nu(x)dx.\end{aligned}$$

□

Example 7.2.3 (Continuation of Example 7.2.2). We decompose the integral of $e^{u+\oslash}$ over the external interval $[0, \pounds]$ into an integral of an internal function and an integral of a neutrix function over $[0, \pounds]$. Generalizing (7.7) we obtain

$$\int_0^{\pounds} e^{-u+\oslash}du = \operatorname*{zup}_{v\in\pounds^+} \int_0^v e^{-u+\oslash}du = \operatorname*{zup}_{v\in\pounds^+} \{1 - e^{-v} + \oslash\} = 1 + \oslash.$$

Exercises

Exercise 7.2.1. *Let $\varepsilon \simeq 0$, $\varepsilon > 0$. A flexible function $\phi : [0, 3] \to \mathbb{E}$ is defined for $x \in [0, 1 + \oslash]$ by $\phi(x) = 0$, and for $x \in (1 + \oslash, 2]$ by $\phi(x) = \varepsilon$. Define ϕ on $(2, 3]$ in such a way that ϕ is inner-integrable.*

Exercise 7.2.2. *Let $f : [0, 1] \to \mathbb{R}$ be defined by*

$$f(x) = \begin{cases} 0, & \exists^{\text{st}} q \in \mathbb{Q} \cup [0, 1](x \simeq q) \\ 1, & \text{otherwise.} \end{cases}$$

Determine $\overline{\int_0^1} f(x)dx$ and $\underline{\int_0^1} f(x)dx$. Is f outer-integrable?

Exercise 7.2.3. *Let $\omega \simeq +\infty$. Determine each of the following integrals:*

1. $\int_0^{\omega} \oslash x \, dx$.
2. $\int_0^{\omega+\pounds} \oslash x^2 dx$.
3. $\int_0^{\omega+\pounds} (1+\oslash)e^x dx$.
4. $\int_0^{\oslash} (1+\oslash)\sin(\omega x)dx$.

Exercise 7.2.4. *Let $\varepsilon \simeq 0$, $\varepsilon > 0$ and $\omega \simeq \infty$, $\omega \in \mathbb{N}$. Determine each of the following sums:*

1. $\sum_{n=1}^{\pounds} \frac{1}{n^2}$.
2. $\sum_{n=1}^{\pounds} \frac{1}{n}$.
3. $\int_1^{\pounds} \frac{1}{x} dx - \sum_{n=1}^{\pounds} \frac{1}{n}$.

7.3 Elementary properties of integrals

We start by considering extensions of the Chasles Relation for dividing the interval of integration into two subintervals. Then we deal with sums and scalar multiplication.

Theorem 7.3.1. *Let $a, b \in \mathbb{R}$ with $a < b$. Let ν be an integrable neutrix function on $[a, b]$. Let $\gamma \in \mathbb{E}$ be such that $a < \gamma < b$. Then*

$$\int_a^b \nu(x)dx = \int_a^{\gamma} \nu(x)dx + \int_{\overline{\gamma}}^b \nu(x)dx = \int_a^{\gamma} \nu(x)dx + \int_{\gamma}^b \nu(x)dx. \tag{7.8}$$

Proof. For s such that $a < s \leq b$, let $N(s) := \int_a^s \nu(x)dx$. Let $N := N(b)$. Let $M(\gamma) := \int_{\gamma}^b \nu(x)dx$ and $K(\gamma) := \int_{\overline{\gamma}}^b \nu(x)dx$.

Assume first that $\gamma = c \in \mathbb{R}$. It follows directly from the Chasles Relation for integrable representatives of ν that

$$N(c) + M(c) = N. \tag{7.9}$$

For the general case let $c \in \gamma$. We prove first the second equality of (7.8). Because $N(c), N(\gamma), M(c), M(\gamma)$ are neutrices, it follows from Proposition 7.2.5 that $N(c) \leq N(\gamma)$ and $M(c) \leq M(\gamma)$. Hence $N \leq N(\gamma) + M(\gamma)$ by (7.9). Conversely, both the neutrices $N(\gamma)$ and $M(\gamma)$ are contained in N. Then $N(\gamma)+M(\gamma) = \max\{N(\gamma), M(\gamma)\} \leq N$. We conclude that $N = N(\gamma)+M(\gamma)$.

As for the first equality of (7.8) one shows as above that $N(\gamma)+K(\gamma) \leq N$.

Let f be a non-negative integrable representative of ν. By Theorem 7.1.1 we have that $\int_a^b f(x)dx \in \int_a^\gamma f(x)dx + \int_{\overline{\gamma}}^b f(x)dx$. Let $F(c) := \int_a^c f(x)dx$. Then $\int_a^\gamma f(x)dx = \sup_{c \leq \gamma} F(c)$. For all $c \leq \gamma$ we have $F(c) \in N(c)$. Then $\sup_{c \leq \gamma} F(c) \leq \sup_{c \leq \gamma} N(c)$. Hence $\int_a^\gamma f(x)dx \leq N(\gamma)$. In fact, $\int_a^\gamma f(x)dx$ is an external number contained in $N(\gamma)$. In the same way one shows that $\int_{\overline{\gamma}}^b f(x)dx \subseteq K(\gamma)$. Hence $\int_a^b f(x)dx \in N(\gamma) + K(\gamma)$. We conclude that $N \leq N(\gamma) + K(\gamma)$. Combining we obtain that $N = N(\gamma) + K(\gamma)$. □

The next theorem gives the Chasles Relation for flexible functions.

Theorem 7.3.2. *Let $a, b \in \mathbb{R}$ with $a < b$. Let ϕ be a non-negative integrable flexible function on $[a, b]$. Let $\gamma \in \mathbb{E}$ be such that $a < \gamma < b$. Then*

$$\int_a^b \phi(x)dx \subseteq \int_a^\gamma \phi(x)dx + \int_{\overline{\gamma}}^b \phi(x)dx \leq \int_a^\gamma \phi(x)dx + \int_\gamma^b \phi(x)dx. \tag{7.10}$$

Proof. Let f be an integrable representative of ϕ and ν be the integrable neutrix function defined for $x \in [a, b]$ by $\nu(x) = \phi(x) - f(x)$. Using Theorems 7.2.6, 7.1.1 and 7.3.1 we find

$$\begin{aligned} \int_a^b \phi(x)dx &= \int_a^b f(x)dx + \int_a^b \nu(x)dx \\ &\subseteq \int_a^\gamma f(x)dx + \int_{\overline{\gamma}}^b f(x)dx + \int_a^\gamma \nu(x)dx + \int_{\overline{\gamma}}^b \nu(x)dx \\ &= \int_a^\gamma f(x)dx + \int_a^\gamma \nu(x)dx + \int_{\overline{\gamma}}^b f(x)dx + \int_{\overline{\gamma}}^b \nu(x)dx \\ &= \int_a^\gamma \phi(x)dx + \int_{\overline{\gamma}}^b \phi(x)dx. \end{aligned} \tag{7.11}$$

In a similar way one shows that

$$\int_a^b \phi(x)dx \leq \int_a^\gamma \phi(x)dx + \int_\gamma^b \phi(x)dx. \tag{7.12}$$

From (7.11) and (7.12) we derive (7.10). □

We show that the integral of the sum of two non-negative flexible functions is equal to the sum of the integrals of the respective functions. First we show that the sum of two integrable flexible functions is integrable.

Proposition 7.3.3. *Let $a, b \in \mathbb{R}$, $a < b$. Let $\phi, \psi : [a, b] \to \mathbb{E}$ be integrable non-negative flexible functions. Then $\phi + \psi$ is integrable over all external intervals included in $[a, b]$.*

Proof. Observe that $\Gamma(\phi) = \bigcup\{f \subseteq \Gamma(\phi) : f \text{ internal and integrable}\}$ and $\Gamma(\psi) = \bigcup\{g \subseteq \Gamma(\psi) : g \text{ internal integrable}\}$. Let $x \in [a,b]$ and $y \in \phi(x)+\psi(x)$. Then there exists an internal integrable function $\overline{f} \subseteq \Gamma(\phi)$ and an internal integrable function $\overline{g} \subseteq \Gamma(\psi)$ such that $\overline{f}(x) + \overline{g}(x) = y$. Then

$$\Gamma(\phi+\psi) \subseteq \bigcup\{f+g : f \subseteq \Gamma(\phi) \wedge g \subseteq \Gamma(\psi) \wedge \ f,g \text{ internal integrable}\}.$$

The other inclusion is obvious, so

$$\Gamma(\phi+\psi) = \bigcup\{f+g : f \subseteq \Gamma(\phi) \wedge g \subseteq \Gamma(\psi) \wedge \ f,g \text{ internal integrable}\}.$$

Hence $\phi + \psi$ is integrable over $[a,b]$, hence also integrable over all external subintervals. □

With respect to integration over external intervals we show now the sum property for internal integrable functions and for neutrix functions.

Proposition 7.3.4. *Let $a,b \in \mathbb{R}$, $a < b$. Let $f,g : [a,b] \to \mathbb{R}$ be internal, non-negative and integrable and let $I \subseteq [a,b]$ be an external interval. Then*

$$\int_I f(x) + g(x)dx = \int_I f(x)dx + \int_I g(x)dx.$$

Proof. We have

$$\begin{aligned}\int_I f(x) + g(x)dx &= \operatorname*{zup}_{\substack{a<b\\ a,b\in I}} \left\{\int_a^b f(x) + g(x)dx\right\}\\ &= \operatorname*{zup}_{\substack{a<b\\ a,b\in I}} \left\{\int_a^b f(x)dx + \int_a^b g(x)dx\right\}\\ &= \operatorname*{zup}_{\substack{a<b\\ a,b\in I}} \left\{\int_a^b f(x)dx\right\} + \operatorname*{zup}_{\substack{a<b\\ a,b\in I}} \left\{\int_a^b g(x)dx\right\}\\ &= \int_I f(x)dx + \int_I g(x)dx.\end{aligned}$$

□

Proposition 7.3.5. *Let $a,b \in \mathbb{R}$, $a < b$. Let $\mu,\nu : [a,b] \to \mathbb{E}$ be integrable neutrix functions and $I \subseteq [a,b]$ be an external interval. Then*

$$\int_I \mu(x) + \nu(x)dx = \int_I \mu(x)dx + \int_I \nu(x)dx. \tag{7.13}$$

Proof. It follows from Proposition 7.3.3 that $\mu + \nu$ is integrable over I. Also $\mu + \nu$ is a neutrix function, so

$$\int_I \mu(x) + \nu(x)dx = \bigcup_{\substack{s<t\\ s,t\in I}} \int_s^t \mu(x) + \nu(x)dx \tag{7.14}$$

by Proposition 7.2.5. For $s, t \in I, s < t$, let $y \in \int_s^t \mu(x) + \nu(x)dx$. Let r be a representative of $\mu + \nu$ such that $\int_s^t r(x)dx = y$. Then there exist representatives m of μ and n of ν such that $m + n = r$. Then

$$\begin{aligned}\int_s^t r(x)dx &= \int_s^t m(x) + \int_s^t n(x)dx \\ &\in \int_s^t \mu(x)dx + \int_s^t \nu(x)dx \\ &\subseteq \int_I \mu(x)dx + \int_I \nu(x)dx.\end{aligned}$$

Hence $\int_s^t \mu(x)+\nu(x)dx \subseteq \int_I \mu(x)dx+\int_I \nu(x)dx$; hence also $\int_I \mu(x)+\nu(x)dx \subseteq \int_I \mu(x)dx + \int_I \nu(x)dx$ by (7.14).

Conversely, let $z \in \int_I \mu(x)dx + \int_I \nu(x)dx$; then z is an element of the maximum of these neutrices, say $z \in \int_I \nu(x)dx$. Then there exist $s, t \in I, s < t$ such that $z \in \int_s^t \nu(x)dx$. Then there exists a representative n of ν such that $z = \int_s^t n(x)dx$. Because n is also a representative of $\mu + \nu$, also $z \in \int_s^t \mu(x) + \nu(x)dx \subseteq \int_I \mu(x) + \nu(x)dx$. Hence $\int_I \mu(x)dx + \int_I \nu(x)dx \subseteq \int_I \mu(x) + \nu(x)dx$. Combining, we obtain (7.13). □

Theorem 7.3.6. *Let $a, b \in \mathbb{R}$, $a < b$. Let $\phi, \psi : [a, b] \to \mathbb{E}$ be non-negative integrable flexible functions and $I \subseteq [a, b]$ be an external interval. Then*

$$\int_I \phi(x) + \psi(x)dx = \int_I \phi(x)dx + \int_I \psi(x)dx.$$

Proof. By Proposition 7.3.3 the flexible function $\phi+\psi$ is integrable over I. Let f, g be integrable internal functions, and μ, ν be integrable neutrix functions such that $\phi = f + \mu$ and $\psi = g + \nu$. Then by Theorem 7.2.6 and Propositions 7.3.4 and 7.3.5

$$\begin{aligned}\int_I \phi(x) + \psi(x)dx &= \int_I (f(x) + \mu(x)) + (g(x) + \nu(x))dx \\ &= \int_I f(x) + g(x)dx + \int_I \mu(x) + \nu(x)dx \\ &= \int_I f(x)dx + \int_I g(x)dx + \int_I \mu(x)dx + \int_I \nu(x)dx \\ &= \int_I f(x)dx + \int_I \nu(x)dx + \int_I g(x)dx + \int_I \nu(x)dx \\ &= \int_I \phi(x)dx + \int_I \psi(x)dx.\end{aligned}$$ □

We consider now the behaviour of integrals under multiplication by an external number. If c is a real number, obviously

$$\int_I c\phi(x)dx = c\int_I \phi(x)dx. \tag{7.15}$$

Due to the existence of non-constant representative functions, the analogous formula on multiplication by an external number is less immediate. It will be established progressively.

Lemma 7.3.7. *Let $a, b \in \mathbb{R}$, $a < b$. Let $f : [a, b] \to \mathbb{R}$ be internal, non-negative and integrable and N be an external neutrix. Then*

$$\int_a^b Nf(x)dx = \left\{ \int_a^b pf(x)dx : p \in N \right\} = N \int_a^b f(x)dx. \tag{7.16}$$

Proof. The second equality is obvious. To prove the first equality, observe that $\left\{ \int_a^b pf(x)dx : p \in N \right\} \subseteq \int_a^b Nf(x)dx$. Conversely, let n be an internal integrable representative of N, when seen as a constant neutrix function. By Cauchy's Principle there exists $q \in \mathbb{R}, 0 \leq q \leq N$ such that $|n(x)| \leq q$ for all $x \in [a, b]$. Then $\left| \int_a^b n(x)f(x)dx \right| \leq q \int_a^b f(x)dx \in N \int_a^b f(x)dx$. Hence $\int_a^b Nf(x)dx \subseteq \left\{ \int_a^b pf(x)dx : p \in N \right\}$. Combining, we obtain (7.16). □

Proposition 7.3.8. *Let $a, b \in \mathbb{R}$, $a < b$. Let $f : [a, b] \to \mathbb{R}$ be internal, non-negative and integrable. Let N be an external neutrix and $I \subseteq [a, b]$ be an external interval. Then*

$$\int_I Nf(x)dx = \left\{ \int_I pf(x)dx : p \in N \right\} = N \int_I f(x)dx.$$

Proof. The second equality is obvious. To prove the first equality, applying Lemma 7.3.7, Definition 7.2.3 and Proposition 4.4.3 we find

$$\begin{aligned} \int_I Nf(x)dx &= \operatorname*{zup}_{\substack{a<b \\ a,b \in I}} \left\{ \int_a^b Nf(x)dx \right\} = \operatorname*{zup}_{\substack{a<b \\ a,b \in I}} \left\{ N \int_a^b f(x)dx \right\} \\ &= N \operatorname*{zup}_{\substack{a<b \\ a,b \in I}} \left\{ \int_a^b f(x)dx \right\} = N \int_I f(x)dx. \end{aligned}$$ □

Lemma 7.3.9. *Let ν be an integrable neutrix function on the interval $[a, b]$. Let N be an external neutrix and $I \subseteq [a, b]$ be an external interval. Then*

$$\int_I N\nu(x)dx = N \int_I \nu(x)dx.$$

Proof. See Exercise 7.3.2. □

Theorem 7.3.10. *Let $a, b \in \mathbb{R}$, $a < b$. Let $\phi : [a, b] \to \mathbb{E}$ be a non-negative integrable flexible function and N be an external neutrix. Let $I \subseteq [a, b]$ be an external interval. Then*

$$\int_I N\phi(x)dx = N \int_I \phi(x)dx. \tag{7.17}$$

Proof. Let f be an internal integrable function and ν an integrable neutrix function such that $\phi = f + \nu$. Then, using distributivity and Proposition 7.3.5

$$\begin{aligned}\int_I N\phi(x)dx &= \int_I N(f(x)+\nu(x))dx = \int_I Nf(x) + N\nu(x)dx \\ &= \int_I Nf(x)dx + \int_I N\nu(x)dx = N\int_I f(x)dx + N\int_I \nu(x)dx \\ &= N\left(\int_I f(x)dx + \int_I \nu(x)dx\right) \\ &= N\int_I f(x)+\nu(x)dx = N\int_I \phi(x)dx.\end{aligned}$$

□

Theorem 7.3.11. *Let $a, b \in \mathbb{R}$, $a < b$. Let $\phi : [a, b] \to \mathbb{E}$ be a non-negative integrable flexible function and $\gamma = c + C$ be an external number. Let $I \subseteq [a, b]$ be an external interval. Then*

$$\int_I \gamma\phi(x)dx = \gamma\int_I \phi(x)dx.$$

Proof. Using distributivity, Proposition 7.3.6, formula (7.15) and Theorem 7.3.10 we obtain

$$\begin{aligned}\int_I \gamma\phi(x)dx &= \int_I (c+C)\phi(x)dx = \int_I c\phi(x) + C\phi(x)dx \\ &= \int_I c\phi(x)dx + \int_I C\phi(x)dx = c\int_I \phi(x)dx + C\int_I \phi(x)dx \\ &= (c+C)\int_I \phi(x)dx = \gamma\int_I \phi(x)dx.\end{aligned}$$

□

We end by showing that inner continuous functions are integrable.

Proposition 7.3.12. *Let $\nu : [a, b] \to \mathbb{E}$ be an inner continuous neutrix function. Then ν is integrable over $[a, b]$; in fact $I_\nu := \int_a^b \nu(x)dx$ is a neutrix.*

Proof. Let F be the set of internal integrable functions $f : [a, b] \to \mathbb{R}$. Because an internal continuous function is integrable, one has $\Gamma(\nu) = \Gamma(\nu) \cap \bigcup F$. By symmetry $\operatorname{zup} I_\nu = \operatorname{winf} I_\nu$. Hence ν is inner-integrable. Then ν is integrable by Theorem 7.2.1 and I_ν is a neutrix. □

Proposition 7.3.13. *Let $\phi : [a, b] \to \mathbb{E}$ be a non-negative inner continuous flexible function; let f be a continuous representative of ϕ and $\nu = \phi - f$. Then ϕ is integrable over $[a, b]$ and*

$$\int_a^b \phi(x)dx = \int_a^b f(x)dx + \int_a^b \nu(x)dx. \tag{7.18}$$

Proof. Let F be the set of internal integrable functions $f : [a, b] \to \mathbb{R}$. Because an internal continuous function is integrable, one has $\Gamma(\phi) = \Gamma(\phi) \cap \bigcup F$. Then ϕ is integrable by Theorem 7.2.4, with integral given by (7.18), noting that ν is integrable by Proposition 7.3.12. □

Exercises

Exercise 7.3.1. *Let $a, b \in \mathbb{R}$, $a < b$. Let $n \in \mathbb{N}$ be standard, $\alpha_0, ..., \alpha_k \in \mathbb{E}$ be bounded and p the polynomial given by $p(x) = \alpha_0 + \alpha_0 x + \cdots + \alpha_n x^n$. Determine $\int_a^b p(x)dx$.*

Exercise 7.3.2. *Prove Lemma 7.3.9.*

7.4 Special integrals and applications

In order to approximate integrals depending on parameters it is important to know the domain where the main contribution to its value takes place. Then usually one needs only to find a local approximation of the integrand, i.e. valid on this domain. To this end we recall in Subsection 7.4.1 the notions of mass and tail introduced in [131, 50]. In Subsection 7.4.2 we consider the problem of local averaging of a rapidly oscillating function, which reduces to assigning a value to an external integral with respect to the mass of a density. In Subsection 7.4.3 we consider integrals of Dirac-like functions, with mass contained in the halo $a + \oslash$ of some standard number a. These are related to asymptotic expansions, in particular to the Laplace method.

7.4.1 Mass and tail of probabilities and integrals

Definition 7.4.1. *Let (Ω, pr) be an internal probability space and $x : \Omega \to \mathbb{R}$ be a random variable. The* upper tail *is defined by $C_x^+ := \{c \in \mathbb{R} : \mathrm{pr}\{x \geq c\} \simeq 0\}$, the* lower tail *is defined by $C_x^- := \{c \in \mathbb{R} : \mathrm{pr}\{x \leq c\} \simeq 0\}$ and the* tail *is defined by $C_x = C_x^- \cup C_x^+$. The* mass *is defined by $M_x := \mathbb{R} \setminus C_x$. If x has a density function f, we may also talk about the mass and tail of f, and write M_f instead of M_x.*

The Mass Concentration Lemma (Theorem 2.2.2) expresses the fact that the mass of a probability distribution with mean $\mu \in \mathbb{R}$ and standard deviation $\sigma > 0$ is contained in $\mu + \pounds\sigma$. Here we present a version for densities.

Lemma 7.4.1 (Mass Concentration Lemma for densities). *Let f be an integrable density function of an internal probability distribution. Assume that the mean $\mu \in \mathbb{R}$ and standard deviation $\sigma > 0$ are well-defined. Then $M_f \subseteq \mu + \pounds\sigma$, otherwise said*

$$\int_{\mu+\pounds\sigma} f(x)dx \subseteq 1 + \oslash. \tag{7.19}$$

The Mass Concentration Lemma for densities is a consequence of the following result.

Lemma 7.4.2. *Let f be an integrable density function of an internal probability distribution taking only non-negative values. Assume that the mean $\mu \in \mathbb{R}$ is well-defined. Then $M_f \subseteq \pounds\mu$; otherwise said*

$$\int_{\pounds\mu} f(x)dx \subseteq 1 + \oslash. \tag{7.20}$$

Proof. By Chebyshev's Lemma $\int_{t\mu}^{\infty} f(x)dx \leq \frac{\mu}{t\mu} = \frac{1}{t}$, so every positive unlimited t is an element of the tail of f. Hence $M_f \subseteq \pounds\mu$, and then (7.20) is a consequence. □

Proof of the Mass Concentration Lemma for densities. Let y be the centralized non-negative random variable defined by $y = (x-\mu)^2$. Then $\mathrm{E(y)} = \mathrm{Var(y)} = \sigma^2$. By Lemma 7.4.2 we have $M_y \subseteq \pounds\mathrm{E(y)} = \pounds\sigma^2$. Then $M_{x-\mu} \subseteq \pounds\sigma$; hence $M_x \subseteq \mu + \pounds\sigma$. Then (7.19) is a consequence. □

Properties of concentration also hold for other integrals. Let $f : \mathbb{R} \to \mathbb{R}^+$ be internal and integrable. Assume $I := \int_{-\infty}^{\infty} f(x)dx > 0$. Put $g = \frac{f}{I}$; then g is a density function. We can extend the notion of tail and mass to f by defining

$$\begin{aligned}
C_f^+ &:= \left\{c \in \mathbb{R} : \int_c^{\infty} f(x)dx \in \oslash I\right\} \\
C_f^- &:= \left\{c \in \mathbb{R} : \int_{-\infty}^{c} f(x)dx \in \oslash I\right\} \\
C_f &:= C_f^- \cup C_f^+ \\
M_f &:= \mathbb{R} \setminus C_f.
\end{aligned}$$

Then $M_f = M_g$, with M_g given by Definition 7.4.1. The following is a direct consequence of the Mass Concentration Lemma.

Theorem 7.4.3. *Let f be an internal integrable non-negative function. Assume that $I := \int_{-\infty}^{\infty} f(x)dx > 0$. Then*

$$\int_{M_f} f(x)dx \subseteq (1 + \oslash)I.$$

By the next theorem, approximation of the integrands on the mass suffices for approximation of the integrals.

Theorem 7.4.4. *Let f, g be internal integrable non-negative functions with $M_f = M_g$. Assume $\int_{-\infty}^{\infty} f(x)dx > 0$ and $g(x) \subseteq (1+\oslash)f(x)$ for $x \in M_f$. Then*

$$\int_{-\infty}^{\infty} g(x)dx \subseteq (1+\oslash)\int_{-\infty}^{\infty} f(x)dx$$

Proof. We have

$$\begin{aligned}\int_{-\infty}^{\infty} g(x)dx &\subseteq (1+\oslash)\int_{M_g} g(x)dx \\ &\subseteq (1+\oslash)\int_{M_g} (1+\oslash)f(x)dx \\ &= (1+\oslash)\int_{M_f} f(x)dx \\ &= (1+\oslash)\int_{-\infty}^{\infty} f(x)dx.\end{aligned}$$

□

7.4.2 On local averaging

Let D be a density function, with mass M. Assume that f is an internal integrable function such that $\int_M f(x)D(x)dx$ is well-defined. Then this integral acts as a local mean of f on M; this is especially relevant in the case where f is rapidly oscillating on M. Below we give precise definitions, in a more general context.

Definition 7.4.2. *Let $f : \mathbb{R} \to \mathbb{R}$ be an internal continuous density function. Let $S \subseteq \mathbb{R}$ be an interval (internal or external). We say that f is* an almost density on *S if $\int_S f(x)dx \in 1+\oslash$.*

Densities are almost densities on their mass, but also on sets containing the mass; by the Mass Concentration Lemma for densities, if the mean μ and the standard deviation σ are well defined, a density is also an almost density on $\mu + \pounds\sigma$. As an example, the *standard normal density* $\frac{e^{-\frac{x^2}{2}}}{\sqrt{2\pi}}$ is an almost density on $\pounds$. Also, if $\varepsilon > 0$, $\varepsilon \simeq 0$, the density $\frac{e^{-\frac{x^2}{2\varepsilon}}}{\sqrt{2\pi\varepsilon}}$ is an almost density on $\pounds\sqrt{\varepsilon}$ and on $\oslash$.

Definition 7.4.3. *Let $S \subseteq \mathbb{R}$ be an external interval. Let $f : \mathbb{R} \to \mathbb{R}^+$ be internal and $D : \mathbb{R} \to \mathbb{R}$ be an internal density function on S. We say that f is* averageable *on S* with respect to *D if fD is integrable on S. Then the external number $M_S(f) := \int_S f(x)\phi(x)dx$ will be called the* mean on *S* with respect to *D.*

Here are some examples concerning the standard normal distribution (see Exercise 1.4.6).

Example 7.4.1. The constant function $f : \mathbb{R} \to \mathbb{R}$ defined by $f(x) = 1$ has average $1 + \oslash$ on $\pounds$, since

$$\int_{\pounds} 1 \cdot \frac{e^{-\frac{x^2}{2}}}{\sqrt{2\pi}}dx = \lim_{\substack{y \to -\not\infty \\ z \to \not\infty}} \int_y^z \frac{e^{-\frac{x^2}{2}}}{\sqrt{2\pi}}dx = 1 + \oslash.$$

The Heaviside function H is also averageable on $\pounds$, because

$$\lim_{\substack{y \to -\not\infty \\ z \to \not\infty}} \int_y^z H(x)\frac{e^{-\frac{x^2}{2}}}{\sqrt{2\pi}}dx = \lim_{z \to \not\infty} \int_0^z H(x)\frac{e^{-\frac{x^2}{2}}}{\sqrt{2\pi}}dx = \frac{1}{2} + \oslash.$$

Theorem 7.4.5. *Let $S \subseteq \mathbb{R}$ be bounded and convex. Let $f : \mathbb{R} \to \mathbb{R}^+$ be internal and Lebesgue integrable, and $D : \mathbb{R} \to \mathbb{R}$ be an internal continuous density function on S. Then f is averageable on S with respect to D.*

Proof. Because f is Lebesgue integrable and D is continuous, the product fD is a Lebesgue integrable non-negative function. Because S is bounded and convex, the possibly external integral $\int_S f(x)D(x)dx$ is well-defined. Hence f is averageable on S with respect to D. □

It is also natural to try to assign a local mean to functions which are fluctuating around 0, like the sine function. To do this we will adapt the notion of limit of functions. For simplicity we restrict ourselves to the case where S is a neutrix.

Definition 7.4.4. *Assume that S is an external neutrix. Let f be an internal function, which is at least integrable on every internal interval $[y, z] \subseteq S$. Let $s \in S$, $s > 0$. We define $\int_S f(x)dx = \alpha$, where $\alpha = a + A$ and $A = N(\alpha)$ is minimal such that for every $\varepsilon \in \mathbb{R}$, $\varepsilon > A$ there exists $s \in S$, $s > 0$ such that $\left|\int_y^z f(x)dx - \alpha\right| < \varepsilon$ for all $y, z \in S$ with $y < -s$, $s < z$.*

Observe that

$$\int_{\pounds} \sin(x)\frac{e^{-\frac{x^2}{2}}}{\sqrt{2\pi}}dx = \oslash.$$

Hence the sine function is averageable on $\pounds$ with average $\oslash$.

From a nonstandard point-of-view, the problem of averaging for rapidly oscillating functions has been intensively studied by C. Reder in [109], and in the context of differential equations by T. Sari in [115]. Here we consider only a simple special case, which deals with the integral of a highly frequent sine-function with respect to a one-side density.

Proposition 7.4.6. *Let $\varepsilon \simeq 0$, $\varepsilon > 0$ and D be a standard continuously differentiable density function. Then $x \mapsto \sin\left(\frac{\pi x}{\varepsilon}\right)$ is averageable on $\pounds$ with respect to D, with average $(1 + \oslash)\frac{\varepsilon}{\pi}$.*

Proof. Let $k \in \mathbb{N}$. Then, using the fact that D' is S-continuous and limited,

$$\begin{aligned}
g(k\varepsilon) &:= \int_{2k\varepsilon}^{(2k+2)\varepsilon} \sin\left(\frac{\pi x}{\varepsilon}\right) D(x)dx \\
&= \int_{2k\varepsilon}^{(2k+1)\varepsilon} \sin\left(\frac{\pi x}{\varepsilon}\right) D(x)dx + \int_{(2k+1)\varepsilon}^{(2k+2)\varepsilon} \sin\left(\frac{\pi x}{\varepsilon}\right) D(x)dx \\
&= \int_{2k\varepsilon}^{(2k+1)\varepsilon} \sin\left(\frac{\pi x}{\varepsilon}\right) (D(x) - D(x+\varepsilon))dx \\
&\in -(D'(2k\varepsilon) + \oslash)\varepsilon \int_{2k\varepsilon}^{(2k+1)\varepsilon} \sin\left(\frac{\pi x}{\varepsilon}\right) dx \\
&= -(D'(2k\varepsilon) + \oslash)\varepsilon \int_{0}^{\varepsilon} \sin\left(\frac{\pi x}{\varepsilon}\right) dx \\
&= -(D'(2k\varepsilon) + \oslash)\frac{2\varepsilon^2}{\pi}.
\end{aligned}$$

Then

$$\begin{aligned}
\int_0^{\pounds} \sin\left(\frac{\pi x}{\varepsilon}\right) D(x)dx = \sum_0^{\pounds/\varepsilon} g(k\varepsilon) &\in -\varepsilon \sum_0^{\pounds/\varepsilon} (D'(2k\varepsilon) + \oslash)\frac{2\varepsilon}{\pi} \\
&= -\frac{\varepsilon}{\pi} \sum_0^{\pounds/\varepsilon} D'(2k\varepsilon)2\varepsilon - \varepsilon \sum_0^{\pounds/\varepsilon} \oslash\varepsilon \\
&= -\frac{\varepsilon}{\pi} \int_0^{\pounds} D'(x)dx + \varepsilon^2 \oslash \frac{\pounds}{\varepsilon} \\
&= (1+\oslash)\frac{\varepsilon}{\pi} + \oslash\varepsilon = (1+\oslash)\frac{\varepsilon}{\pi}.
\end{aligned}$$

□

7.4.3 The concentration lemma and the Laplace method

This section deals with both classical and nonstandard asymptotics. Let f, g be real functions. We use the common notation $f(x) \sim g(x)$ for $x \to \infty$ to say that $\lim_{x\to+\infty} \frac{f(x)}{g(x)} = 1$. The *Laplace method* deals with asymptotic estimates of integrals of the form

$$\int_0^{\infty} e^{-\omega x^p} g(x)dx \tag{7.21}$$

with, in a nonstandard setting, $\omega > 0$ unlimited and $p > 0$ appreciable. Two-sided integrals of the form

$$\int_{-\infty}^{\infty} e^{-\omega |x|^p} g(x)dx$$

also occur.

For $p = 1$ the integral (7.21) reduces to a Laplace-transform. The contribution to the value of the integrals is very concentrated, with mass strictly contained in the infinitesimals. Integrals of such type exhibit Dirac-like behaviour. For instance if g is S-continuous and is of polynomial growth at infinity, only the value of g at zero counts, and then, say, for $p = 2$ we have the simplification

$$\int_{-\infty}^{\infty} e^{-\omega x^2} g(x)dx \in (1+\oslash)g(0)\int_{-\infty}^{\infty} e^{-\omega x^2}dx = (1+\oslash)g(0)\sqrt{\frac{\pi}{\omega}}.$$

In fact, for such integrals it does not suffice to locate the mass, for as the example shows we need sharper approximations than just an arbitrary infinitesimal order; also the integrals give rise to expansions in powers of $\frac{1}{\omega}$, and then an estimate better than all the standard powers of $\frac{1}{\omega}$ is useful. The next lemma expresses that outside the infinitesimals the contribution is in fact exponentially small.

Lemma 7.4.7 (Concentration Lemma). *Let $\omega \simeq +\infty$. Then*

$$\int_{\overline{\oslash}}^{\infty} \pounds e^{-\omega x^{@}}dx = \pounds e^{-@\omega}. \tag{7.22}$$

Proof. Observe that $x \mapsto \pounds e^{-\omega x^{@}}$ is a neutrix function whose graph is the union of the internal integrable functions f, defined by $f(x) = Ke^{-\omega x^p}$ for $x \geq a$, with $a, p > 0$ appreciable and $K > 0$ limited. If $p \geq 1$ we have $f(x) \leq Ke^{-\omega x}$, hence

$$\int_a^{\infty} f(x)dx \leq \int_a^{\infty} Ke^{-\omega x}dx = \frac{K}{\omega}e^{-\omega a} \subseteq \pounds e^{-@\omega}.$$

For $p < 1$, put $g(x) = x^p - a^p$. Now $g(x) \geq (x-a)\min_{a\leq t\leq x} g'(t)$. Hence for $x > 2a$,

$$g(x) \geq (x-a)\min_{a\leq t\leq x} g'(t) = (x-a)px^{p-1} = p\frac{x-a}{x}x^p \geq \frac{p}{2}x^p.$$

For $x \leq 2a$ we have $g(x) \geq 0$. So

$$\begin{aligned}\int_a^{\infty} f(x)dx &= K\int_a^{\infty} e^{-\omega x^p}dx \\ &= Ke^{-\omega a^p}\int_a^{\infty} e^{-\omega g(x)}dx \\ &= Ke^{-\omega a^p}\int_a^{2a} e^{-\omega g(x)}dx + Ke^{-\omega a^p}\int_{2a}^{\infty} e^{-\omega g(x)}dx \\ &\leq Ke^{-\omega a^p}\int_a^{2a} dx + Ke^{-\omega a^p}\int_{2a}^{\infty} e^{-\frac{p}{2}\omega x^p}dx \\ &\subseteq \pounds e^{-\omega a^p} + \pounds e^{-\omega a^p} \\ &\subseteq \pounds e^{-@\omega}.\end{aligned}$$

Hence again $\int_a^{\infty} \pounds e^{-\omega x^{@}}dx \subseteq \pounds e^{-@\omega}$. Taking suprema we obtain (7.22). □

A special case of the Concentration Lemma is given by Lemma 7.4.8. Note that that if $p > 0$ is appreciable, for appreciable $x > 0$ it holds that $@x^p \subseteq x^@$.

Lemma 7.4.8. *Let $\omega \simeq +\infty$ and $p > 0$ be appreciable. Then*

$$\int_{\overline{\oslash}}^{\infty} \pounds e^{-@\omega x^p} dx = \pounds e^{-@\omega}.$$

One may verify that functions of the form $e^{-\omega f(x)} g(x)$, with f standard, having an isolated minimum at $x = 0$, and such that ultimately $f \geq mx^p$, and g internal with $g(x) \leq Ke^{Cf(x)}$ are of the form $\pounds e^{-@\omega x^p}$.

Lemma 7.4.9 (Local Laplace method)**.** *Let $\omega \simeq +\infty$ and $a, b, r, s > 0$ be standard. Then*

$$\int_0^{\oslash} e^{-(1+\oslash)\omega a x^r} (1+\oslash) b x^s dx = (1+\oslash) \frac{b\Gamma\left(\frac{s+1}{r}\right)}{ra^{\frac{s+1}{r}}} \frac{1}{\omega^{\frac{s+1}{r}}}. \tag{7.23}$$

Proof. Put $x := \left(\frac{t}{a\omega}\right)^{\frac{1}{r}}$. Then

$$\int_0^{\oslash} e^{-(1+\oslash)\omega a x^r} (1+\oslash) b x^s dx = \frac{b}{ra^{\frac{s+1}{r}}} \cdot \frac{1}{\omega^{\frac{s+1}{r}}} \int_0^{\oslash a\omega} e^{-(1+\oslash)t} (1+\oslash) t^{\frac{s+1}{r}-1} dt.$$

Now $\frac{s+1}{r} > 1$, so $\int_0^\infty e^{-t} t^{\frac{s+1}{r}-1} dt$ is convergent and we have $\int_0^\infty e^{-t} t^{\frac{s+1}{r}-1} dt = \Gamma\left(\frac{s+1}{r}\right)$. Because $t \mapsto e^{-t} t^{\frac{s+1}{r}-1}$ is standard, its mass is @. The mass of an internal integrable representative of $e^{-t(1+\oslash)}(1+\oslash)t^{\frac{s+1}{r}-1}$ is also equal to @, for it has a standard convergent majorant. On the mass, $e^{-t(1+\oslash)}(1+\oslash)t^{\frac{s+1}{r}-1} \subseteq (1+\oslash)e^{-t}t^{\frac{s+1}{r}-1}$.

Let $\nu \in \oslash a\omega$ be limited. Then, with the help of Theorem 7.4.4

$$\begin{aligned} \int_0^{\nu} e^{-t(1+\oslash)}(1+\oslash)t^{\frac{s+1}{r}-1} dt &\subseteq (1+\oslash) \int_0^{\pounds} e^{-t} t^{\frac{s+1}{r}-1} dt \\ &= (1+\oslash) \int_0^{\infty} e^{-t} t^{\frac{s+1}{r}-1} dt \\ &= (1+\oslash)\Gamma\left(\frac{s+1}{r}\right). \end{aligned}$$

Then (7.23) follows from Definition 7.2.1. □

Theorem 7.4.10 (Laplace method)**.** *Let $\omega \simeq +\infty$ and $a, r, m, q, K, C > 0$, $b \neq 0$ and $s > -1$ be standard. Let $f : [0, +\infty) \to \mathbb{R}^+$ be standard, continuous and increasing, asymptotic to ax^r for $x \to 0$, such that there exists $y \geq 0$ with $f(x) \geq mx^q$ for $x \geq y$. Let $g : (0, +\infty) \to \mathbb{R}$ be internal, continuous, and such that $g(x) \in (1+\oslash)bx^s$ for all $s \simeq 0$, $g(x)$ is limited for appreciable x and $|g(x)| \leq Ke^{Cf(x)}$ for all unlimited x. Then*

$$\int_0^{\infty} e^{-\omega f(x)} g(x) dx \in (1+\oslash) \frac{b\Gamma\left(\frac{s+1}{r}\right)}{ra^{\frac{s+1}{r}}} \frac{1}{\omega^{\frac{s+1}{r}}}.$$

Proof. We have $f(x) \in (1+\oslash)ax^r$ for all $x \simeq 0$ and $e^{-\omega f(x)}g(x) \in e^{-\omega x^{@}} \pounds e^{@x^{@}} = \pounds e^{-\omega x^{@}}$ for all $x \gtrsim 0$. Then, applying the Chasles Relation, the Local Laplace method and the Concentration Lemma

$$\begin{aligned}
\int_0^{\infty} e^{-\omega f(x)} g(x)dx &\subseteq \int_0^{\oslash} e^{-\omega f(x)} g(x)dx + \int_{\overline{\oslash}}^{\infty} e^{-\omega f(x)} g(x)dx \\
&\subseteq \int_0^{\oslash} e^{-(1+\oslash)\omega a x^r}(1+\oslash)bx^r dx + \int_{\overline{\oslash}}^{\infty} \pounds e^{-\omega x^{@}} dx \\
&\subseteq (1+\oslash)\frac{b\Gamma\left(\frac{s+1}{r}\right)}{ra^{\frac{s+1}{r}}}\frac{1}{\omega^{\frac{s+1}{r}}} + \pounds e^{-@\omega} \\
&\subseteq (1+\oslash)\frac{b\Gamma\left(\frac{s+1}{r}\right)}{ra^{\frac{s+1}{r}}}\frac{1}{\omega^{\frac{s+1}{r}}}.
\end{aligned}$$

□

Remark 7.4.11. If in Theorem 7.4.10 the function g is standard, the condition that $g(x)$ is limited for appreciable x is automatically satisfied.

Using Remark 7.4.11, we formulate the special case $r = 1$ in terms of an ordinary Laplace transform. We recall that a function g is of *exponential order* if there exists $d, K, C > 0$ such that for all $x \geq d$ it holds that $|g(x)| \leq Ke^{Cx}$.

Corollary 7.4.12. *Let $\omega \simeq +\infty$ and $K, C > 0$, $b \neq 0$ and $n \in \mathbb{N}$ be standard. Let $g : (0, +\infty) \to \mathbb{R}$ be standard, continuous, of exponential order and such that $g(x) \sim bx^n$ for $x \to 0$. Then*

$$\int_0^{\infty} e^{-\omega x} g(x)dx \in (1+\oslash)\frac{bn!}{\omega^{n+1}}. \tag{7.24}$$

Stated in standard terms, (7.24) may be reformulated into

$$\int_0^{\infty} e^{-\omega x} g(x)dx \sim \frac{bn!}{\omega^{n+1}}, \text{ for } \omega \to \infty. \tag{7.25}$$

The case $r = 2$ of the Laplace method is used in [130] to obtain global asymptotic approximations of the remainders of divergent asymptotic expansions of Laplace transforms or Borel transforms. Let $\varepsilon \simeq 0$, $\varepsilon > 0$. Suppose the real function f is defined at least on $[0, +\infty)$, where it has at most exponential growth, and has the asymptotic expansion $f(x) \sim \sum_{n=0} c_n x^n$ at 0. Let ϕ be given by the Borel transform $\phi(x) = \int_0^{\infty} e^{-t} f(xt)dt$. Watson's Lemma [97] (for a nonstandard proof see [130] or Exercise 7.4.2) states that the function ϕ has at zero the asymptotic expansion

$$\phi(x) \sim \sum_{n=0} c_n n! x^n. \tag{7.26}$$

If f is meromorphic, the expansion (7.26) is certainly divergent. If f itself is the result of applying a Borel transform to an entire function, and if the coefficients c_n have sufficient regularity, i.e.

$$\frac{c_{n+1}}{c_n} = -c + o\left(\sqrt{\frac{1}{n}}\right), \tag{7.27}$$

we have the uniform expression for the remainder (we take $c = 1$ for simplicity)

$$R_{n-1}(\varepsilon) := \phi(\varepsilon) - c_k k! \varepsilon^k \in (1+\oslash)\frac{c_n n! \varepsilon^n}{1+n\varepsilon}, \tag{7.28}$$

(see [130, Theorem 3.13]). Let $[x]$ denote the integer part of x. Formula (7.28) permits us to show easily that, $R_{\left[\frac{1}{\varepsilon}\right]}(\varepsilon)$ is *nearly minimal*; i.e.

$$\left|R_{\left[\frac{1}{\varepsilon}\right]}(\varepsilon)\right| \lesssim |R_n(\varepsilon)|,$$

for all $n \in \mathbb{N}$ (see Exercise 7.4.3). This confirms the common practice (for instance as observed by Poincaré [101]) that the best results in approximation by diverging series are obtained when summing up to the smallest term.

We prove now a general approximation result based on the Laplace method with $s = 2$, and then apply it to formula (7.28).

Proposition 7.4.13. *Let $\varepsilon \simeq 0$, $\varepsilon > 0$. Let $\omega \in \mathbb{N}$ be unlimited such that $\omega\varepsilon$ is limited. Let g be internal, continuous and S-continuous, appreciable for limited arguments and such that there exist standard $K, C > 0$ with $|g(x)| \leq Ke^{Cf(x)}$ for all unlimited x. Then*

$$\int_0^\infty e^{-t} t^\omega g(\varepsilon t) dt \in (1+\oslash)\omega! g(\varepsilon\omega). \tag{7.29}$$

Proof. The function $t \mapsto e^{-t} t^\omega$ takes its maximum at $t = \omega$. We rescale around $t = \omega$ in order to be able to apply the Laplace method; i.e. we put $t = \omega(1+u)$. Then

$$\int_0^\infty e^{-t} t^\omega g(\varepsilon t) dt = \omega^\omega e^{-\omega} \omega \int_{-1}^\infty e^{-\omega(u - \log(1+u))} g(\varepsilon\omega + \varepsilon\omega u) du.$$

In fact we have simultaneously two applications of the Laplace method, on $[0, +\infty)$, and backwards on $(-1, 0]$. Now $u - \log(1+u) \sim -\frac{u^2}{2}$ for $u \to 0$. As for the growth conditions on $[0, +\infty)$, one has $u - \log(1+u) \sim u$ for $u \to +\infty$, while clearly $|g(\varepsilon\omega + \varepsilon\omega u)| \leq Ke^{Cu}$. With respect to the growth conditions on $(-1, 0]$ it suffices to observe that, going backwards, the function $u \mapsto u - \log(1+u)$ is majorized by u, while the interval $(-1, 0]$ is bounded, so $g(\varepsilon\omega + \varepsilon\omega u)$ is limited on this interval. So by Theorem 7.4.10, with $a = \frac{1}{2}$, $r = 2$ and $s = 0$, using the S-continuity of g at $\varepsilon\omega$, we have

$$\begin{aligned} \omega^\omega e^{-\omega} \omega \int_{-1}^\infty e^{-\omega(u - \log(1+u))} g(\varepsilon\omega + \varepsilon\omega u) du &\in (1+\oslash)\omega^\omega e^{-\omega} \omega \frac{\sqrt{\pi}}{\sqrt{\frac{1}{2}}\sqrt{\omega}} g(\varepsilon\omega) \\ &= (1+\oslash)\omega^\omega e^{-\omega} \sqrt{\omega}\sqrt{2\pi} g(\varepsilon\omega). \end{aligned}$$

Then (7.29) follows from Stirling's formula (Theorem 2.2.3). □

A well-known special case concerns the Euler expansion $\sum_{n=0}(-1)^n n!\varepsilon^n$ of the (modified) exponential integral given by $\phi(\varepsilon) := \int_0^\infty \frac{e^{-t}}{1+\varepsilon t}dt$. In this case formula (7.28) becomes

$$R_{n-1}(\varepsilon) \in (1+\oslash)\frac{(-1)^n n!\varepsilon^n}{1+n\varepsilon}, \tag{7.30}$$

valid for all $n \in \mathbb{N}$. For $n \in \mathbb{N}$ such that $n\varepsilon$ is limited, formula (7.30) follows from Proposition 7.4.13. Indeed, ϕ is the Borel transform of the function f defined by $f(x) = \frac{1}{1+x}$. Its asymptotic expansion coincides with the geometric series $\sum_0^\infty c^n x^n = \sum_0^\infty \frac{(-1)^n}{x^n}$. It satisfies the condition of regularity of the coefficients with equality; i.e. we always have $\frac{c_{n+1}}{c_n} = -1$. Also the error term is defined for all $x \neq -1$, in fact with the exact expression $R_n(x) := \frac{x^{n+1}}{1+x}$, with $\frac{1}{1+x}$ S-continuous and even bounded for $x \geq 0$. Then the conditions of Proposition 7.4.13 are indeed satisfied. As shown in [130], it is not difficult to verify the validity of (7.30) in the extreme cases $n\varepsilon \simeq 0$ and $n\varepsilon \simeq \infty$ (see also Exercise 7.4.3), which extends the validity of formula (7.29) to all indices.

Many expansions satisfy condition (7.27); in fact typically the coefficients c_n allow from an expansion starting with $c_{n+1} = (-1)c_n = c + \frac{a}{n} + \cdots$. In [130] examples are given like hypergeometric functions, Bessel and Hankel functions, with asymptotic expansions given by Borel or Laplace transforms, which admit an error analysis similar to the methods described in this section.

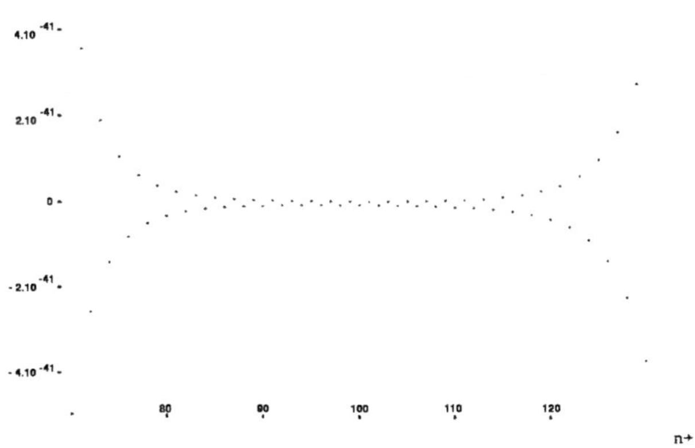

FIGURE 7.1: Remainders $R_n(\varepsilon)$ of the Euler expansion of the exponential integral $\int_0^\infty \frac{e^{-t}}{1+\varepsilon t}dt$ [130].

Exercises

Exercise 7.4.1. *Let $f : \mathbb{R}^+ \to \mathbb{R}$ be standard, positive, continuous and decreasing, and $\varepsilon \simeq 0$, $\varepsilon > 0$.*

1. *Using the Fresnel integral $\int_0^\infty \sin(x^2)dx = \sqrt{\frac{\pi}{8}}$, show that*

$$\int_0^\infty \sin\left(\frac{x^2}{\varepsilon}\right) f(x)dx = (1+\oslash)\sqrt{\frac{\pi}{8}}f(0).$$

2. *Using the Dirichlet integral $\int_0^\infty \frac{\sin(x)}{x}dx = \frac{\pi}{2}$, show that*

$$\int_0^\infty \frac{\sin\left(\frac{x^2}{\varepsilon}\right)}{\sqrt{x}} f(x)dx = (1+\oslash)\sqrt{\varepsilon}\frac{\pi}{2}f(0).$$

Exercise 7.4.2. *Give a nonstandard proof of Watson's Lemma. One may use* (7.24) *and its standard version* (7.25).

Exercise 7.4.3. *Consider the Euler expansion $\sum_{n=0}(-1)^n n!\varepsilon^n$ of the exponential integral $\int_0^\infty \frac{e^{-t}}{1+\varepsilon t}dt$ with remainder $R_n(\varepsilon)$ (see Figure 7.1).*

1. *Verify formula* (7.30) *for all standard $n \in \mathbb{N}$.*
2. *Verify formula* (7.30) *for $n \in \mathbb{N}, n \simeq \infty$ such that $n\varepsilon \simeq 0$.*
3. *Verify formula* (7.30) *for $n \in \mathbb{N}$ such that $n\varepsilon \simeq \infty$.*
4. *Derive the property of "summation to the smallest term": For all $n \in \mathbb{N}$*

$$\left|\frac{R_n(\varepsilon)}{R_{\left[\frac{1}{\varepsilon}\right]}(\varepsilon)}\right| \gtrsim 1.$$

Chapter 8

Flexible systems of linear equations

8.1 Flexible systems 189
8.2 Determinants 192
8.3 On Gauss-Jordan elimination 196
8.4 Parameter method 200
8.4.1 Non-singular systems 202
8.4.2 Singular systems with strict rank equal to the number of equations 204
8.4.3 Singular systems with strict rank less than the number of equations 206

8.1 Flexible systems

In this chapter we study systems of linear equations with external numbers. Following [68, 146] we call these systems *flexible systems of linear equations*, or simply *flexible systems*. The coefficients of flexible systems are external numbers each with its own, individual neutrix. Such a neutrix may be seen to represent imprecise data. These may have several kinds of origins, like measurement problems, model deficiencies or error propagation in calculations. Let m, n be standard. If we have n equations with m unknowns, and taking into account uncertainties in the second member, we thus get $(m + 1)n$ possibly different neutrices. The solutions are given by external vectors, in the form of an internal vector surrounded by a neutrix in higher dimension. There are many approaches to deal with such imprecisions, for example based on interval calculus or statistical methods (see e.g. [90, 89, 96, 63, 121]). Such approaches have a more obvious numerical implementation; in contrast flexible systems deal more easily with individual uncertainties, obey many more algebraic laws, and have explicit solutions under fairly general conditions. We describe some of the main results of studies by J. Justino and Nam Van Tran, partly with the second author.

Definition 8.1.1. *A* flexible system *is a system of linear equations of the form:*

$$\begin{cases} \alpha_{11}\xi_1+ & \alpha_{12}\xi_2+ & \cdots & +\alpha_{1n}\xi_n & \subseteq \beta_1 \\ \vdots & \vdots & \ddots & \vdots & \vdots \\ \alpha_{m1}\xi_1+ & \alpha_{m2}\xi_2+ & \cdots & +\alpha_{mn}\xi_n & \subseteq \beta_m, \end{cases} \tag{8.1}$$

where α_{ij}, β_i are external numbers and m, n are two standard natural numbers. We write $\alpha_{ij} = a_{ij} + A_{ij}, \beta_i = b_i + B_i$, where A_{ij} and B_i are neutrices and a_{ij} and b_i are representatives. The variables ξ_j designate external numbers, and we denote their representatives by x_j.

An admissible solution *is a (possibly external) set of vectors $x := (x_1, ..., x_n)^T$ of $\mathbb{R}^n$ satisfying (8.1) (using $\in$ instead of $\subseteq$). The* maximal solution *of the system is the set of all vectors of $\mathbb{R}^m$ which are admissible solutions. The maximal solution is* exact *if it satisfies (8.1) with equality.*

In the case where $m = n$ we say that the flexible system is square.

As in the definition of flexible system, in this chapter we will aways assume that m, n are standard natural numbers. We will make use of the following notation.

Notation 8.1.1. *For $1 \leq i \leq m, 1 \leq j \leq n$, let $\alpha_{ij} = a_{ij} + A_{ij}$, with $a_{ij} \in \mathbb{R}$ and A_{ij} neutrices. We write*

1. $\mathcal{A} = [\alpha_{ij}]_{m \times n}$, *an $m \times n$ matrix.*
2. $\overline{\alpha} = \max\limits_{\substack{1 \leq i \leq m \\ 1 \leq j \leq n}} |\alpha_{ij}|$.
3. $\overline{a} = \max\limits_{\substack{1 \leq i \leq m \\ 1 \leq j \leq n}} |a_{ij}|$.
4. $\overline{A} = \max\limits_{\substack{1 \leq i \leq m \\ 1 \leq j \leq n}} A_{ij}$.
5. $\underline{A} = \min\limits_{\substack{1 \leq i \leq m \\ 1 \leq j \leq n}} A_{ij}$.

In particular, for a column vector $\mathcal{B} = [\beta_i]$, with $\beta_i = b_i + B_i \in \mathbb{E}$ for $1 \leq i \leq m$, we write $\overline{\beta} = \max\limits_{1 \leq i \leq m} |\beta_i|$, $\overline{b} = \max\limits_{1 \leq i \leq m} |b_i|$, $\overline{B} = \max\limits_{1 \leq i \leq m} B_i$ and $\underline{B} = \min\limits_{1 \leq i \leq m} B_i$.

Definition 8.1.2. *Let $\mathcal{A} = [\alpha_{ij}]_{m \times n}$ be a matrix. We say that $\mathcal{A}$ is* reduced *if $\overline{\alpha} = \alpha_{11} = 1 + A_{11}$ with $A_{11} \subseteq \oslash$.*

In [67, 68] conditions that guarantee the existence, in terms of inclusion, of maximal solutions were given. The components of the maximal solution are external numbers. So the maximal solution is the sum of an internal vector, in fact the solution of a system of representatives, and a direct sum of neutrices. This is in line with an earlier theoretical result. The definition of neutrices in $\mathbb{R}^1$ (scalar neutrices) is easily carried over to convex groups in $\mathbb{R}^d$, with $d > 1$. The following decomposition theorem permits us to attribute a notion of dimension to neutrices in $\mathbb{R}^d$.

Theorem 8.1.2 ([134]). *Let* $\operatorname{st}(d) \in \mathbb{N} \setminus \{0\}$. *Let* $N \subseteq \mathbb{R}^d$ *be a neutrix. Then there exist* d *scalar neutrices* $N_1, ..., N_d$ *and orthonormal vectors* $u_1, ..., u_d$ *such that* $N = N_1 u_1 \oplus \cdots \oplus N_d u_d$; *the scalar neutrices are unique, if ordered by inclusion.*

In analogy to linear algebra, the maximal solution of a non-singular non-homogeneous flexible system of linear equations is given by Cramer's rule. However some restrictions need to be added which are induced by the size of the uncertainties of the system. They are stated in terms of determinants. Determinants of square matrices are defined as usual by signed sums of products along appropriate paths through the matrix (see Section 8.2). For instance, the condition that the determinant of the matrix of coefficients is non-zero is replaced by a condition stating that the determinant of the matrix should not be too small. This can be achieved in terms of absorbers of neutrices. In some cases, by adapting Cramer's rule it is still possible to produce admissible solutions even in the case where not all of those restrictions are satisfied.

The maximal solution produced by Cramer's rule is related to the procedure of Gauss-Jordan elimination which is the basis of numerical methods for solving systems of linear equations. In fact, under reasonable conditions, both methods lead to the same solution. The conditions are stated in Section 8.3. For a full account with proofs we refer to [67, 146]; here we illustrate the solution strategy by a typical example.

One of the conditions for effective application of the procedure of Gauss-Jordan elimination is that all neutrices in the right-hand side are equal. This is of course restrictive. In practice this means that the smallest neutrix $\underline{B}$ in the right-hand side prescribes our approach, for to have an admissible solution we should replace every neutrix of the right-hand side of the system by $\underline{B}$. In Section 8.4 we present a second method, called the *parameter method*, which is applicable with arbitrary neutrices in the second member. Under fairly general conditions it even yields the exact solution [146, 147]. Instead of the vector representation of the solution, with liberties in terms of subspaces, we have now restricted liberties given by neutrices. This comes in the form of an orthogonal decomposition within a higher-dimensional neutrix, called *feasibility space*. In order to determine the length of the neutrices of the orthogonal decomposition within the feasibility space, sometimes it is needed to solve a new linear system, and then Gauss-Jordan elimination comes again into view.

Exercises

Exercise 8.1.1. *Let* $\varepsilon \simeq 0$, $\varepsilon > 0$. *By, if necessary, interchanging rows and columns, write the following matrices as the product of a real number and a reduced matrix.*

1. $\begin{bmatrix} 1+\varepsilon\pounds & 2 & 1+\varepsilon\pounds \\ 2+\varepsilon\pounds & 2+\varepsilon & 2 \\ 1 & \oslash & \varepsilon\pounds \end{bmatrix}.$

2. $\begin{bmatrix} 1+\varepsilon\oslash & 1 & 1 & 1+\oslash \\ 6+\sqrt{\varepsilon}+\varepsilon\oslash & 4 & 5\oslash & 1 \\ \varepsilon\pounds & 1+\oslash & 2+\varepsilon^2\pounds & 3+\varepsilon\pounds \\ \varepsilon\oslash & 1+\oslash & -3+\varepsilon\pounds & 6+\sqrt{\varepsilon}\pounds \end{bmatrix}.$

Exercise 8.1.2. *Let $\varepsilon \simeq 0$, $\varepsilon > 0$. Write the neutrix $\{(x, y, z) \in \mathbb{R}^3 : x$ limited $\wedge\, y \simeq x \wedge z + y \in \varepsilon\pounds\}$ as an orthogonal decomposition.*

Exercise 8.1.3. *A flexible system is* homogeneous *if the components of the right-hand side β are neutrices, and* non-homogeneous *if $\overline{\beta}$ is zeroless. Show, by giving an explicit example, that there exists a third category, of a flexible system which is* upper homogeneous *if $\overline{\beta}$ is a neutrix* [146].

8.2 Determinants

Determinants of matrices with external numbers are defined as sums of signed products of entries.

Definition 8.2.1. *Let $\mathcal{A} = [\alpha_{ij}]_{n\times n}$ be a square matrix over the external numbers. We will denote by S_n the set of all permutations of $\{1, \ldots, n\}$. The* determinant *of $\mathcal{A}$, denoted by* det($\mathcal{A}$) *or* $|\mathcal{A}|$, *is defined by*

$$\det(\mathcal{A}) = \sum_{\sigma\in S_n} \operatorname{sgn}(\sigma)\alpha_{1\sigma(1)}\alpha_{2\sigma(2)}\cdots\alpha_{n\sigma(n)}.$$

Common techniques for calculation with determinants need to be applied with care, since they often use distributivity which, as seen in Chapter 3, may change the size of the neutrix part. Also some care is needed with the notion of rank, for it is easy to construct matrices with determinant equal to a neutrix, but for which at least some matrices of representatives have non-zero rank. This problem is avoided by the notion of strict rank introduced in the last part of this section.

As shown in the following example, the Laplace expansion of a determinant along a column or a row may not be equal to the determinant. In fact, it gives an approximation of the exact value; in particular if the determinant is zeroless, the result is again zeroless, with a common representative. Indeed we prove that the Laplace expansion possibly reduces the neutrix part (see [67]).

Example 8.2.1. We have that $\begin{vmatrix} \oslash & 0 & 0 \\ 0 & 1 & 1 \\ 0 & 1 & 1 \end{vmatrix} = \oslash$, while if we expand the determinant along the first column we obtain that $\oslash \begin{vmatrix} 1 & 1 \\ 1 & 1 \end{vmatrix} - 0 \begin{vmatrix} 0 & 0 \\ 1 & 1 \end{vmatrix} + 0 \begin{vmatrix} 0 & 0 \\ 1 & 1 \end{vmatrix} = 0 \in \oslash$.

Definition 8.2.2. *Let $\mathcal{A} = [\alpha_{ij}]_{n\times n}$ be a matrix. If $k = \{1, \ldots, n\}$ and $1 \le i_1 \le \cdots \le i_k \le n$, $1 \le j_1 \le \cdots \le j_k \le n$, we denote the* minor *with elements from the rows $i_1 < \cdots < i_k$ and columns $j_1 < \cdots < j_k$ by $M_{i_1,\ldots,i_k,j_1,\ldots,j_k}$. If there is no ambiguity we write simply $M_{I,J}$.*

Proposition 8.2.1 (Laplace formula)**.** *Let $\mathcal{A} = [\alpha_{ij}]_{n\times n}$ be a matrix over the external numbers. Then for all $j \in \{1, \ldots, n\}$*

$$\sum_{k=1}^{n} (-1)^{j+k} \alpha_{kj} |M_{k,j}| \subseteq |\mathcal{A}|.$$

Proof. We only consider the case $j = 1$; the other cases are similar. Let $\sigma \in S_n$. The Laplace expansion along the first column and subdistributivity yield

$$\begin{aligned}
&\alpha_{11}|M_{1,1}| - \alpha_{21}|M_{2,1}| + \cdots + (-1)^{1+n}|M_{n,1}| \\
&= \alpha_{11} \sum_{\substack{\sigma\in S_n \\ \sigma(1)=1}} \operatorname{sgn}(\sigma)\alpha_{\sigma(2)2} \cdots \alpha_{\sigma(n)n} + \cdots \\
&\qquad\qquad + \alpha_{n1} \sum_{\substack{\sigma\in S_n \\ \sigma(1)=n}} \operatorname{sgn}(\sigma)\alpha_{\sigma(1)1} \cdots \alpha_{\sigma(n-1)n-1} \\
&\subseteq \sum_{\substack{\sigma\in S_n \\ \sigma(1)=1}} \alpha_{11} \left(\operatorname{sgn}(\sigma)\alpha_{\sigma(2)2} \cdots \alpha_{\sigma(n)n}\right) + \cdots \\
&\qquad\qquad + \sum_{\substack{\sigma\in S_n \\ \sigma(1)=n}} \alpha_{n1}(\operatorname{sgn}(\sigma)\alpha_{\sigma(1)1} \cdots \alpha_{\sigma(n-1)n-1}) \\
&= \sum_{\sigma\in S_n} \operatorname{sgn}(\sigma)\alpha_{\sigma(1)1} \cdots \alpha_{\sigma(n)n} = |\mathcal{A}|.
\end{aligned}$$

□

As shown in the next proposition, the determinant and all minors of a reduced matrix are limited, with neutrix part no bigger than $\overline{A} \subseteq \oslash$.

Proposition 8.2.2. *Let $\mathcal{A} = [\alpha_{ij}]_{n\times n}$ be a reduced matrix. Let $k = \{1, \ldots, n\}$ and $1 \le i_1 \le \cdots \le i_k \le n$, $1 \le j_1 \le \cdots \le j_k \le n$. Then*

1. $|\det(M_{I,J})| \le \pounds$.

2. $N(M_{I,J}) \subseteq \overline{A} \subseteq \oslash$.

Proof. Because $\mathcal{A}$ is a reduced matrix, it follows that $|\alpha_{ij}| \leq 1 + \oslash$ for all $1 \leq i, j \leq n$.

1. We have

$$\begin{aligned} |\det(M_{I,J})| &= \left| \sum_{\sigma \in S_k} \operatorname{sgn}(\sigma) \alpha_{i_1\sigma(i_1)} \cdots \alpha_{i_k\sigma(i_k)} \right| \\ &\leq \sum_{\sigma \in S_k} \left|\alpha_{i_1\sigma(i_1)}\right| \cdots \left|\alpha_{i_k\sigma(i_k)}\right| \\ &\leq \sum_{\sigma \in S_k} (1+\oslash)^k = k!(1+\oslash). \end{aligned}$$

Since n is standard and $k \leq n$, it follows by External Induction that $k!$ is standard. So $k!(1+\oslash) \leq \pounds$. Hence $|\det(M_{I,J})| \leq \pounds$.

2. Note that $\overline{A} \subseteq \oslash$, otherwise $\overline{A} \supseteq \pounds$ and $\mathcal{A}$ would not be reduced. It follows from Exercise 3.3.8 that $(1+\overline{A})^k = 1 + \overline{A}$. So

$$\begin{aligned} N\left(|\det(M_{I,J})|\right) &= N\left(\sum_{\sigma \in S_k} \operatorname{sgn}(\sigma) \alpha_{i_1\sigma(i_1)} \cdots \alpha_{i_k\sigma(i_k)} \right) \\ &= \sum_{\sigma \in S_k} N\left(\alpha_{i_1\sigma(i_1)} \cdots \alpha_{i_k\sigma(i_k)}\right) \subseteq \sum_{\sigma \in S_k} N\left((1+\overline{A})^k\right) \\ &= \sum_{\sigma \in S_k} k\overline{A} = kk!\overline{A} = \overline{A}. \end{aligned}$$

For $k = n$ we obtain that $N(\det(\mathcal{A})) \subseteq \overline{A}$. □

The following proposition gives a useful relation between the determinant of a square matrix $\mathcal{A}$ and the determinant of the matrix obtained by adding c times the tth row to the kth row of $\mathcal{A}$, where $k \neq t$.

Proposition 8.2.3. *Let $c \in \mathbb{R}$, with $|c| \leq 1$, and $k, t \in \{1, ..., n\}$ with $k \neq t$. Let $\mathcal{A} = [\alpha_{ij}]_{n \times n}$ be a reduced non-singular matrix over the external numbers and $\Delta = \det(\mathcal{A})$. Let $\mathcal{A}' = [\alpha'_{ij}]_{n \times n}$ be such that $\alpha'_{ij} = \begin{cases} \alpha_{ij}, & i \neq k \\ \alpha_{ij} + c\alpha_{tj}, & i = k \end{cases}$ and $\Delta' = \det(\mathcal{A}')$, for $1 \leq i, j \leq n$. Then $\Delta' \subseteq (1+\oslash)\Delta$.*

Proof. One has

$$\mathcal{A}' = \begin{bmatrix} \alpha_{11} & \cdots & \alpha_{1n} \\ \vdots & & \vdots \\ \alpha_{t1} & \cdots & \alpha_{tn} \\ \vdots & & \vdots \\ \alpha_{k1} + c\alpha_{t1} & \cdots & \alpha_{kn} + c\alpha_{tn} \\ \vdots & & \vdots \\ \alpha_{n1} & \cdots & \alpha_{nn} \end{bmatrix}.$$

Let $\sigma = (p_1, ..., p_n) \in S_n$. So, using subdistributivity,

$$\begin{aligned}\Delta' &= \sum_{\sigma \in S_n} \operatorname{sgn}(\sigma)\alpha_{1p_1} \cdots \alpha_{tp_t} \cdots (\alpha_{kp_k} + c\alpha_{tp_k}) \cdots \alpha_{np_n} \\ &\subseteq \sum_{\sigma \in S_n} \operatorname{sgn}(\sigma)\alpha_{1p_1} \cdots \alpha_{tp_t} \cdots \alpha_{kp_k} \cdots \alpha_{np_n} + \\ &\qquad c\sum_{\sigma \in S_n} \operatorname{sgn}(\sigma)\alpha_{1p_1} \cdots \alpha_{tp_t} \cdots \alpha_{tp_k} \cdots \alpha_{np_n}.\end{aligned}$$

Now $\alpha_{1p_1} \cdots \alpha_{tp_t} \cdots \alpha_{tp_k} \cdots \alpha_{np_n} = \alpha_{1p_1} \cdots \alpha_{tp_k} \cdots \alpha_{tp_t} \cdots \alpha_{np_n}$ and they appear with opposite signs in the sum of permutations. So

$$\sum_{\sigma \in S_n} \operatorname{sgn}(\sigma)\, \alpha_{1p_1} \cdots \alpha_{tp_t} \cdots \alpha_{tp_k} \cdots \alpha_{np_n} = N(\alpha_{1p_1} \cdots \alpha_{tp_t} \cdots \alpha_{tp_k} \cdots \alpha_{np_n}).$$

Since $|\alpha_{ij}| \leq 1 + \oslash$ and $N(\alpha_{ij}) \subseteq \overline{A}$, by Exercise 3.3.8,

$$N(\alpha_{1p_1} \cdots \alpha_{tp_t} \cdots \alpha_{tp_k} \cdots \alpha_{np_n}) \subseteq N\left(\left(1 + \overline{A}\right)^n\right) = \overline{A}.$$

Hence, $\Delta' \subseteq \Delta + cN(\alpha_{1p_1} \cdots \alpha_{tp_t} \cdots \alpha_{tp_k} \cdots \alpha_{np_n}) \subseteq \Delta + c\overline{A} \subseteq \Delta + \oslash\Delta = (1 + \oslash)\Delta$. □

Exercises

Exercise 8.2.1. *Let $\varepsilon \simeq 0$. Calculate the determinant of the following matrices.*

1. $\begin{bmatrix} 1 + \varepsilon\oslash & -1 \\ \varepsilon\pounds & 2 + \varepsilon^2\pounds \end{bmatrix}$.

2. $\begin{bmatrix} -1 + \varepsilon\oslash & 1 + \varepsilon\pounds \\ \varepsilon\oslash & 2 + \varepsilon \end{bmatrix}$.

3. $\begin{bmatrix} 3 + \varepsilon\oslash & -1 + \oslash \\ 2 + \varepsilon\pounds & 1 + \varepsilon\oslash \end{bmatrix}$.

4. $\begin{bmatrix} 1 + \oslash & -1 + \oslash & \varepsilon^2\pounds \\ \varepsilon\pounds & 2 + \varepsilon\pounds & \varepsilon\oslash \\ 2 & 3 + \varepsilon & \oslash \end{bmatrix}$.

5. $\begin{bmatrix} 1 + \varepsilon^2\oslash & 1 & 1 + \varepsilon^3\pounds \\ 2 + \varepsilon^3\pounds & -1 + \varepsilon^2\oslash & -1 \\ \varepsilon + \varepsilon^3\oslash & 1 & 2 + \varepsilon^2\oslash \end{bmatrix}$.

Exercise 8.2.2. *Solve, if possible, the equations given by the determinants*

1. $\begin{vmatrix} 1+\oslash & -1 \\ \varepsilon\pounds & x \end{vmatrix} = 0.$

2. $\begin{vmatrix} 1+\oslash & -1 \\ \varepsilon\pounds & x \end{vmatrix} = \oslash.$

Exercise 8.2.3. *Let $\gamma \in \mathbb{E}$ and $\mathcal{A}, \mathcal{B}$ be $n \times n$ matrices over the external numbers. Show that the following equalities are false in general.*

1. $\det(\gamma\mathcal{A}) = \gamma^n \det(\mathcal{A})$.

2. $\det(\mathcal{A}\mathcal{B}) = \det(\mathcal{A})\det(\mathcal{B})$.

3. $\det((\mathcal{A}+\mathcal{B})^2) = (\det(\mathcal{A}+\mathcal{B}))^2$.

8.3 On Gauss-Jordan elimination

The determinant of a square matrix, being an external number, can either be a neutrix or zeroless. We say that the matrix is *singular* in the former case and *non-singular* in the latter. This definition has an obvious extension to systems. Under the following conditions non-singular systems are solved by Cramer's rule, and then the solutions are the same as given by Gauss-Jordan elimination.

Definition 8.3.1. [146, slightly modified] *The flexible system* (8.1) *is called* Gauss-Jordan eliminable *if*

1. $m = n$.

2. *The matrix $\mathcal{A}$ is reduced.*

3. Systematic Gauss-Jordan elimination *is possible in the following sense. After Gauss-Jordan elimination of the first column with pivot α_{11} (creating neutrices instead of zeros), there appears a new zeroless element which is maximal in absolute value, which without restriction of generality may be assumed to occur at place 2,2; then after Gauss-Jordan elimination of the second column with this pivot appears a new zeroless element which is maximal in absolute value etc., finally obtaining a matrix with zeroless elements with representative 1 on the diagonal, and neutrices less than $\oslash$ outside the diagonal.*

4. *All neutrices of the second member are equal to some scalar neutrix B.*

5. *We have $R(\mathcal{A}) \subseteq P(B)$. Here*

$$R(\mathcal{A}) = \frac{\overline{A}}{\Delta}, \tag{8.2}$$

with $\overline{A}$ *the maximum of the neutrices* A_{ij}*, is the* relative uncertainty *of* $\mathcal{A}$ *and*

$$P(B) = \begin{cases} \frac{B}{\overline{\beta}}, & \overline{\beta} \text{ zeroless} \\ B : B, & \overline{\beta} = B, \end{cases} \tag{8.3}$$

where $\overline{\beta}$ *is the largest in absolute value of the* β_i *is the* relative uncertainty *of the vector* β.

6. Δ *is not an absorber of* B.

Part 2 of Definition 8.3.1 concerns an obvious generalization of classical reduced matrices and Part 6 is a weakening of the condition of singularity in terms of zero determinants. Part 5 ensures that during the process of Gauss-Jordan elimination there is no inadmissible blow-up of uncertainties in the matrix of coefficients. Part 4 is of course restrictive; in practice we must take $B = \underline{B}$, with $\underline{B}$ the smallest of the neutrices in the second member.

Theorem 8.3.1. *Assume the flexible system* (8.1) *is Gauss-Jordan eliminable. Then*

1. [68] *the application of Cramer's rule yields the maximal solution.*

2. [67, 146] *the solution found by Cramer's rule equals the solution by systematic Gauss-Jordan elimination.*

To some extend, the solution by Gauss-Jordan elimination and Cramer's rule still applies to singular flexible systems, though additional conditions are required. We do not present a formal theorem, but instead illustrate the method by an example, adapted from [146].

Example 8.3.1. 1. Given a flexible system, by appropriate multiplication and rearranging, we put the coefficient matrix in reduced form. This is always possible, but we must assume that once in reduced form, all entries of the second member have the same neutrix part.

Let $\varepsilon > 0$ be an infinitesimal. The following system is already in reduced form:

$$\begin{cases} (1+\varepsilon £)x + (-1+4\varepsilon+\varepsilon\oslash)y + (-1+\varepsilon £)z \subseteq 1+\oslash \\ \varepsilon £ x + (-4\varepsilon+\varepsilon\oslash)y + \oslash z \subseteq \oslash \\ \varepsilon £ x + (-1+4\varepsilon+\varepsilon\oslash)y + (-1+\varepsilon\oslash)z \subseteq 3+\oslash. \end{cases} \tag{8.4}$$

2. Applying some row operations, it may be supposed that in all columns the coefficients have the same neutrix part. This can be done column by column; in (8.4) the two first columns are already in this form. As for the third column, consider the line with the biggest neutrix, and add it to the other lines. We get

$$\begin{cases} (1+\varepsilon £)x + (-1+\varepsilon\oslash)y + (-1+\oslash)z \subseteq 1+\oslash \\ \varepsilon £ x + (-4\varepsilon+\varepsilon\oslash)y + \oslash z \subseteq \oslash \\ \varepsilon £ x + (-1+\varepsilon\oslash)y + (1+\oslash)z \subseteq 3+\oslash. \end{cases} \tag{8.5}$$

3. The system (8.5) does not have full rank since the determinant equals $\oslash$; however we recognize non-singular 2×2-minors. A 2×2-minor which is not an absorber of $\oslash$ should be situated above-left. We must try to choose the minor in such a way that the neutrices appearing in the columns with index superior to the rank of the non-singular minor are bigger than any neutrix with index less than this rank. This is of course not always possible. Exchanging the second and third line we obtain

$$\begin{cases} (1+\varepsilon\pounds)x + (-1+\varepsilon\oslash)y + (-1+\oslash)z \subseteq 1+\oslash \\ \varepsilon\pounds x + (-1+\varepsilon\oslash)y + (1+\oslash)z \subseteq 3+\oslash \\ \varepsilon\pounds x + (-4\varepsilon+\varepsilon\oslash)y + \oslash z \subseteq \oslash. \end{cases} \tag{8.6}$$

We see that in (8.6) the neutrices of the first and second column are indeed smaller than the neutrices of the last column.

4. In contrast to common matrices of rank 2, it is not possible to transform the coefficients of the third line into zeros, still they can be transformed into neutrices, by adding -4ε times the second line. Then we get

$$\begin{cases} (1+\varepsilon\pounds)x + (-1+\varepsilon\oslash)y + (-1+\oslash)z \subseteq 1+\oslash \\ \varepsilon\pounds x + (-1+\varepsilon\oslash)y + (1+\oslash)z \subseteq 3+\oslash \\ \varepsilon\pounds x + \varepsilon\oslash y + \oslash z \subseteq \oslash. \end{cases} \tag{8.7}$$

5. In a common matrix a line of zeros can be neglected if the second member is also zero. Here we may neglect a line of neutrices if the second member is also a neutrix; in fact we recognize the last line of (8.7) as part of both the first and the second lines. We obtain

$$\begin{cases} (1+\varepsilon\pounds)x + (-1+\varepsilon\oslash)y + (-1+\oslash)z \subseteq 1+\oslash \\ \varepsilon\pounds x + (-1+\varepsilon\oslash)y + (1+\oslash)z \subseteq 3+\oslash. \end{cases} \tag{8.8}$$

6. We verify the conditions on relative uncertainty and absence of absorption of Definition 8.3.1, with $\mathcal{A} := \begin{bmatrix} 1+\varepsilon\pounds & -1+\varepsilon\oslash \\ \varepsilon\pounds & -1+\varepsilon\oslash \end{bmatrix}$, $\mathcal{\beta} := \begin{bmatrix} 1+\oslash \\ 3+\oslash \end{bmatrix}$, $B := \begin{bmatrix} \oslash \\ \oslash \end{bmatrix}$. We have $\overline{A} = \varepsilon\pounds$, $\Delta = -1+\varepsilon\pounds$ and $\overline{\beta} = 3+\oslash$. Hence

$$R(\mathcal{A}) = \frac{\overline{A}}{\Delta} = \varepsilon\pounds \subset \oslash = \frac{\oslash}{3+\oslash} = \frac{B}{\overline{\beta}} = P(B).$$

We also see that $\Delta = -1+\varepsilon\pounds$ is not an absorber of $B = \oslash$.

7. We know now that the system (8.8) is solvable and apply Gauss-Jordan elimination. We multiply the second line by -1 and add it to the first line. Then we get

$$\begin{cases} (1+\varepsilon £)x + \varepsilon \oslash y + (-2+\oslash)z \subseteq -2+\oslash \\ \varepsilon £ x + (1+\varepsilon\oslash)y + (-1+\oslash)z \subseteq -3+\oslash. \end{cases} \tag{8.9}$$

8. We now consider z as a parameter. It will not range over the whole of $\mathbb{R}$, because from (8.9) we infer that $\oslash z \subseteq \oslash$, i.e. z ranges over $£$. Putting z at the right-hand side we get

$$\begin{cases} (1+\varepsilon £)x + \varepsilon \oslash y \subseteq (2+\oslash)z - 2+\oslash \\ \varepsilon £ x + (1+\varepsilon\oslash)y \subseteq (1+\oslash)z - 3+\oslash. \end{cases} \tag{8.10}$$

9. The unperturbed coefficient matrix of system (8.10) is equal to the identity matrix. With this we write the solution S of (8.4) explicitly, and obtain

$$S = \{(-2+2z+\oslash, -3+z+\oslash, z) : z \in £\}. \tag{8.11}$$

Exercises

Exercise 8.3.1. *Let $\varepsilon \simeq 0$, $\varepsilon > 0$. Solve the following flexible systems using Gauss-Jordan elimination; solve the first equation also by Cramer's rule.*

1. $\begin{cases} \oslash x + (1+\varepsilon £)y = 3+\oslash \\ 3x + (1+\oslash)y = \oslash. \end{cases}$

2. $\begin{cases} (2+\varepsilon\oslash)x - (1+\varepsilon £)y = 1+\varepsilon £ \\ (3+\varepsilon £)x + (1+\varepsilon\oslash)y = \varepsilon £. \end{cases}$

3. $\begin{cases} (1+\varepsilon^2\oslash)x + y + (1+\varepsilon^3 £)z = \frac{1}{\varepsilon} + \varepsilon\oslash \\ (2+\varepsilon^3 £)x + (3+\varepsilon^2\oslash)y + z = \varepsilon\oslash \\ (\varepsilon+\varepsilon^3\oslash)x - y + (2+\varepsilon^2\oslash)z = 1+\varepsilon\oslash. \end{cases}$

Exercise 8.3.2. *Let $\varepsilon \simeq 0, \varepsilon > 0$. Verify that Cramer's rule and Gauss-Jordan elimination do not work for the following flexible systems.*

1. $\begin{cases} 3x - (1+\varepsilon £)y = 3+\varepsilon\oslash \\ (2+\varepsilon £)x + y = \varepsilon\oslash. \end{cases}$

2. $\begin{cases} 3x - (1+\varepsilon^2\oslash)y = 1+\oslash \\ 2\varepsilon x + \varepsilon y = \varepsilon^2 £. \end{cases}$

8.4 Parameter method

In this section we present the parameter method introduced in [146, 147]. This method consists in replacing external numbers in the second member of each equation of a flexible system by a parameter which is a representative of the external number in case. The replaced system may be solved using the usual means, e.g. the Gauss-Jordan method. Finally the parameters are replaced by the corresponding external numbers; sometimes truncations are needed to respect feasibility. If the strict rank r is equal to the number of equations m, the solution thus obtained corresponds to the set of all possible solutions and is therefore maximal. If the strict rank is less than m, additional conditions are needed. In that case, it is natural to expect that after Gauss-Jordan elimination $m-r$ equations will have relatively small coefficients, for they generate neutrix-valued minors. The additional conditions enable us to neglect these $m-r$ equations indeed.

We start by solving non-singular systems, then singular systems where the number of equations equals the strict rank, and finally singular systems where the number of equations is greater than the strict rank. In fact, the class of equations introduced at each successive step may be reduced to the former, at the last stage in the presence of additional conditions.

Definition 8.4.1. Let $\mathcal{A} = [\alpha_{ij}]_{m\times n}$ be a matrix over the external numbers. A natural number r is called the *strict rank* of the matrix $\mathcal{A}$, written $\mathrm{sr}(\mathcal{A}) = r$, if it satisfies the following conditions:

1. There is a zeroless minor of order r of $\mathcal{A}$.
2. There is a matrix $\hat{A} = [a_{ij}]_{m\times n}$ of representatives of the α_{ij} such that the rank of $\hat{A}$ is r.

It follows from the definition that if a matrix has strict rank r, then there exists a submatrix S of order $r \times r$ such that all choices of its representatives give rise to a non-zero minor; i.e. all matrices of representatives of S are of rank r, and it is possible to extend S to a matrix of representatives $\hat{A}$ of $\mathcal{A}$ without augmenting the rank. Then the rank of a submatrix T of order $(r+1)\times(r+1)$ may depend on the choice of representatives; if $\det(T)$ is a neutrix, then for at least one of these choices the submatrix has rank r, as seen in the next example.

Example 8.4.1. Let $\varepsilon > 0$, $\varepsilon \simeq 0$. Consider the matrix

$$M = \begin{bmatrix} 1+\oslash & 1+\varepsilon\oslash & 3 \\ 1+\varepsilon\pounds & 1+\varepsilon\oslash & 2 \\ 2\varepsilon+\varepsilon\oslash & \varepsilon\pounds & 5\varepsilon \end{bmatrix}.$$

It is obvious that the strict rank of M is at least 2. It is also easy to find a matrix of representatives of rank 3 with determinant equal to ε. For example

$$\begin{bmatrix} 1 & 1 & 3 \\ 1 & 1 & 2 \\ 2\varepsilon & \varepsilon & 5\varepsilon \end{bmatrix},$$

However,

$$\begin{bmatrix} 1 & 1 & 3 \\ 1 & 1 & 2 \\ 2\varepsilon & 2\varepsilon & 5\varepsilon \end{bmatrix}$$

is a matrix of representatives with rank 2. Notice that $\det(M) = \varepsilon \pounds$, which is a neutrix. So $\mathrm{sr}(M) = 2$.

Definition 8.4.2. *Consider the flexible system* (8.1)*. For each i with $1 \leq i \leq n$ we write*

$$K_i := \min_{1 \leq j \leq m} (B_i : A_{ij}). \tag{8.12}$$

We adopt the usual notation e_i for the ith element of the unit basis of $\mathbb{R}^n$, where $1 \leq i \leq n$. We define the feasibility space *F by $F := \oplus_{i=1}^n K_i e_i$.*

Algebraically, the feasibility space is a neutrix with at most dimension n, and as such a module over $\pounds$.

Example 8.4.2. For the system (8.4) of Example 8.3.1 we have $K_1 = \oslash : \varepsilon\pounds = \frac{\oslash}{\varepsilon}, K_2 = \oslash : \varepsilon\oslash = \frac{\pounds}{\varepsilon}, K_3 = \oslash : \oslash = \pounds$. So $F = (\frac{\pounds}{\varepsilon}, \frac{\pounds}{\varepsilon}, \pounds)$. Hence the solution S given by (8.11) is feasible indeed, and the truncation of the variable z corresponds to the truncation with respect to F.

Theorem 8.4.1 transforms the system (8.1) into a system with real coefficient matrix, while the variables range over the feasibility space instead of $\mathbb{R}^n$.

Theorem 8.4.1. *The system* (8.1) *is equivalent to*

$$\left\{ \begin{array}{ccccccccc} a_{11}x_1 & + & \cdots & + & a_{1n}x_n & \in & b_1 & + & B_1 \\ \vdots & & \ddots & & \vdots & & \vdots & & \\ a_{m1}x_1 & + & \cdots & + & a_{mn}x_n & \in & b_m & + & B_m, \\ \multicolumn{9}{l}{x_j \in K_j, j \in \{1, \ldots, n\},} \end{array} \right. \tag{8.13}$$

where $a_{ij} \in \alpha_{ij}, 1 \leq i, j \leq n$.

Proof. The vector $\xi = (\xi_1, \ldots, \xi_n)^T \in \mathbb{R}^n$ is a solution of the system (8.1) if and only if

$$\left\{ \begin{array}{ccccccccc} (a_{11} + A_{11})\xi_1 & + & \cdots & + & (a_{1n} + A_{1n})\xi_n & \subseteq & b_1 & + & B_1 \\ \vdots & & \ddots & & \vdots & & \vdots & & \\ (a_{m1} + A_{m1})\xi_1 & + & \cdots & + & (a_{mn} + A_{mn})\xi_n & \subseteq & b_m & + & B_m. \end{array} \right.$$

This is equivalent to

$$\begin{cases} a_{11}x_1 & + & a_{12}x_2 & + & \cdots & + & a_{1n}x_n & \in & b_1 & + & B_1 \\ \vdots & & \vdots & & \ddots & & \vdots & & \vdots & & \\ a_{m1}x_1 & + & a_{m2}x_2 & + & \cdots & + & a_{mn}x_n & \in & b_m & + & B_m, \end{cases}$$

and $A_{ij}\xi_j \subseteq B_i$ for all $1 \leq i \leq m$ and $1 \leq j \leq n$; the latter is equivalent to $x_j \in K_j$ for all $1 \leq j \leq n$. □

Not all ordinary systems of linear equations have a solution. Likewise, there exist flexible systems without solutions. However, it is possible for a flexible system to have no solution even if the system is non-singular. Indeed, let $\varepsilon \simeq 0$, $\varepsilon > 0$. Consider the equation

$$(1 + \oslash)x \subseteq 1 + \varepsilon\pounds. \tag{8.14}$$

Putting the equation in the form of Theorem 8.4.1 we get the contradictory system

$$\begin{cases} x \in 1 + \varepsilon\pounds \\ x \in \varepsilon\pounds. \end{cases}$$

8.4.1 Non-singular systems

Theorem 8.4.2. *Assume that the system*

$$\begin{cases} \alpha_{11}\xi_1 + & \alpha_{12}\xi_2 + & \cdots & +\alpha_{1n}\xi_n & \subseteq \beta_1 \\ \vdots & \vdots & \ddots & \vdots & \vdots \\ \alpha_{n1}\xi_1 + & \alpha_{n2}\xi_2 + & \cdots & +\alpha_{nn}\xi_n & \subseteq \beta_n, \end{cases} \tag{8.15}$$

is non-singular. Then the solution is given by

$$\xi = \left(P^{-1}b + \sum_{i=1}^{n} B_i P^{-1} e_i\right) \cap \left(\oplus_{i=1}^{n} K_i e_i\right). \tag{8.16}$$

Proof. By Theorem 8.4.1, a vector $x = (x_1, \ldots, x_n)^T \in \mathbb{R}^n$ is a solution of the system (8.15) if and only if its representatives satisfy

$$\begin{cases} a_{11}x_1 & + & a_{12}x_2 & + & \cdots & + & a_{1n}x_n & \in & b_1 + B_1 \\ \vdots & & \vdots & & \ddots & & \vdots & & \vdots \\ a_{n1}x_1 & + & a_{n2}x_2 & + & \cdots & + & a_{nn}x_n & \in & b_n + B_n \end{cases} \tag{8.17}$$

with the additional conditions $x_j \in K_j, 1 \leq j \leq n$.

On the other hand, a vector $x = (x_1, \ldots, x_n)^T \in \mathbb{R}^n$ is a solution of the system (8.17) if and only if for all $i \in \{1, \ldots, n\}$ there exists $\varepsilon_i \in B_i$ such that

$$\begin{cases} a_{11}x_1 & + & a_{12}x_2 & + & \cdots & + & a_{1n}x_n & = & b_1 + \varepsilon_1 \\ \vdots & & \vdots & & \ddots & & \vdots & & \vdots \\ a_{n1}x_1 & + & a_{n2}x_2 & + & \cdots & + & a_{nn}x_n & = & b_n + \varepsilon_n. \end{cases}$$

This implies that such solutions have the form

$$\begin{aligned} x &= P^{-1}b + \sum_{i=1}^{n} \varepsilon_i P^{-1} e_i, \quad \varepsilon_i \in B_i \\ &= P^{-1}b + \sum_{i=1}^{n} B_i P^{-1} e_i. \end{aligned}$$

With the additional condition $x \in \oplus_{i=1}^{n} K_i e_i$, we obtain that

$$\xi = \left(P^{-1}b + \sum_{i=1}^{n} B_i P^{-1} e_i \right) \cap (\oplus_{i=1}^{n} K_i e_i)$$

solves the system (8.15). □

We see that the neutrix part of each component of the solution of (8.15) is determined by the maximum of the corresponding component of $B_1 P^{-1} e_1, ..., B_n P^{-1} e_n$. If necessary, the neutrix parts need to be restricted further according to the additional bounds of the variables originating from the neutrix parts of the coefficients α_{ij}, for they should take values in the feasibility space.

We give a simple example, where the solution is contained in the feasibility space. At the end of the next section we give an example where the feasibility space interacts, for the additional restrictions will reduce the validity of the domain of the solutions indeed.

Example 8.4.3. Let $\varepsilon > 0$, $\varepsilon \simeq 0$. Consider the flexible system

$$\begin{cases} (1 + \oslash)\xi + (1 + \varepsilon\oslash)\eta \subseteq \oslash \\ (1 + \varepsilon \pounds)\xi - (1 + \varepsilon \pounds)\eta \subseteq \varepsilon \pounds. \end{cases} \tag{8.18}$$

Though the system is homogeneous it is not trivial. For instance if we try naively $\xi = \oslash, \eta = \oslash$, which obviously satisfies the first equation, it does not satisfy the second equation, for $\oslash - \oslash = \oslash \supset \varepsilon \pounds$. Observe that Cramer's rule and Gauss-Jordan elimination *a priori* do not apply, because of the presence of different neutrices in the right-hand side. The parameter method works with representatives $x \in \xi$ and $y \in \eta$, and starts by introducing a parameter ε_1 ranging over $\oslash$ and a parameter ε_2 ranging over $\varepsilon \pounds$. This transforms (8.18) into

$$\begin{cases} x + y = \varepsilon_1 \\ x - y = \varepsilon_2 \end{cases}$$

with

$$\begin{cases} \varepsilon_1 \in \oslash \\ \varepsilon_2 \in \varepsilon \pounds. \end{cases}$$

Gauss-Jordan elimination shows that $x = \frac{\varepsilon_1+\varepsilon_2}{2}$ and $y = \frac{\varepsilon_1-\varepsilon_2}{2}$. So the solution set is

$$\begin{bmatrix} \xi \\ \eta \end{bmatrix} = \oslash \begin{bmatrix} \frac{1}{2} \\ \frac{1}{2} \end{bmatrix} + \varepsilon \pounds \begin{bmatrix} \frac{1}{2} \\ -\frac{1}{2} \end{bmatrix}. \tag{8.19}$$

The additional constraints coming from the neutrix parts of the coefficients are $\oslash x \subseteq \oslash$, $\varepsilon \oslash y \subseteq \oslash$, $\varepsilon \pounds x \subseteq \varepsilon \pounds$, $\varepsilon \pounds y \subseteq \varepsilon \pounds$. Taking the narrowest restrictions, these reduce to $\begin{cases} x \in \oslash : \oslash = \pounds \\ y \in \varepsilon \pounds : \varepsilon \pounds = \pounds \end{cases}$. They obviously do not intervene, so the solution set remains (8.19). We show that Cramer's rule and Gauss-Jordan elimination fall short on (8.18). Applying direct calculation of determinants (Cramer's rule) we obtain

$$\begin{bmatrix} \xi \\ \eta \end{bmatrix} = \begin{bmatrix} \oslash \\ \oslash \end{bmatrix}. \tag{8.20}$$

It was already seen that the second equation of (8.18) is not satisfied, for $\oslash - \oslash = \oslash \supset \varepsilon \pounds$.

As for Gauss-Jordan elimination, subtracting the first equation of (8.18) from the second equation yields

$$\begin{cases} (1+\oslash)x + (1+\varepsilon\oslash)y \subseteq \oslash \\ \oslash x - (2+\varepsilon\pounds)y \subseteq \oslash. \end{cases}$$

This has the obvious solution (8.20), which as we already saw was not feasible. Direct verification along the above lines shows that both Cramer's rule and Gauss-Jordan elimination applied to the more restrictive system

$$\begin{cases} (1+\oslash)x + (1+\varepsilon\oslash)y \subseteq \varepsilon\pounds \\ (1+\varepsilon\pounds)x - (1+\varepsilon\pounds)y \subseteq \varepsilon\pounds \end{cases} \tag{8.21}$$

yield the solution

$$\begin{bmatrix} \xi \\ \eta \end{bmatrix} = \begin{bmatrix} \varepsilon\pounds \\ \varepsilon\pounds \end{bmatrix}. \tag{8.22}$$

This solution is feasible indeed, but is only part of the exact solution (8.19).

In [67, 146] general conditions for the application of Cramer's rule and Gauss-Jordan elimination were given. As it turns out, the system (8.18) does not verify one of these conditions but the system (8.21) does indeed.

The additional constraints do not give rise to additional restrictions. In some cases the intersection given by (8.16) is empty. For example, the contradictory equation (8.14) yields $(1+\varepsilon\pounds) \cap \varepsilon\pounds = \emptyset$.

8.4.2 Singular systems with strict rank equal to the number of equations

Theorem 8.4.3. *Consider the flexible system (8.1) with $n > m$. If the strict rank of the system is equal to the number of equations and $\Delta :=$*

$\begin{vmatrix} \alpha_{11} & \cdots & \alpha_{1m} \\ \vdots & \ddots & \vdots \\ \alpha_{m1} & \cdots & \alpha_{mm} \end{vmatrix}$ *is zeroless, the set of solutions is given by*

$$\begin{bmatrix} \xi_1 \\ \vdots \\ \xi_m \\ \xi_{m+1} \\ \vdots \\ \xi_n \end{bmatrix} = \left(\begin{bmatrix} M^{-1}b \\ 0 \end{bmatrix} + \sum_{i=1}^{m} \begin{bmatrix} B_i M^{-1} e_i \\ 0 \end{bmatrix} + \sum_{k=m+1}^{n} \mathbb{R} \begin{bmatrix} -M^{-1} a_k^T \\ e_{k-m}^{(n-m)} \end{bmatrix} \right) \cap \left(\oplus_{i=1}^{n} K_i e_i \right),$$

where $e_i^{(n-m)}$ *is the* i^{th} *unit vector in* $\mathbb{R}^{n-m}$ *and* $M := \begin{bmatrix} a_{11} & \cdots & a_{1m} \\ \vdots & \ddots & \vdots \\ a_{m1} & \cdots & a_{mm} \end{bmatrix}$,

with $a_{ij} \in \alpha_{ij}, 1 \leq i, j \leq m$.

A proof of Theorem 8.4.3 is given in [146]. As in (8.17) we take representatives, and similarly to Example 8.3.1, we put the variables not corresponding to the matrix M at the right-hand side. The rest of the proof is along the lines of the proof of Theorem 8.4.2.

We illustrate this procedure by the following system with two equations, with strict rank also equal to 2. This example also shows how the feasibility conditions reduce the solution set. Let again $\varepsilon > 0$ be an infinitesimal. Consider the flexible system

$$\begin{cases} (2 + \varepsilon\oslash)\xi + (1 + \varepsilon\oslash)\eta + (1 + \oslash)\zeta \subseteq 1 + \oslash \\ (1 + \varepsilon\oslash)\xi + (1 + \varepsilon\pounds)\eta + (3 + \varepsilon\oslash)\zeta \subseteq 2 + \varepsilon\oslash . \end{cases} \tag{8.23}$$

The determinant of the first 2×2-minor being equal to $1 + \varepsilon\pounds$, the strict rank of the system is obviously equal to 2. Taking representatives, and considering the restriction due to the neutrix parts, the system is equivalent to

$$\begin{cases} 2x + y + z \in 1 + \oslash \\ x + y + 3z \in 2 + \varepsilon\oslash \\ x \in \pounds, y \in \oslash, z \in \pounds. \end{cases} \tag{8.24}$$

We introduce parameters $\varepsilon_1 \in \oslash$ and $\varepsilon_2 \in \varepsilon\oslash$ and solve the system

$$\begin{cases} 2x + y + z = 1 + \varepsilon_1 \\ x + y + 3z = 2 + \varepsilon\varepsilon_2 \end{cases} \tag{8.25}$$

by Gauss-Jordan elimination. With z as a parameter ranging over $\mathbb{R}$, and substituting ε_1, ε_2 and z by their range, the solution of (8.25) takes the form

$$\begin{bmatrix} \xi \\ \eta \\ \zeta \end{bmatrix} = \begin{bmatrix} -1 \\ 3 \\ 0 \end{bmatrix} + \oslash \begin{bmatrix} 1 \\ -1 \\ 0 \end{bmatrix} + \varepsilon\oslash \begin{bmatrix} -1 \\ 2 \\ 0 \end{bmatrix} + \mathbb{R} \begin{bmatrix} 2 \\ -5 \\ 1 \end{bmatrix} .$$

By the restrictions of (8.24), the feasibility neutrix F becomes

$$F := \pounds \begin{bmatrix} 1 \\ 0 \\ 0 \end{bmatrix} + \oslash \begin{bmatrix} 0 \\ 1 \\ 0 \end{bmatrix} + \pounds \begin{bmatrix} 0 \\ 0 \\ 1 \end{bmatrix}.$$

So the solution of the original system (8.23) is

$$\begin{bmatrix} \xi \\ \eta \\ \zeta \end{bmatrix} = \left(\begin{bmatrix} -1 \\ 3 \\ 0 \end{bmatrix} + \oslash \begin{bmatrix} 1 \\ -1 \\ 0 \end{bmatrix} + \varepsilon \oslash \begin{bmatrix} -1 \\ 2 \\ 0 \end{bmatrix} + \mathbb{R} \begin{bmatrix} 2 \\ -5 \\ 1 \end{bmatrix} \right) \cap F. \tag{8.26}$$

We will see that the infinitesimal second coordinate induces an additional restriction. Obviously

$$\oslash \begin{bmatrix} 1 \\ -1 \\ 0 \end{bmatrix} + \varepsilon \oslash \begin{bmatrix} -1 \\ 2 \\ 0 \end{bmatrix} \subseteq F,$$

so we only need to introduce a new parameter λ ranging over $\mathbb{R}$. This leads to the system

$$\begin{cases} -1 + \lambda \in \pounds \\ 3 - 5\lambda \in \oslash \\ \lambda \in \pounds. \end{cases}$$

It follows from the second line that $\lambda \in \frac{3}{5} + \oslash$; hence the solution (8.26) takes the final form

$$\begin{bmatrix} \xi \\ \eta \\ \zeta \end{bmatrix} = \begin{bmatrix} \frac{1}{5} \\ 0 \\ \frac{3}{5} \end{bmatrix} + \oslash \begin{bmatrix} 1 \\ -1 \\ 0 \end{bmatrix} + \varepsilon \oslash \begin{bmatrix} -1 \\ 2 \\ 0 \end{bmatrix} + \oslash \begin{bmatrix} 2 \\ -5 \\ 1 \end{bmatrix}.$$

8.4.3 Singular systems with strict rank less than the number of equations

As already mentioned, in case the strict rank of the system is less than the number of equations we need some extra conditions for the parameter method to be applied.

Consider the flexible system (8.1) with strict rank $r < \min\{m, n\}$. Then we may assume that $\Delta := \det \begin{bmatrix} \alpha_{11} & \cdots & \alpha_{1r} \\ \vdots & \ddots & \vdots \\ \alpha_{r1} & \cdots & \alpha_{rr} \end{bmatrix}$ is zeroless and the matrix of representatives $M := \begin{bmatrix} a_{11} & \cdots & a_{1r} \\ \vdots & \ddots & \vdots \\ a_{r1} & \cdots & a_{rr} \end{bmatrix}$ has non-zero determinant. For $r+1 \leq k \leq m$ define

$$M_{rk}^{(i)} = \begin{bmatrix} a_{11} & \cdots & a_{1r} \\ \vdots & \ddots & \vdots \\ a_{(i-1)1} & \cdots & a_{(i-1)r} \\ a_{k1} & \cdots & a_{kr} \\ a_{(i+1)1} & \cdots & a_{(i+1)r} \\ \vdots & \ddots & \vdots \\ a_{r1} & \cdots & a_{rr} \end{bmatrix}.$$

Observe that the row i of M has been substituted by the first r elements of the k-th row of $[P|b]$. Because the rank of M is r, this row is a linear combination of the first r rows of $[P|b]$, and, due to the non-singularity of M, its first r elements can be calculated from $\det\left(M_{rk}^{(i)}\right), 1 \leq i \leq r$, and $\det(M)$ by Cramer's rule. This gives the coefficients, say $t_1, \dots, t_r$, of the linear combination. For this operation to have no consequences for the neutrix of the k-th second member, these coefficients should not be too large; in fact $\det\left(M_{rk}^{(i)}\right)$ should not be too large with respect to $\det(M)$. It is reasonable to expect this when the entries of the last $m-r$ rows do not exceed too much the entries of the first r rows, which in turn should correspond to relatively small neutrices in the constant term. The next lemma gives precise conditions for this to happen.

Lemma 8.4.4. *Consider the flexible system* (8.1). *Assume that the strict rank is* $r < \min\{m, n\}$ *and*

1. $\frac{\det\left(M_{rk}^{(i)}\right)}{\det(M)}$ *is not an exploder of* B_k *for all* $1 \leq i \leq r$, $r+1 \leq k \leq m$.
2. $B_i \subseteq B_k$ *for* $r+1 \leq k \leq m, 1 \leq i \leq r$.

Then the system (8.1) *is equivalent to*

$$\left\{ \begin{array}{cccccccccc} a_{11}x_1 & + & a_{12}x_2 & + & \cdots & + & a_{1n}x_n & \in & b_1 + B_1 \\ \vdots & & \vdots & & \ddots & & \vdots & & \vdots \\ a_{r1}x_1 & + & a_{r2}x_2 & + & \cdots & + & a_{rn}x_n & \in & b_r + B_r \\ \multicolumn{9}{l}{x_i \in K_i, 1 \leq i \leq n.} \end{array} \right. \tag{8.27}$$

A solution $x = (x_1, \dots, x_n) \in \mathbb{R}^n$ of the system (8.1) obviously satisfies (8.27). For the converse we refer to [146].

Theorem 8.4.5. *Consider the flexible system* (8.1). *Assume that the strict rank of the system is* $r < \min\{m, n\}$ *and that assumptions* (1) *and* (2) *of*

Lemma 8.4.4 are satisfied. Then the solution of the system is given by

$$\begin{bmatrix} \xi_1 \\ \vdots \\ \xi_r \\ \xi_{r+1} \\ \vdots \\ \xi_n \end{bmatrix} = \left(\begin{bmatrix} M^{-1}b \\ 0 \end{bmatrix} + \sum_{i=1}^{r} \begin{bmatrix} B_i M^{-1} e_i \\ 0 \end{bmatrix} + \sum_{k=r+1}^{n} \mathbb{R} \begin{bmatrix} -M^{-1}a_k^T \\ e_{k-r}^{(n-r)} \end{bmatrix} \right) \cap (\oplus_{i=1}^{n} K_i e_i), \tag{8.28}$$

where $e_i^{(n-r)}$ is the ith unit vector in $\mathbb{R}^{(n-r)}$.

Proof. By Lemma 8.4.4, the system is equivalent to system (8.27). Applying Theorem 8.4.3 to this system, the set of solutions is then given by (8.28). □

Remark 8.4.6. In case Δ and $M_{rk}^{(i)}$ are both appreciable, we can always rearrange the system such that the neutrix part of the second member is increasing, so that the assumptions (1) and (2) of Lemma 8.4.4 are satisfied.

In general, we can interchange the rows of P such that always $|\det(M)| \geq \left|\det(M_{rk}^{(i)})\right|$, implying assumption (1). Then it depends on the nature of the neutrices of the right-hand side whether the constant term assumption (2) is also satisfied.

Example 8.4.4. In system (8.5) we changed rows, to obtain system (8.6), for which assumption (1) of Lemma 8.4.4 is satisfied, for we would have $\det(M)$ appreciable and $\det\left(M_{rk}^{(1)}\right), \det\left(M_{rk}^{(2)}\right)$ infinitesimal. Assumption (2) is also satisfied, because all neutrices at the right-hand side are equal. Then the transition of the system (8.6) to the system (8.8) by neglecting the last equation is in line with Lemma 8.4.4.

Example 8.4.5. Let $\varepsilon > 0$ be an infinitesimal. Consider the system

$$\begin{cases} (1+\oslash)\xi + (1+\varepsilon\oslash)\eta + 3\zeta \subseteq 1+\oslash \\ (1+\varepsilon \pounds)\xi - (1+\varepsilon\oslash)\eta + 2\zeta \subseteq 2+\varepsilon\pounds \\ (2\varepsilon+\varepsilon\oslash)\xi + \varepsilon\pounds\eta + 5\varepsilon\zeta \subseteq 3\varepsilon+\oslash. \end{cases}$$

One verifies that the system has strict rank 2 and that it is equivalent to

$$\begin{cases} x+y+3z \in 1+\oslash \\ x-y+2z \in 2+\varepsilon\pounds \end{cases} \tag{8.29}$$

with $x \in \pounds, y \in \pounds, z \in \mathbb{R}$.

Using z as a parameter the solutions of the system (8.29) are given by

$$\begin{bmatrix} x \\ y \\ z \end{bmatrix} = \begin{bmatrix} \frac{3}{2} \\ -\frac{1}{2} \\ 0 \end{bmatrix} + \oslash \begin{bmatrix} \frac{1}{2} \\ \frac{1}{2} \\ 0 \end{bmatrix} + \varepsilon\pounds \begin{bmatrix} \frac{1}{2} \\ -\frac{1}{2} \\ 0 \end{bmatrix} + \mathbb{R} \begin{bmatrix} -\frac{5}{2} \\ -\frac{1}{2} \\ 1 \end{bmatrix}.$$

Noting that x and z must be limited, the solution of the original system is given by

$$\begin{bmatrix}\xi\\ \eta\\ \zeta\end{bmatrix} = \left(\begin{bmatrix}\frac{3}{2}\\ -\frac{1}{2}\\ 0\end{bmatrix} + \oslash\begin{bmatrix}\frac{1}{2}\\ \frac{1}{2}\\ 0\end{bmatrix} + \varepsilon £\begin{bmatrix}\frac{1}{2}\\ -\frac{1}{2}\\ 0\end{bmatrix} + \mathbb{R}\begin{bmatrix}-\frac{5}{2}\\ -\frac{1}{2}\\ 1\end{bmatrix}\right)$$
$$\cap\left(£\begin{bmatrix}1\\ 0\\ 0\end{bmatrix} + £\begin{bmatrix}0\\ 1\\ 0\end{bmatrix} + \mathbb{R}\begin{bmatrix}0\\ 0\\ 1\end{bmatrix}\right)$$
$$= \begin{bmatrix}\frac{3}{2}\\ -\frac{1}{2}\\ 0\end{bmatrix} + \oslash\begin{bmatrix}\frac{1}{2}\\ \frac{1}{2}\\ 0\end{bmatrix} + \varepsilon £\begin{bmatrix}\frac{1}{2}\\ -\frac{1}{2}\\ 0\end{bmatrix} + £\begin{bmatrix}-\frac{5}{2}\\ -\frac{1}{2}\\ 1\end{bmatrix}.$$

Exercises

Exercise 8.4.1. *Let $\varepsilon \simeq 0$, $\varepsilon > 0$. Solve the following flexible systems using the parameter method.*

1. $\begin{cases}(2+\varepsilon\oslash)x - (1+\varepsilon £)y \in 1+\varepsilon £\\ (3+\varepsilon £)x + (1+\varepsilon\oslash)y \in \oslash.\end{cases}$

2. $\begin{cases}3x - (1+\varepsilon\oslash)y \in 3+\varepsilon £\\ 2x + y \in \varepsilon\oslash.\end{cases}$

3. $\begin{cases}3x - (1+\varepsilon^2\oslash)y \in 1+\oslash\\ 2\varepsilon x + \varepsilon y \in \varepsilon £.\end{cases}$

4. $\begin{cases}(1+\varepsilon^2\oslash)x + y + (1+\varepsilon^3 £)z \in \frac{1}{\varepsilon} + £\\ (2+\varepsilon^3 £)x + (3+\varepsilon^2\oslash)y + z \in \oslash\\ (\varepsilon+\varepsilon^3\oslash)x - y + (2+\varepsilon^2\oslash)z \in 1+\varepsilon\oslash.\end{cases}$

Exercise 8.4.2. *Solve*

$$\begin{cases}(1+\varepsilon\oslash)\xi + (\varepsilon+\varepsilon^2\oslash)\eta \subseteq \varepsilon\oslash\\ (1+\varepsilon\oslash)\xi + (1+\varepsilon^2 £)\eta \subseteq \varepsilon\oslash.\end{cases}$$

Exercise 8.4.3. *Apply Gauss-Jordan elimination towards neglecting the last line of the next system, and then solve*

$$\begin{cases}(1+\varepsilon\oslash)\xi + (-1+\varepsilon £)\eta + \oslash\zeta \subseteq 2+\oslash\\ (1+\varepsilon\oslash)\xi + \varepsilon £\eta + (2+\oslash)\zeta \subseteq -1+\oslash\\ (-1+\varepsilon\oslash)\xi + (2+\varepsilon £)\eta + (2+\oslash)\zeta \subseteq -5+\oslash.\end{cases}$$

Exercise 8.4.4. *Solve the homogeneous system*

$$\begin{cases} (1+\varepsilon\oslash)\xi + (\varepsilon^2 \pounds)\eta \subseteq \varepsilon\pounds \\ (-2+\varepsilon\oslash)\xi + (1+\varepsilon\pounds)\eta \subseteq \varepsilon\oslash \,. \end{cases}$$

Chapter 9

Applications in asymptotics

9.1	Nonstandard Borel-Ritt Theorem	211
9.2	Tools for solution of external equations	213
9.3	Matching principles	225
9.4	An external singular perturbation with canard solutions	230
9.4.1	External differentiable equations and their solutions	230
9.4.2	The external Riccati-Hermite equation	231
9.4.3	Solving the external Riccati-Hermite equation	232
9.4.4	Description of the canard behaviour	236
9.4.5	Influence of the singular point on the localization of canards	237

9.1 Nonstandard Borel-Ritt Theorem

Expansions in ε-shadow of numbers have been introduced by Francine Diener as a nonstandard alternative to asymptotic expansions of functions; in fact the two notions coincide for standard functions (see [35]). Given an asymptotic expansion, the classical Borel-Ritt theorem guarantees the existence of a real function having this expansion. This theorem is particularly useful for divergent series: they are not empty, in the sense that they are always the expansion of some function. Proofs of the existence of real numbers having a prescribed expansion in ε-shadow are given in [35, 130], they form a sort of nonstandard Borel-Ritt theorems.

For ε infinitesimal, we interpret an expansion in ε-shadow in terms of an N-Cauchy sequence. Here N is the set of all real numbers having shadow expansion identically zero, i.e. the *microhalo* $M_\varepsilon := \bigcap_{\mathrm{st}(n)\in\mathbb{N}}[-\varepsilon^n, \varepsilon^n]$ introduced in Definition 1.4.4. We recall that the microhalo is a neutrix built from unlimited powers of ε, and may be denoted by $M_\varepsilon = \pounds\varepsilon^{\not\infty}$. By Theorem 5.3.8 an N-Cauchy sequence has a limit. This yields a new proof of such a nonstandard Borel-Ritt theorem; in fact it defines the set of real numbers having a given expansion in ε-shadow in the form of an external number.

We recall that the *shadow* of a limited real number x, denoted by $^\circ x$, is the unique standard real number infinitely close to x.

Definition 9.1.1. *Let ε be a positive infinitesimal real number, (a_n) be a standard sequence of real numbers and $b \in \mathbb{R}$. Then the formal series $\sum_{n=0}^{\infty} a_n \varepsilon^n$ is called an* expansion in ε-shadow *of b if for all standard $n \in \mathbb{N}$*

$$^{\circ}\left(\frac{b - \sum_{k=0}^{n} a_k \varepsilon^k}{\varepsilon^{n+1}}\right) = a_{n+1}. \tag{9.1}$$

Theorem 9.1.1 (Nonstandard Borel-Ritt Theorem [45]). *Let ε be a positive infinitesimal real number and (a_n) be a standard sequence of real numbers. Then there exists $b \in \mathbb{R}$ such that $\sum_{n=0}^{\infty} a_n \varepsilon^n$ is an expansion in ε-shadow of b.*

Proof. Put $s_n = \sum_{k=0}^{n} a_k \varepsilon^k$ for $n \in \mathbb{N}$. Let $\eta > \pounds\varepsilon^{\not\infty}$. Then there exists a standard $m \in \mathbb{N}$ such that $\varepsilon^m < \eta$. Then for standard $n \in \mathbb{N}$ with $n \geq m+1$ we have

$$|s_n - s_m| = \left|\sum_{k=m+1}^{n} a_k \varepsilon^k\right| \subset \pounds\varepsilon^{m+1} < \eta.$$

Hence s is an $\pounds\varepsilon^{\not\infty}$-Cauchy sequence with respect to ${}^{\sigma}\mathbb{N}$. By Theorem 5.3.8 there exists $b \in \mathbb{R}$ such that $\lim_{n\to\not\infty} s_n = b + \pounds\varepsilon^{\not\infty}$. Let $n \in \mathbb{N}$ be standard. Then there exists standard $p \in \mathbb{N}, p > n$ such that $b - s_p \in \oslash\varepsilon^{n+1}$. Then

$$\begin{aligned}
\frac{b - \sum_{k=0}^{n} a_k \varepsilon^k}{\varepsilon^{n+1}} &= \frac{b - \sum_{k=0}^{p} a_k \varepsilon^k + \sum_{k=n+1}^{p} a_k \varepsilon^k}{\varepsilon^{n+1}} \\
&\in \frac{\oslash\varepsilon^{n+1} + a_{n+1}\varepsilon^{n+1} + \oslash\varepsilon^{n+1}}{\varepsilon^{n+1}} \\
&\subseteq a_{n+1} + \oslash.
\end{aligned}$$

Hence $^{\circ}\left(\frac{b-\sum_{k=0}^{n} a_k \varepsilon^k}{\varepsilon^{n+1}}\right) = a_{n+1}$, meaning that $\sum_{n=0}^{\infty} a_n \varepsilon^n$ is an expansion in ε-shadow of b. □

Exercises

Exercise 9.1.1. *An* order scale *is an internal sequence of positive real numbers (u_n) such that $\frac{u_{n+1}}{u_n} \simeq 0$ for all $n \in \mathbb{N}$. If we generalize Definition 9.1.1 to order scales, using*

$$^{\circ}\left(\frac{b - \sum_{k=0}^{n} a_k u_k}{u_{n+1}}\right) = a_{n+1}$$

instead of (9.1) *for all standard $n \in \mathbb{N}$, we obtain a* shadow-expansion. *In analogy to the Nonstandard Borel-Ritt Theorem, state and prove an existence theorem for shadow-expansions with respect to an order scale.*

9.2 Tools for solution of external equations

We study equations of the type

$$f(x) \in \alpha, \tag{9.2}$$

where $f : \mathbb{R} \to \mathbb{R}$ is internal and $\alpha = a + A$ is an external number. The difference with respect to ordinary equations lies in the fact that the second member allows for some imprecision, due to the appearance of $N(\alpha) = A$. So, in a sense, instead of determining all x such that $f(x) = a$, we try to determine all x such that the value of $f(x)$ is equal to a, up to an imprecision with size A. A classical theory of approximate equations is given by asymptotic equations. Typically, these are of the form $f(x) \sim \omega$ for $\omega \to \infty$, and one searches for an asymptotic expansion for x in terms of ω. Such an expansion tends to be infinite, even divergent, but in the case of (9.2), if the neutrix A is large enough, we can hope that it will be truncated and restricted to a standard finite number of terms, the last one being a neutrix related to A.

In this chapter we present some tools for solving such equations, guided by some examples of which some are rather simple and others are more elaborate; the latter deal with the domain of validity of asymptotic approximations. The main strategy in solving the external equation (9.2) consists in trying to replace this equation by a simpler equation, with the same solutions. Indeed, as a consequence of the flexibility arising from the invariance of external numbers under some additions and multiplications we have some flexibility in the functions defining the equation. In general (9.2) is solved stepwise, giving rise to a common asymptotic expansion of the solutions.

We already saw that the neutrix $N(\alpha) = A$ is the set of all real numbers leaving α invariant by addition, and since $-A = A$ its elements leave α also invariant by subtraction. We now consider the set of real numbers leaving α invariant by multiplication.

Definition 9.2.1. *The* module $M(\alpha)$ *of an external number* α *is defined by* $M(\alpha) := \{x \in \mathbb{R} : \alpha x = \alpha\}$.

If α is zeroless, we have $M(\alpha) = R(\alpha)$, with $R(\alpha) = 1 + \frac{A}{a}$. If $\alpha = A$, by Theorem 4.5.1 we may write $A = pJ$, with $p \in \mathbb{R}$ and J idempotent for multiplication, i.e. $J = JJ$. If $1 < J$, by Theorem 4.5.3 the neutrix J has a maximal ideal I, and $M(\alpha) = J \setminus I$. If $J < 1$, it is the maximal ideal of an idempotent neutrix with $1 < K$, and $M(\alpha) = K \setminus J$. Due to the inclusion $R(\alpha) \subseteq \oslash$ and the idempotency of J in both cases $M(\alpha) = \frac{1}{M(\alpha)}$: its elements also leave α invariant by division.

The next theorem gives conditions enabling us to replace the external equation (9.2) by a (simpler) equation with the same set of solutions.

Theorem 9.2.1 (Substitution theorem). *Consider equation (9.2). Let $S \subseteq \mathbb{R}$.*

1. *Let $g : \mathbb{R} \to \mathbb{R}$ be internal and such that $f(x) - g(x) \in A$ for all $x \in S$. Then on S equation (9.2) and the equation $g(x) \in \alpha$ have the same solutions.*

2. *Assume that f is non-zero on S. Let $h : S \to \mathbb{R} \setminus \{0\}$ be internal and such that $\frac{f(x)}{h(x)} \in M(\alpha)$ for all $x \in S$. Then on S equation (9.2) and the equation $h(x) \in \alpha$ have the same solutions.*

Proof. 1. Let $x \in S$. If $g(x) \in \alpha$, then $f(x) \in g(x)+A \subseteq \alpha+A = \alpha$. Conversely, if $f(x) \in \alpha$, then $g(x) \in f(x) - A \subseteq \alpha + A = \alpha$. Hence $f(x) \in \alpha \Leftrightarrow g(x) \in \alpha$.

2. Let $x \in S$. If $h(x) \in \alpha$, then $f(x) \in \alpha M(\alpha) = \alpha$. Conversely, if $f(x) \in \alpha$, then $h(x) \in \frac{f(x)}{M(\alpha)} \subseteq \alpha M(\alpha) = \alpha$. Hence $f(x) \in \alpha \Leftrightarrow h(x) \in \alpha$. □

Example 9.2.1. Let $\varepsilon \simeq 0, \varepsilon > 0$. Consider the equation

$$x^2 + \varepsilon x^5 \in 1 + \oslash. \tag{9.3}$$

We distinguish the domains

(i) $S_1 := \left\{x \in \mathbb{R} : \frac{x^2}{\varepsilon x^5} \simeq 0\right\}$.

(ii) $S_2 := \left\{x \in \mathbb{R} : \frac{x^2}{\varepsilon x^5} \simeq \infty\right\}$.

(iii) $S_3 := \left\{x \in \mathbb{R} : \frac{x^2}{\varepsilon x^5} \text{ appreciable}\right\}$.

Note that $M(1 + \oslash) = 1 + \oslash$.

(i) On S_1 it holds that $\frac{x^2+\varepsilon x^5}{\varepsilon x^5} \in 1 + \oslash = M(1 + \oslash)$. By Part 1 of Theorem 9.2.1 one may as well solve $\varepsilon x^5 \in 1+\oslash$. Its solution is $\frac{1+\oslash}{\varepsilon^{1/5}}$. However this solution must be rejected for it is not in S_1.

(ii) On S_2 one has $\frac{x^2+\varepsilon x^5}{x^2} \in 1+\oslash = M(1+\oslash)$. By Part 1 of Theorem 9.2.1 the equation $x^2 \in 1 + \oslash$ has the same set of solutions. These are $x_1 = 1 + \oslash$ and $x_2 = -1 + \oslash$. Both are admissible, for they are subsets of S_2.

(iii) On S_3 Theorem 9.2.1 provides no simplification. We put $a =: \frac{\varepsilon x^5}{x^2} = \varepsilon x^3$. Then a is appreciable and satisfies $a^{2/3}+a^{5/3} = a^{2/3}(a+1) \in (1+\oslash)\varepsilon^{2/3}$. It has the obvious solutions $a_1 := (1 + \oslash)\varepsilon$ and $a_2 := -1 + (1 + \oslash)\varepsilon^{2/3}$; only a_2 is appreciable, hence admissible. Then

$$x_3 = \frac{a_2^{1/3}}{\varepsilon^{1/3}} = \frac{(-1 + (1 + \oslash)\varepsilon^{2/3})^{1/3}}{\varepsilon^{1/3}} = -\frac{1}{\varepsilon^{1/3}} + (1 + \oslash)\frac{1}{3}\varepsilon^{1/3}$$

is the third and last root of (9.3).

Example 9.2.2. In Chapter 2 the approximation of the exponential function e^{-x} by the partial sums of its Taylor series $\sum_{n=0}^{\infty} \frac{(-1)^n x^n}{n!}$ was considered. In particular "the approximation by a Taylor polynomial is good" was modelled by "the remainder is infinitesimal" and the domain was determined where the remainder is infinitesimal by solving an external equation. At certain points the equation was simplified; here we wish to justify the simplifications by the Substitution Theorem.

We recall that by convergence, after truncating at an unlimited index $\omega-1$ the remainder $R_{\omega-1}(x)$, takes the form

$$R_{\omega-1}(x)=\sum_{n=\omega}^{\infty}(-1)^n\frac{x^n}{n!}\in\oslash.$$

This corresponds to

$$\frac{x^\omega}{\omega!}A_\omega(x)\in\oslash, \tag{9.4}$$

with the approximation factor $A_\omega(x)$ given by

$$A_\omega(x)=\sum_{k=0}^{\infty}(-1)^n\frac{x^k}{(\omega+1)\cdots(\omega+k)}. \tag{9.5}$$

It was shown in Theorem 2.1.1 that we may confine to $S:=\left\{x\in\mathbb{R}:\left|\frac{x}{\omega}\right|\lessapprox 1\right\}$, and then $A_\omega(x)\in\frac{1+\oslash}{1+\frac{x}{\omega}}$, by formula (2.9). Now $M(\oslash)=@$, and $A_\omega(x)\in @$ for $x\in S$. Then by the Substitution Theorem equation (9.4) and $\frac{x^\omega}{\omega!}\in\oslash$ have the same set of solutions. By Stirling's formula $\frac{\omega^\omega}{\omega!}\in\omega^\omega e^{-\omega}\sqrt{\omega}\sqrt{2\pi(1+\oslash)}$. Now $\sqrt{2\pi(1+\oslash)}$ is appreciable, so again by the Substitution Theorem it is equivalent to solve either

$$x^\omega\in\omega^\omega e^{-\omega}\sqrt{\omega}\oslash,$$

or

$$x\in\frac{\omega}{e}\omega^{\frac{1}{2\omega}}\oslash^{\frac{1}{\omega}},$$

which, as seen in Part 2 of Theorem 2.1.1, amounts to

$$x\in\left(-\frac{\omega}{e}-\frac{\log(\omega)}{2e}+£,\frac{\omega}{e}+\frac{\log(\omega)}{2e}+£\right). \tag{9.6}$$

We now turn to *quasi-linear* equations. This is a natural class of equations, for which a strategy for determining an expansion and/or solving can be given. In principle a solution is given by the corresponding linear equation. If it is not admissible, it will determine the first term of an expansion, and a method and conditions will be given to determine further terms of the expansion.

Definition 9.2.2. *Let $\alpha=a+A$ be a zeroless external number. Let f be an internal continuous real function, at least defined at α. Assume $f(\alpha)\subseteq\oslash\alpha$. Then the equation*

$$x+f(x)\in\alpha \tag{9.7}$$

is called quasi-linear.

Example 9.2.3. The equation $x + \sqrt{x} \in \omega + \pounds$ is obviously quasi-linear.

A typical class of quasi-linear equations arises from Taylor expansions, in the natural case where the principal term of the expansion is linear.

Example 9.2.4. Consider the equation

$$\sqrt[3]{1+x} - 1 \in \varepsilon + \pounds\varepsilon^2. \tag{9.8}$$

Now $\sqrt[3]{1+x} - 1 = \frac{1}{3}(x + f(x))$, with $f(x) = 3(\sqrt[3]{1+x} - 1) - x$ and $f(\varepsilon + \pounds\varepsilon^2) \subseteq \pounds\varepsilon^2 \subseteq \oslash\varepsilon$. So (9.8) corresponds to the quasi-linear equation $x + f(x) \in 3\varepsilon + \pounds\varepsilon^2$.

We present now a more elaborate example of an equation which leads to a quasi-linear equation. The example deals with the approximation of the exponential function by Euler's formula and is taken from [134].

Example 9.2.5. Let $\omega \in \mathbb{N}$ be infinitely large. Consider the approximation of the exponential function by Euler's formula $\left(1 + \frac{x}{\omega}\right)^\omega$. Put

$$R_\omega(x) := e^x - \left(1 + \frac{x}{\omega}\right)^\omega.$$

Suppose we wish to know the set of positive values x where the error $R_\omega(x)$ is infinitesimal. This leads to the external equation

$$e^x - \left(1 + \frac{x}{\omega}\right)^\omega \in \oslash. \tag{9.9}$$

We show that (9.9) may be substituted by the simpler equation

$$e^x \frac{x^2}{\omega} \in \oslash. \tag{9.10}$$

This will be done by reducing the domain in two steps, each one leading to a sharper asymptotic expansion. Indeed, (9.9) may be rewritten as

$$e^x \left(1 - e^{\omega\left(\log\left(1+\frac{x}{\omega}\right) - \frac{x}{\omega}\right)}\right) \in \oslash. \tag{9.11}$$

Since $\log(1+a) - a$ is non-infinitesimal for appreciable a, equation (9.11) can only have solutions on $S_1 := \left\{x \in \mathbb{R} : \frac{x}{\omega} \simeq 0\right\}$. Then

$$e^{\omega\left(\log\left(1+\frac{x}{\omega}\right) - \frac{x}{\omega}\right)} \subseteq e^{\frac{x^2}{\omega}(1+\oslash)}.$$

Hence $e^x \left(1 - e^{\omega\left(\log\left(1+\frac{x}{\omega}\right) - \frac{x}{\omega}\right)}\right) \subseteq e^x \left(1 - e^{\frac{x^2}{\omega}(1+\oslash)}\right)$.

We will further reduce the domain of possible solutions of the equation (9.11) to $S_2 := \{x \in \mathbb{R} : x \in \oslash\sqrt{\omega}\}$. Indeed, if $x \in S_1$, $x > \oslash\sqrt{\omega}$, the value of $1 - e^{\frac{x^2}{\omega}}$ is appreciable; hence $e^x \left(1 - e^{\frac{x^2}{\omega}(1+\oslash)}\right)$ is even infinitely large.

For $x \in S_2$ we have the approximation

$$1 - e^{\frac{x^2}{\omega}(1+\oslash)} \subseteq \frac{x^2}{\omega}(1+\oslash). \tag{9.12}$$

Because $(1+\oslash) \subseteq M(\oslash) = @$, equation (9.9), which through equation (9.11) leads to (9.12), and equation (9.10) have on S_2 the same solutions.

We reduce now equation (9.10) to a quasi-linear equation. Taking logarithms we obtain

$$x + 2\log(x) < \log(\omega) + \pounds. \tag{9.13}$$

By monotony we might as well solve

$$x + 2\log(x) \in \log(\omega) + \pounds, \tag{9.14}$$

for the solutions of (9.13) precede the solutions of (9.14). The latter equation is quasi-linear, because $2\log(\log(\omega) + \pounds) \subseteq \oslash(\log(\omega) + \pounds)$.

In order to solve quasi-linear equations we introduce some notions and notation.

Definition 9.2.3. *Let $N > 0$ be a neutrix. Let $S, T \subseteq \mathbb{R}$ be external sets. We say that S is* bounded *in T, and write $S \sqsubset T$, if there exists an internal set I such that $S \subseteq I \subseteq T$.*

Clearly an internal set is always bounded in any set. By the Fehrele Principle, galaxies contained in halos are bounded in those halos.

Observe that if N is a neutrix such that $0 < N < \mathbb{R}$, and an external set S satisfies $S \sqsubset N$, there exists $n \in N$ such that $S \subseteq [-n, n]$.

We consider now some notions which express that functions are growing so slowly, that their images have locally a property of boundedness in the sense of Definition 9.2.3. These are "S-notions", i.e. are equivalent to classical notions when applied to standard functions, but we do not establish the equivalences here.

Definition 9.2.4. *Let $\alpha = a + A$ be a zeroless external number and f be an internal and measurable function, defined at least at α. Assume that $f(\alpha) \in \mathbb{E}$ is also zeroless. The function f is called* S-selfneglecting at α *if $\frac{f(\alpha)}{\alpha} \subseteq \oslash$ and $f(\alpha + \pounds f(\alpha)) \subseteq (1+\oslash)f(\alpha)$.*

If f is S-selfneglecting clearly $f(\alpha + \pounds f(\alpha))$ is bounded in $\pounds f(\alpha)$.

Example 9.2.6. The logarithm is S-selfneglecting at every positive infinitely large external number $\xi = x + X$, for

$$\begin{aligned}
\log(\xi + \pounds\log(\xi)) &= \log\left(\xi\left(1 + \frac{\pounds\log(\xi)}{\xi}\right)\right) \\
&= \log(\xi) + \log\left(1 + \frac{\pounds\log(\xi)}{\xi}\right) \\
&= \log(\xi) + \frac{\pounds\log(\xi)}{\xi} \\
&\sqsubset (1+\oslash)\log(\xi).
\end{aligned}$$

Other functions which are S-selfneglecting at ξ are the function $x \mapsto \frac{1}{x^s}$, with $\mathrm{st}(s) > -1$. Indeed,

$$\frac{1}{\left(\xi + \frac{\pounds}{\xi^s}\right)^s} = \frac{1}{\xi^s \left(1 + \frac{\pounds}{\xi^{s+1}}\right)^s} = \frac{1}{\xi^s \left(1 + \frac{s\pounds}{\xi^{s+1}}\right)} \sqsubset (1+\oslash)\frac{1}{\xi^s}.$$

Definition 9.2.5. *Let $f : \mathbb{R} \to \mathbb{R}$ be internal and differentiable, and α be a zeroless external number. The function f is called S-*asymptotically differentiable *at α if $f'(\alpha)$ is a zeroless external number and for $a, y \in \mathbb{R}$ such that $\alpha = a + A$ and $y \in \oslash\alpha$ it holds that $\frac{f(a+y)-f(a)}{y} \in (1+\oslash)f'(\alpha)$.*

Proposition 9.2.2. *S-Asymptotically differentiable functions f are S-selfneglecting at points α where $\frac{f(\alpha)}{\alpha}$, $f'(\alpha) \subseteq \oslash$.*

Proof. Let $f : \mathbb{R} \to \mathbb{R}$ be internal and S-asymptotically differentiable at some zeroless external number $\alpha = a + A$. Then $f'(a) \neq 0$, $f'(a) \simeq 0$ and $f(a) \in \oslash a$. Then f is S-selfneglecting at a, for if $y \in \pounds f(a)$

$$\begin{aligned} f(a+y) &= \frac{f(a+y)-f(a)}{y}y + f(a) \\ &\in (1+\oslash)yf'(a) + f(a) \\ &\subseteq (1+\oslash)\pounds f(a) \oslash + f(\alpha) \\ &= (1+\oslash)f(\alpha). \end{aligned}$$

□

As a consequence, $f(\alpha + \pounds f(\alpha))$ is bounded in $\pounds f(\alpha)$; in fact $f(\alpha + N) \subseteq (1+\oslash)f(\alpha) \sqsubset \pounds f(\alpha)$ for all neutrices $N \subseteq \oslash f(\alpha)$. Because $\alpha + f(\alpha) \subseteq \alpha + \oslash\alpha = (1+\oslash)\alpha$, locally the solutions of (9.7) are contained in the external number $(1+\oslash)\alpha$. The next localization theorems state that within $(1+\oslash)\alpha$ solutions exist indeed, and at the same time restrict (locally) the domain on which solutions may occur to an external number of the form $\alpha + N$, with $A \subseteq N \subseteq \oslash\alpha$.

First we present a lemma of "permanence" relating functions and neutrices.

Lemma 9.2.3. *Assume that α and β are external numbers, and M, N are external neutrices such that $N(\alpha) \leq M$. Let f be an internal real function such that $f(\alpha + M) \sqsubset \beta + N$. Then there exists $p \in \mathbb{R}, p > M$ such that $f[\alpha - p, \alpha + p] \sqsubset \beta + N$.*

Proof. Assume that $B \subseteq N$; the proof for $N \subset B$ is analogous, working with B instead of N. Let $a \in \alpha$ and $b \in \beta$. Then there exists $n \in N$ such that $f(x) \in [b-n, b+n]$ for all $x \in \alpha + M$. Then $f[a-m, a+m] \subset [b-n, b+n]$ for all $m \in M$. Because M is external, by Cauchy's Principle there exists $q \in \mathbb{R}, q > M$ such that still $f[a-q, a+q] \subset [b-n, b+n]$. Then $f[\alpha - p, \alpha + p] \sqsubset N$, for $p = \frac{q}{2}$. □

Theorem 9.2.4 (Localization Theorem). *Consider the quasi-linear equation (9.7). Assume there exists a neutrix $N > 0$ such that $N(\alpha) \subseteq N \subseteq \oslash\alpha$ and*

$f(\alpha+N) \sqsubset N$. Then (9.7) has a solution in $\alpha+N$. Moreover, there exists $p \in \mathbb{R}, p > N$ such that all solutions of (9.7) on $[\alpha-p, \alpha+p]$ are contained in $\alpha+N$.

Proof. With $\alpha = a + A$, let $n \in N$ be such that $f(x) \in [-n, n]$ for all $x \in a + A+N$. By Lemma 9.2.3 there exists $p \in \mathbb{R}, p > A+N$ such that $f(x) \in [-n, n]$ for all $x \in [\alpha-p, \alpha+p]$. Observe that

$$a+p+f(a+p) \geq a+p-n > a+\frac{p}{2} > a+A = \alpha$$

and

$$a-p+f(a-p) \leq a-p+n < a-\frac{p}{2} < a+A = \alpha.$$

Hence by the continuity of f there exists $x \in [a-p, a+p] \subseteq [\alpha-p, \alpha+p]$ such that $x + f(x) \in \alpha$. By Theorem 9.2.1 the equations $x + f(x) \in \alpha + N$ and $x \in \alpha + N$ have the same solutions on $[\alpha-p, \alpha+p]$. Hence if $x + f(x) \in \alpha$ and $x \in [\alpha-p, \alpha+p]$, then in fact $x \in \alpha + N$. □

If the perturbation function f is increasing, the condition of boundedness of Theorem 9.2.4 can be omitted.

Theorem 9.2.5 (Localization theorem for monotonous functions). *Consider the quasi-linear equation (9.7). Assume there exists a neutrix $N > 0$ such that $A \subseteq N \subseteq \oslash\alpha$, f is increasing on $\alpha+N$ and $f(\alpha+N) = N$. Then (9.7) has a solution in $\alpha+N$. Moreover, there exists $p \in \mathbb{R}, p > N$ such that all solutions of (9.7) on $[\alpha-p, \alpha+p]$ are contained in $\alpha+N$.*

Proof. By Cauchy's Principle there exists $p \in \mathbb{R}$, $p > N$ such that f is increasing on $[a-p, a+p]$. Note that $f(x) < N$ for $x \in [a-p, a+N)$ and $f(x) > N$ for $x \in (a+N, a+p]$. So

$$a+p+f(a+p) > a+N \geq a+A = \alpha$$

and

$$a-p+f(a-p) < a-p+N < a-\frac{p}{2} < a+A = \alpha.$$

Then by continuity there exists $x \in \alpha+N$ such that $x + f(x) \in \alpha$, satisfying equation (9.7). Also $x+f(x) < a+N+N = \alpha+N$ for $x \in [a-p, a+N)$ and $x + f(x) > a+N+N = \alpha+N$ for $x \in (a+N, a+p]$. Hence all solutions of (9.7) on $[\alpha-p, \alpha+p]$ are contained in $\alpha+N$. □

As a consequence, we obtain that if $f(\alpha) \sqsubset A$, the right-hand side α itself solves the quasi-linear equation. In particular, if $A \subset f(\alpha)$ additional steps are needed, giving rise to an expansion.

Corollary 9.2.6 (Direct solution for quasi-linear equations). *Let $S \subseteq \mathbb{R}$ and $\alpha = a + A \subseteq S$. Let f be an internal real continuous function at least defined on S. Consider the quasi-linear equation (9.7). If $f(\alpha) \sqsubset A$, the set of solutions of (9.7) on S equals α.*

Proof. For $x \in \alpha$ it holds that $x + f(x) \in \alpha + A = \alpha$. By Theorem 9.2.4 all solutions of (9.7) are contained in α. Hence α is the set of solutions of (9.7) on S. □

Example 9.2.7 (Continuation of Example 9.2.4). With $f(x) = 3(\sqrt[3]{1+x} - 1) - x$ equation (9.8) was reduced to the quasi-linear equation $x + f(x) \in 3\varepsilon + \pounds\varepsilon^2$. By the Taylor expansion up to second order

$$f(3\varepsilon + \pounds\varepsilon^2) \subset -\frac{2}{3}(3\varepsilon + \pounds\varepsilon^2)^2(1 + \oslash) \sqsubset \pounds\varepsilon^2.$$

Hence $3\varepsilon + \pounds\varepsilon^2$ solves (9.8).

Example 9.2.8. Consider the equation $x + \frac{1}{x} = \omega + \pounds$, with $x \geq 1$. Put $f(x) = \frac{1}{x}$. Then $f\ (\omega + \pounds) = \frac{1}{\omega+\pounds} = \frac{1}{\omega} + \frac{\pounds}{\omega^2} \sqsubset \pounds$. Hence the equation is solved by $\omega + \pounds$.

An important special case of the Localization Theorem, expressed in the next result, deals with S-selfneglecting functions, with $N = \pounds f(\alpha)$, for then $f(\alpha + \pounds f(\alpha)) \subseteq (1 + \oslash)f(\alpha) \sqsubset \pounds f(\alpha)$.

Corollary 9.2.7 (Localization theorem for S-selfneglecting functions). *Consider the quasi-linear equation* (9.7), *with f S-selfneglecting at α, and $A < f(\alpha)$. Then* (9.7) *has a solution in $\alpha + \pounds f(\alpha)$. Moreover there exists $p \in \mathbb{R}$, $p > \pounds f(\alpha)$ such that $f[\alpha - p, \alpha + p] \subseteq (1 + \oslash)f(\alpha)$ and all solutions of* (9.7) *on $[\alpha - p, \alpha + p]$ are contained in $\alpha + \pounds f(\alpha)$.*

Proof. One has $f(\alpha + \pounds f(\alpha)) \subseteq (1 + \oslash)f(\alpha) \sqsubset \pounds f(\alpha)$. By the Fehrele Principle, there exists $p \in \mathbb{R}$, $p > \pounds f(\alpha)$ such that still $f[\alpha - p, \alpha + p] \subseteq (1 + \oslash)f(\alpha) \sqsubset \pounds f(\alpha)$. Then the result follows from Theorem 9.2.4. □

If $\pounds f(\alpha) \subseteq A$, by Corollary 9.2.6 the solution of the quasi-linear equation is α itself, for $f(\alpha) = f(\alpha + \pounds f(\alpha)) \sqsubset \pounds f(\alpha) \subseteq A$. If $A \subseteq \pounds f(\alpha)$, the importance of the theorem lies in the fact that the possible domain for solutions is restricted from $a + \oslash\alpha$ to $a + \pounds f(\alpha) \sqsubset a + \oslash\alpha$. As such, it suggests the first step of an expansion; i.e. the solutions x will be of the form $x = a + y$, with $y \in \pounds f(\alpha)$. This is illustrated by Example 9.2.9, and formalized in Theorem 9.2.8.

Example 9.2.9 (Continuation of Example 9.2.5). As seen in Example 9.2.5 the equation (9.9) leads to the quasi-linear equation

$$x + 2\log(x) \in \xi + \pounds, \tag{9.15}$$

with $\xi = \log(\omega)$. Put $f(x) := 2\log(x)$. Because the logarithm is S-selfneglecting and $\pounds < f(\xi)$, it follows from Corollary 9.2.7 that all solutions of (9.15) must be contained in $\xi + \pounds + \pounds\log(\xi) = \xi + \pounds\log(\xi)$. Otherwise said, they must be of the form

$$x = \xi + y, \tag{9.16}$$

with $\frac{y}{\log(\xi)}$ limited. Formula (9.16) gives the first term of an expansion, and delimits the domain where to look for the second term.

Substituting (9.16) in equation (9.15) will yield a new quasi-linear equation in the variable y. This is a consequence of the next theorem which gives a condition enabling us to continue the expansion.

Theorem 9.2.8 (Expansion theorem). *Let f be internal and defined at least on $\alpha + £f(a)$. Consider the quasi-linear equation (9.7). Let $\alpha = a + A$. Assume that $A < |f(\alpha)|$ and that f is S-selfneglecting. There exists an interval $I := [a-p, a+p]$, with $p > £f(a)$ such that all solutions of (9.7) on I are contained in $a + £f(a)$. Let $x = a + y$ be a solution of (9.7). Then y satisfies on $[-p, p]$ the quasi-linear equation*

$$y + g(y) \in \beta, \tag{9.17}$$

with $g(y) = f(a+y) - f(a)$ and $\beta = -f(a) + A$.

Proof. By Corollary 9.2.7 there exists an interval $I := [a-p, a+p]$, with $p > £f(a)$, such that all solutions of (9.7) on I are contained in $a + £f(a)$, and $f[a-p, a+p] \subseteq (1+\oslash)f(\alpha)$. We have for $y \in [-p, p]$

$$y + g(y) = x - a + f(x) - f(a) \in a - a - f(a) + A = \beta.$$

Because f is S-selfneglecting, one has

$$g(\beta) = f(a - f(a) + A) - f(a) \subseteq \oslash(-f(a) + A) \subseteq \oslash\beta.$$

Also, β is zeroless, because $A < |f(\alpha)|$. Hence the equation (9.17) is quasi-linear on $[-p, p]$. □

If $g(\beta) \sqsubset A$, we do not need to make further steps, for by Corollary 9.2.6 the solution of (9.7) is β itself; hence the solution of the original equation (9.7) is given by $\alpha - f(\alpha) + A$. If $A < |g(\beta)|$, more terms are needed.

In practice the function g tends to be more complicated than f. Note however that it is given by a difference, and then we could try a Taylor expansion. If f is sufficiently often asymptotically differentiable, by the Substitution Theorem, we should take as many terms as needed to obtain that the remainder is contained in A.

Example 9.2.10 (Continuation of Examples 9.2.5 and 9.2.9). Consider the equation $x + 2\log(x) \in \xi + £$ of (9.15). The Expansion Theorem suggests the substitution $x = \xi + y$ of (9.16), with $y \in £\log(\xi)$. Then we have $y + 2\log(\xi + y) \in £$. The Expansion Theorem suggests to further reduce this equation to

$$y + 2\log(\xi + y) - 2\log(\xi) \in -2\log(\xi) + £. \tag{9.18}$$

Putting $g(y) := 2\log(\xi + y) - 2\log(\xi)$, we see that the equation (9.18) is quasi-linear indeed, for, the logarithm being S-selfneglecting, we obtain

$$\begin{aligned} g(-2\log(\xi) + £) &= 2\log(\xi - 2\log(\xi) + £) - 2\log(\xi) \\ &\subseteq \oslash(-2\log(\xi) + £). \end{aligned}$$

Applying a Taylor expansion of first order to $g(-2\log(\xi)+\pounds)$ we find a sharper estimate, in fact

$$\begin{aligned} g(-2\log(\xi)+\pounds) &= 2\log(\xi-2\log(\xi)+\pounds)-2\log(\xi) \\ &\subseteq 2\frac{-2\log(\xi)+\pounds}{\xi}(1+\oslash) \\ &\subseteq \oslash \sqsubset \pounds. \end{aligned}$$

Then by Corollary 9.2.6 the solution of (9.18) is given by its right-hand member $-2\log(\xi)+\pounds$. Hence the solution of the equation (9.14) is given by $\log(\omega)-2\log(\log(\omega))+\pounds$. By monotony, the solution of the original equation (9.9) for the remainder of the approximation of the exponential function by Euler's formula at positive values is given by the external interval $[0,\log(\omega)-2\log(\log(\omega))+\pounds)$.

The final theorem of this section gives the first terms of the expansion of the solution in the case of an S-asymptotically differentiable function.

Theorem 9.2.9 (Expansion for S-asymptotically differentiable functions). *Let $\alpha = a+A$ be an external number and f be internal and differentiable, defined at least on $\alpha+\pounds f(\alpha)$, S-asymptotically differentiable with $A<|f(\alpha)|<\alpha$, $f'(\alpha)\simeq 0$. Let g be defined by $g(y)=f(a+y)-f(a)$ and $\beta=-f(a)+A$. Consider the quasi-linear equation $x+f(x)=\alpha$ of (9.7), reduced to the equation $y+g(y)\in\beta$ of (9.17). Then*

1. *there exists an interval $I:=[a-p,a+p]$, with $p>\pounds f(a)$ such that all solutions of (9.7) on I are contained in $a+\pounds f(a)$, and the equation (9.17) is quasi-linear on I.*
2. *all solutions of (9.17) are contained in $-f(a)+\pounds f'(a)f(a)$.*
3. *(a) if $|f'(a)f(a)|<A$ the solution of (9.7) is given by $a-f(a)+A$.*
   (b) if $|f'(a)f(a)|>A$ all solutions of (9.7) are contained in $a-f(a)+f'(a)f(a)+\oslash f'(a)f(a)$.

Proof. 1. By Proposition 9.2.2 the function f is S-selfneglecting. Then by Corollary 9.2.7 there exists an interval $I:=[a-p,a+p]$, with $p>\pounds f(\alpha)$, such that all solutions of (9.7) on I are contained in $a+\pounds f(\alpha)$, and $f[a-p,a+p]\subseteq(1+\oslash)f(\alpha)$. By Theorem 9.2.8 the equation (9.17) is quasi-linear on I.

2. For $y\in -f(a)+\pounds f'(a)f(a)$ it holds that

$$\begin{aligned} g(y) &= f(a+y)-f(a)\subseteq f(a-f(a)+\pounds f'(a)f(a))-f(a) \\ &\subseteq -f'(a)f(a)+\pounds(f'(a))^2f(a)+\oslash f'(a)f(a) \\ &\subseteq -f'(a)f(a)+\oslash f'(a)f(a). \end{aligned} \tag{9.19}$$

Since $-f'(a)f(a)+\oslash f'(a)f(a)\sqsubset \pounds f'(a)f(a)$, by Theorem 9.2.4 all solutions of (9.17) on S are contained in $-f(a)+\pounds f'(a)f(a)$.

3a. We have $\pounds f'(a)f(a) \subseteq A$. Then by Part 2 all solutions of (9.17) on I are contained in $-f(a)+A$. Also $-f(a)+A$ satisfies equation (9.17), for

$$\begin{aligned}
-f(a)+A+g(-f(a)+A) &= -f(a)+A+f(a-f(a)+A)-f(a)\\
&\subseteq -f(a)+A+f(a-(1+\oslash)f(a))-f(a)\\
&\subseteq -f(a)+A+(1+\oslash)f'(a)f(a)\\
&= -f(a)+A.
\end{aligned}$$

Hence the solution of (9.7) is given by $a-f(a)+A$.

3b. Note that $A \subseteq \oslash f'(a)f(a)$. By (9.19) a solution y of (9.17) satisfies

$$y+g(y) = y-f'(a)f(a)+\varepsilon f'(a)f(a) \in -f(a)+A,$$

for some $\varepsilon \simeq 0$. Then

$$\begin{aligned}
y &\in -f(a)+f'(a)f(a)-\varepsilon f'(a)f(a)+A\\
&\subseteq -f(a)+f'(a)f(a)+\oslash f'(a)f(a)+A\\
&= -f(a)+f'(a)f(a)+\oslash f'(a)f(a).
\end{aligned}$$

Hence a solution of (9.7) is contained in $a-f(a)+f'(a)f(a)+\oslash f'(a)f(a)$. □

Example 9.2.11. We found that the solution of equation $x+2\log(x) \in \xi+\pounds$ of (9.15), was given by $x \in \xi-2\log(\xi)+\pounds$. We allow now smaller errors, and ask for the solution of

$$x+2\log(x) \in \xi+\frac{\pounds\log(\xi)}{\xi} \tag{9.20}$$

and

$$x+2\log(x) \in \xi+\frac{\pounds}{\xi}. \tag{9.21}$$

We use again the change of variable $x=\xi+y$ of (9.18). This reduces equation (9.20) to

$$y+2(\log(\xi+y)-\log(y)) \in -2\log(\xi)+\frac{\pounds\log(\xi)}{\xi}. \tag{9.22}$$

We put $f(x) := 2\log(x)$ and apply Theorem 9.2.9. As predicted by Part 1 of the theorem, this is a quasi-linear equation. By Part 2 of the theorem all solutions of (9.22) are contained in $-2\log(\xi)+\frac{\pounds\log(\xi)}{\xi}$, and $\xi-2\log(\xi)+\frac{\pounds\log(\xi)}{\xi}$ is the solution of (9.20) by Part 3a of Theorem 9.2.9.

As regards equation (9.21), because $\frac{\pounds}{\xi} < f'(\xi)f(\xi)$ we need at least one more term, and by Part 3b of Theorem 9.2.9 all solutions are of the form

$$x = \xi-2\log(\xi)+\frac{4\log(\xi)}{\xi}+z,$$

with $z \in \oslash\frac{\log(\xi)}{\xi}$. We put $\eta := \xi - 2\log(\xi) + \frac{4\log(\xi)}{\xi}$. Then

$$\begin{aligned} f(x) &= 2\log(\eta + z) \\ &= 2\log(\xi) + 2\log\left(1 - \frac{2\log(\xi)}{\xi} + \frac{4\log(\xi)}{\xi^2} + \frac{z}{\xi}\right) \\ &\subseteq 2\log(\xi) - \frac{4\log(\xi)}{\xi} + \frac{\oslash}{\xi}. \end{aligned} \tag{9.23}$$

Put $h(x) := f(x) - 2\log(\xi) + \frac{4\log(\xi)}{\xi}$. Then equation (9.21) becomes

$$x + h(x) = x + \left(f(x) - 2\log(\xi) + \frac{4\log(\xi)}{\xi}\right) \in \eta + \frac{\pounds}{\xi}.$$

Because $\frac{\pounds}{\xi} \subseteq \oslash\frac{\log(\xi)}{\xi}$, we derive from (9.23) that $h(\eta + \frac{\pounds}{\xi}) \subseteq \frac{\oslash}{\xi}$. So by the Substitution Theorem $x+h(x) \in \eta+\frac{\pounds}{\xi}$ and $x \in \eta+\frac{\pounds}{\xi}$ have the same solutions. Hence $\eta + \frac{\pounds}{\xi}$ solves $x + h(x) \in \eta + \frac{\pounds}{\xi}$, implying that $\xi - 2\log(\xi) + \frac{4\log(\xi)}{\xi} + \frac{\pounds}{\xi}$ solves (9.21).

Exercises

Exercise 9.2.1. *Assume that equation (9.3) is modified into* $x^2 + \varepsilon x^5 \in \frac{1+\oslash}{\varepsilon^4}$. *Show that all solutions must be positive and unlimited, in fact are contained in* $S := \frac{@}{\varepsilon}$. *Show that the solution is given by* $\xi = \frac{1+\oslash}{\varepsilon}$.

Exercise 9.2.2. *Let* $\omega \simeq +\infty$. *Solve the following equations:*

1. $\sqrt{\omega + x} - \sqrt{\omega} \in \oslash$.
2. $\log(\omega + x) - \log(\omega) \in 1 + \oslash$.

Exercise 9.2.3. *Formula (9.6) suggests that the arithmetic progression in the domain of validity of the Taylor approximation of Example 9.2.2 equals about* $\frac{1}{e}$. *Confirm this by solving*

$$R_{\omega-1}(x) \in \frac{e+\oslash}{e+1}.$$

Exercise 9.2.4. *Verify that standard iterated logarithms are also S-selfneglecting at every positive infinitely large external number* ξ.

Exercise 9.2.5. *Verify that the logarithm is S-asymptotically differentiable at all* $a > 0$ *and all the monomials* x^s, *for* st(s), *are S-asymptotically differentiable at all* $a \neq 0$. *Show that, as a consequence, all the derivatives of standard order of these functions are S-asymptotically differentiable at all* $a \neq 0$.

Exercise 9.2.6. *Solve the following equations:*

1. $x + \frac{1}{\sqrt{x}} = \omega + \frac{\pounds}{\omega^2}$.
2. $x + \sqrt[3]{x} \in \pounds e^{@\omega}$.

9.3 Matching principles

Theorem 5.1.13 on internal sequences may be seen as a form of "matching". Indeed the internal sequence exhibits two approximative behaviours: being outside a limit set, while approaching it, and being inside this limit set. Observe that we have only information on the sequence outside the limit set, and that the theorem adds information on the sequence within the limit set: it enters and stays there for some time. So matching consists of joining two behaviours: being outside the limit set and approaching it, and being inside the limit set and staying there some time. As a consequence the matching property may be expressed by strong convergence theorems: convergence to an external number implies strong convergence.

The property of strong convergence and matching may also be formulated for functions, and as such it is particularly relevant to solutions of singularly perturbed differential equations, which typically exhibit "slow" and "fast" behaviour. Usually one has separate information on slow behaviour of solutions close to a "slow curve", and fast behaviour when approaching it. These separate behaviours have to be matched into an overall description of the behaviour of a solution, showing how the transition from fast behaviour to slow behaviour is made, and vice-versa.

Classical tools are based on the Du Bois-Reymond Lemma [60] on persistence of convergence outside a sequence of convergent functions or Van Dyke's Principle [144] on equality of terms in two locally valid asymptotic expansions. The practice shows that sometimes a general result can be applied, such as the Extension Principle of Eckhaus [46], but mostly matching has to be proved case-by-case, with *ad hoc* methods.

Somewhat surprisingly, Nonstandard Analysis happened to yield a general method which is based on logic, *in casu* the Fehrele Principle mentioned in Chapter 1, expressing that no halo is a galaxy. In [136] it is shown that this principle of permanence enables matching of fast and slow behaviour in a fairly general context. As we will see, the strong convergence theorems of Chapter 5 enable still more general matching theorems.

We start with a matching theorem for sequences. A matching theorem for continuous functions is proved by relating convergence and strong convergence of sequences and functions. Both matching theorems were proved in [45]. We give two applications in the context of singular perturbations, on approaching the slow curve at two different orders of magnitude.

Theorem 9.3.1 (Matching for sequences)**.** *Let S be an initial segment of* $\mathbb{N}$. *Let $u : S \to \mathbb{R}$ be an internal sequence and $\alpha = a + A \in \mathbb{E}$, with $A \neq 0$. Assume $u_0 \notin \alpha$. Let*

$$C := \{n \in S : \forall m (0 \leq m \leq n \Rightarrow u_m \notin \alpha)\}. \tag{9.24}$$

If $u_{\restriction C}$ converges to α with respect to C, there exists $\nu \in S, \nu > C$ such that u converges strongly to α with respect to C on $\{0, ..., \nu\}$.

Proof. Without restriction of generality we may suppose that $\alpha = A$. We put $D = S \setminus C$.

We show first that C is external. Indeed, if C is internal it must be finite as a consequence of Theorem 5.1.11. Let $\mu := \min\{|u_m| : m \in C \wedge u_m \notin A\}$. Then $\mu > A$; hence $u_{\restriction C}$ does not converge to A with respect to C, a contradiction. Hence C is external. As a consequence $C \subset S$ and we conclude that $D \neq \emptyset$.

It follows from (9.24) that if A is a halo, the set C is a complement of a prehalo; hence being external it is a galaxy, and similarly, if A is a galaxy the set C is a halo. Then, knowing that $u_{\restriction C}$ converges to A with respect to C, by Theorem 5.1.13 there exists $\nu \in S, \nu > C$ such that u converges strongly to α with respect to C on $\{0, ..., \nu\}$. □

Let $S \subset \mathbb{R}$ be an initial segment, and $\alpha = a + A$ be an external number. Convergence to α (entering in every ε-neighbourhood of α, with $A < \varepsilon$) and strong convergence to α (becoming element of α) for functions are defined in Definitions 6.1.1 and 6.1.2 of Chapter 6. The next theorem on matching for functions is proved by relating strong convergence of sequences and functions, when the latter are continuous.

Theorem 9.3.2 (Matching for functions)**.** *Let S be an initial segment of $\mathbb{R}^+$. Let $f : S \to \mathbb{R}$ be an internal continuous function and $\alpha = a + A \in \mathbb{E}$, with A external. Assume $f(0) \notin \alpha$. Let*

$$C := \{y \in S : \forall x(0 \leq x \leq y \Rightarrow f(x) \notin \alpha)\}. \tag{9.25}$$

If $f_{\restriction C}$ converges to α with respect to C, there exists $s \in S$, $s > C$ such that f converges strongly to α with respect to C on $[0, s]$.

Proof. Without restriction of generality we may suppose that $\alpha = A$ and $f(0) > A$. We put $D = S \setminus C$.

We show first that $D \neq \emptyset$. If $C = S$, the internal set $f(S) \cap [0, f(0)]$ is contained in the external set $(A, f(0)]$. By Cauchy's Principle $f(S) \subset (A, f(0)]$. Hence there exists $\varepsilon \in \mathbb{R}$, $\varepsilon > A$ such that $f(x) \geq \varepsilon$ for all $x \in S$. Hence f does not converge to A, a contradiction. Hence $C \subset S$ and we conclude that $D \neq \emptyset$. Choose $d \in D$, then f is uniformly continuous on $[0, d]$. Hence there exists $\tau > 0$, $\tau \leq t$ such that $|f(v) - f(w)| \in A$ for all $v, w \in [0, d]$ with $|v - w| < \tau$. As a consequence $C + \tau = C$.

Let $S_\tau := \{n \in \mathbb{N} : n\tau \in S\}$ and define $u_\tau : S_\tau \to \mathbb{R}$ by $u_\tau(n) = f(\tau n)$. By Proposition 6.1.14 the local sequence $u_{\tau\restriction C/\tau}$ converges to A with respect to $\frac{C}{\tau}$. It follows from (9.25) and the continuity of f that C is a galaxy if A is a halo, and C is a halo if A is a galaxy. Then by Theorem 5.1.13 there exists $\nu \in \frac{D}{\tau}$ such that the sequence u_τ converges strongly to A on $\{0, ..., \nu\}$ with respect to $\frac{C}{\tau}$. Let $x \in [0, \nu\tau] \cap D$. Let $n \in \{0, ..., \nu\}$ be minimal such that $x \leq n\tau$. Then $|f(x)| = |f(x) - f(n\tau)| + |u_\tau(n)| \in A + A = A$. Hence f converges strongly to A on $[0, \nu\tau]$ with respect to C. □

As an application, we present here two examples of matching in the setting of singular perturbations. We recall here some notions and properties of Chapter 2 in a more formal setting.

Definition 9.3.1. *Let $D \subseteq \mathbb{R}^2$ be a standard open set. Let $f : D \to \mathbb{R}$ be internal, continuously differentiable (of class C^1) and limited, and S-continuously S-differentiable at near-standard points of D (of class S^1, see [35]). Let $\varepsilon > 0$, $\varepsilon \simeq 0$. Then a differential equation of the form*

$$\varepsilon \frac{dx}{dt} = f(t, x) \tag{9.26}$$

is called a singular perturbation.

Let us write $\frac{dx}{dt} = \frac{f(t,x)}{\varepsilon}$. We observe two types of behaviour of solutions $\lambda(t)$: *fast* behaviour at points where the derivative $\frac{f(t,\lambda(t))}{\varepsilon}$ is unlimited, and *slow* behaviour at points where the derivative $\frac{f(t,\lambda(t))}{\varepsilon}$ is limited.

Definition 9.3.2. *Consider a singular perturbation as given by* (9.26). *We define the* slow curve L *by* $L := {}^{\circ}\{(t,x) \in D : f(t,x) \simeq 0\}$.

Typically the slow curve consists of one or more standard differentiable curves; to fix ideas we consider the case of only one differentiable curve, say g, defined on a standard interval $[a, b]$ with $a < b$.

Definition 9.3.3. *The slow curve g is said to be* attractive *if $f_2'(t, g(t)) \lessapprox 0$ for $a \leq t \leq b$, and* repulsive *if $f_2'(t, g(t)) \gtrapprox 0$ for $a \leq t \leq b$.*

A standard point (t_0, x_0) such that for some $t \in [a, b]$ it holds that $g(t) \simeq x_0$ and $f_2'(t, g(t)) \simeq 0$ is called a singular point.

For simplicity we suppose that g is identically zero on $[a, b]$ as in Section 2.3. If $f_2'(t, g(t)) \lessapprox 0$, the attractivity of the slow curve may be explained as follows. It was shown that there exists standard $c \in \mathbb{R}$, $c > 0$ such that for all t with $a \leq t \leq b$ and $|x| \leq c$

$$\begin{cases} f(t,x) \lessapprox 0, & x \gtrapprox 0 \\ f(t,x) \simeq 0, & x \simeq 0 \\ f(t,x) \gtrapprox 0, & x \lessapprox 0. \end{cases} \tag{9.27}$$

A solution λ at a time t with $a \leq t \leq b$ and $0 \lessapprox \lambda(t) \leq c$ has an infinitely large negative derivative, so it will be moving very fast towards the slow curve. Similarly, if $-c \leq \lambda(t) \lessapprox 0$ has an infinitely large positive derivative, hence again it moves very fast towards the slow curve. Hence trajectories approach g at points at appreciable distance to g, less than c.

If $f_2'(t, g(t)) \gtrapprox 0$, the repulsivity of the slow curve is shown similarly.

On the halo $H := [a, b] \times \oslash$ the derivative of the solution is no longer of order $\frac{1}{\varepsilon}$. Observe that the galaxy $G := \{(t,x) \in [a,b] \times \mathbb{R} : f(t,x) \in \pounds\varepsilon\}$, where the derivative of solutions is limited, is strictly included in H. So $H \setminus G$

may be seen as an intermediate region, where the transition between fast behaviour and slow behaviour takes place.

All these regions are external. To obtain a global description of the behaviour of a solution, which is internal, the behaviours of the solution on those external regions need to be matched.

The next theorem gives a matching result: solutions approaching the slow curve become infinitely close to the slow curve.

Theorem 9.3.3. *Consider the equation (9.26), with $\varepsilon \simeq 0$, $\varepsilon > 0$ and f of class $C^1 \cap S^1$. Let a, b be limited and such that $a \lessapprox b$ and let $g : [a, b] \to \mathbb{R}$, defined by $g(t) = 0$, be an attractive slow curve. Let λ be a solution such that $\lambda(a) \not\simeq 0$. Then there exists $\tau > a$ such that $\lambda(\tau) \simeq 0$.*

Proof. Without loss of generality we may suppose that $a = 0$, that the value of c such that (9.27) holds on $[a, b] \times [-c, c]$ is equal to 1, and that $\lambda(0) = 1$. Let

$$C := \{t \geq 0 : \forall s(0 \leq s \leq t \wedge \lambda(s) \gtrapprox 0)\}.$$

Since possible singular points only may occur at infinitesimal values, as long as t is such that $\lambda(s) > \oslash$ for all s with $0 \leq t$, by the Existence Theorem for Differential Equations $\lambda(s)$ is defined up to t. Then by the Mean Value Theorem there exists $\bar{t}$ such that $0 \leq \bar{t} \leq t$ and

$$\lambda(t) - 1 = \lambda(t) - \lambda(0) = \lambda'\left(\bar{t}\right) t = \frac{f\left(\bar{t}, \lambda\left(\bar{t}\right)\right)}{\varepsilon} t. \tag{9.28}$$

We show first that λ converges to $\oslash$ with respect to C. Let $0 \lessapprox \eta \leq 1$. Suppose that $\lambda(t) > \eta$ for all $t \in [0, b]$. Then it follows from (9.28) that $\lambda(t) - 1$; hence also $\lambda(t)$, is negative unlimited for all appreciable $t \in [0, b]$, a contradiction. Hence there exists $t_\eta \in [0, b]$, such that $\lambda(t_\eta) = \eta$. Because λ remains decreasing for times $t \in C$, one has $\lambda(t) < \eta$ as long as $t \in C, t > t_\eta$. Hence λ converges to $\oslash$ with respect to C.

By Theorem 9.3.2 the solution λ strongly converges to $\oslash$ with respect to C on some interval $[0, \tau]$. □

A well-known Theorem of Tikhonov [126] (see [81] for a nonstandard proof) expresses that a solution remains close to the slow curve, at least until meeting a singular point $(s, 0)$, where $f_2'(s, 0) = 0$. Typically $f_2'(t, 0) > 0$ for $t > s$, so the slow curve turns repulsive, and nearby solutions may leave. Also, solutions become closer to the slow curve; in fact the distance becomes of order $\varepsilon \pounds$; this matching result is a refinement of Theorem 9.3.3, and is also proved by strong convergence.

Theorem 9.3.4. *Consider the equation (9.26), with $\varepsilon \simeq 0$, $\varepsilon > 0$ and f of class $C^1 \cap S^1$. Let a, b be limited with $a \lessapprox b$ and let $g : [a, b] \to \mathbb{R}$, defined by $g(t) = 0$, be an attractive slow curve. Let λ be a solution such that $\lambda(a) \not\simeq 0$. Then there exists $\tau > a$ such that $\lambda(\tau) \in \varepsilon \pounds$.*

Proof. Without loss of generality we may suppose that $a = 0$, that the value of c such that (9.27) holds on $[a, b] \times [-c, c]$ is equal to 1, and that $\lambda(0) = 1$. Let

$$C := \{t \geq 0 : \forall s(0 \leq s \leq t \wedge \lambda(s) > \varepsilon\pounds)\}.$$

Noting that $\frac{f(t,\lambda(t))}{\varepsilon}$ is negative unlimited for all $t \in C$, the proof that λ converges to $\varepsilon\pounds$ with respect to C is similar to the argument used in the proof of Theorem 9.3.3. By Theorem 9.3.2 the convergence is strong. □

Exercises

Exercise 9.3.1. *Let $\varepsilon \simeq 0$, $\varepsilon > 0$. Consider the singular perturbation*

$$\varepsilon \frac{dx}{dt} = x^2 - t.$$

1. *Determine the slow curve, indicating the attractive part and the repulsive part.*

2. *Let λ be a solution such that $\lambda(1) = 0$. Describe the behaviour of λ for positive appreciable t.*

Exercise 9.3.2. *Let $\varepsilon \simeq 0$, $\varepsilon > 0$. Consider the singular perturbation*

$$\varepsilon \frac{dx}{dt} = (x^2 - t)^2.$$

1. *Determine the slow curve, and show that it is neither attractive nor repulsive.*

2. *Show that the set where $\frac{dx}{dt}$ is limited for positive appreciable t is given by two strips around $\sqrt{t}$ of width $\pounds\sqrt{\varepsilon}$.*

3. *Put $z = \sqrt{t} + \sqrt{\varepsilon}z$. Show that this leads to a singular perturbation for z, with attractive slow curve $z = -\frac{1}{t^{3/4}}$ and repulsive slow curve $z = \frac{1}{t^{3/4}}$.*

4. *Show that if we put $z = \sqrt{t} - \sqrt{\varepsilon}z$ we get a singular perturbation without slow curve.*

5. *Let λ be a solution such that $\lambda(1) = 0$. Describe the behaviour of λ for positive appreciable t.*

Exercise 9.3.3. *(See Example 2.3.2 and Exercise 2.3.2) Let $\varepsilon \simeq 0$, $\varepsilon > 0$. Consider the singular perturbation*

$$\varepsilon \frac{dx}{dt} = tx. \tag{9.29}$$

Answer the following questions without solving the equation.

1. *Determine the slow curve, and indicate the attractive part and the repulsive part.*

2. *Show that the constant function $c \equiv 0$ is a solution.*

3. *Let λ be a solution such that $\lambda(-1) = 1$. Describe the behaviour of λ for positive appreciable t.*

4. *Show that for $t \simeq 0$, $t > 0$ the solution λ is decreasing, without ever reaching the value 0.*

5. *Using symmetry, describe the behaviour of λ up to $t = 1$. In particular show that $\lambda(t)$ is infinitely close to the repulsive part of the slow curve for $0 \lessapprox t \lessapprox 1$ (like the zero solution it is a* canard solution, *see next subsection).*

9.4 An external singular perturbation with canard solutions

9.4.1 External differentiable equations and their solutions

Our context is the study of differential equations of the form

$$\frac{dx}{dt} \in F(t,x), \tag{9.30}$$

where F is an internally representable flexible function of class C^1 defined on some appropriate internal domain $D \subseteq \mathbb{R}^2$. This means that $\Gamma(F)$ is the union of the graphs of internal functions of class C^1.

We define first a solution for equation (9.30), also in the form of a union of the graphs of internal solutions.

Definition 9.4.1. *Let F be an internally representable flexible function of class C^1 defined on an internal domain $D \subseteq \mathbb{R}^2$. Let $\alpha_0 \in \mathbb{E}$ such that $(0, \alpha_0) \subset D$. The* solution Φ *of (9.30) with initial condition $(0, \alpha_0)$ is defined by*

$$\Gamma(\Phi) = \bigcup \{\phi : \exists a_0 \in \alpha_0, \exists f \in F, f \text{ of class } C^1 \wedge \phi(0) = a_0$$
$$\wedge \frac{d\phi}{dt}(t,x) = f(t,x), \text{ as long as } (t,x) \in D\}.$$

Flexible singular perturbations *are of the form*

$$\varepsilon \frac{dx}{dt} \in F(t,x), \tag{9.31}$$

where ε is an infinitesimal positive real number and F is an internally representable flexible function which is limited at least on a standard open subset of D. If there is no ambiguity, we may also write equation (9.31) *using the equality symbol instead of* $\in$.

We study a particular flexible singular perturbation for canard solutions, which are a particular kind of solution losing stability, while remaining bounded. This can be seen as a study of how canard solutions behave under small modifications in the differential equation. Already Callot [8] showed that generically these modifications can only be exponentially small. Some stochastic modifications have also been considered by Lobry (unpublished); the latter suggested (private communication) to retry in the context of the most simple singular perturbation with canard solutions, introducing modifications in terms of small external numbers.

9.4.2 The external Riccati-Hermite equation

Definition 9.4.2. *Consider a singular perturbation as given by Definition 9.3.1. Let g be a slow curve, which has an attractive and a repulsive part; we assume that g is a real function defined at least on some interval of appreciable length. Let λ be a solution. Let $S := \{t \in \mathbb{R} : \lambda(t) \simeq g(t),\ g \text{ attractive at } t\}$ and $T := \{t \in \mathbb{R} : \lambda(t) \simeq g(t),\ g \text{ repulsive at } t\}$. Then λ is called a* canard *if S and T are both external intervals which contain internal intervals of appreciable length and whenever $s \in S$ and $t \in T$, it holds that $s < t$.*

An example of canard is the solution λ of the equation $\varepsilon\frac{dx}{dt} = tx$ given in Example 2.3.2. This is a special case of the equation

$$\varepsilon\frac{dx}{dt} = tx + a, \tag{9.32}$$

which was studied from the point-of-view of canards by J.-L. Callot [24]. He called equation (9.32) the *equation of Riccati-Hermite*, for it can be transformed in both a Riccati and a Hermite equation. It is the simplest class of equations exhibiting canard behaviour. This is most obvious for equation (9.29), corresponding to the case where $a = 0$, which is symmetrical with respect to 0, and even is easily solved in terms of elementary functions. Callot showed that the canard phenomenon persists for $a \in \pounds e^{\frac{-@}{\varepsilon}}$, but disappears if a is no longer exponentially small with respect to ε, and also that this is typical for the canard phenomenon ("canards have a short life").

This suggests studying the flexible differential equation

$$\varepsilon\frac{dx}{dt} \in tx + \pounds e^{\frac{-@}{\varepsilon}}. \tag{9.33}$$

In the interpretation of (9.30) this corresponds to the study of the equations

$$\varepsilon\frac{dx}{dt} = tx + \phi(t), \tag{9.34}$$

where the functions ϕ are internal and continuously differentiable such that $\phi(t) \in \pounds e^{\frac{-@}{\varepsilon}}$ for all limited $t \in \mathbb{R}$. Let us denote by Φ the set of all such functions ϕ, and let for each $\phi \in \Phi$ the solution of (9.34) be denoted by x_ϕ. According to Definition 9.4.1 the solution ξ of (9.33) is internally representable and given by

$$\xi = \bigcup_{\phi \in \Phi} x_\phi. \tag{9.35}$$

One has $\pounds e^{\frac{-@}{\varepsilon}} = \pounds \sqrt{\varepsilon} e^{\frac{-@}{\varepsilon}}$. It follows from Callot's arguments that the latter representation is somewhat more convenient, and so we will study

$$\varepsilon \frac{dx}{dt} \in tx + \pounds \sqrt{\varepsilon} e^{-\frac{A}{\varepsilon}}, \tag{9.36}$$

with $A > 0$ fixed appreciable. Otherwise said, for these values of the parameter we study the equations

$$\varepsilon \frac{dx}{dt} = tx + f(t)\sqrt{\varepsilon} e^{-\frac{A}{\varepsilon}}, \tag{9.37}$$

with f a continuously differentiable real function, taking always limited values. The set of such functions f will be denoted by F. We will see that the set of solutions of (9.37) is equal to the set of solutions of the equation

$$\varepsilon \frac{dx}{dt} = tx + L\sqrt{\varepsilon} e^{-\frac{A}{\varepsilon}},$$

with $L \in \mathbb{R}$ limited, so it suffices to consider only constant representatives of the flexible function defining equation (9.36). Still it is interesting to consider representatives of flexible functions of different types. Below we consider S-continuous representatives and continuously differentiable functions which may have more rapid fluctuations.

9.4.3 Solving the external Riccati-Hermite equation

Theorem 9.4.1. *Let $A > 0$ be appreciable. Let $t_0 < 0$ be appreciable. Let ξ be the solution of* (9.36) *with limited initial condition $\xi(0) = x_0 \in \mathbb{R}$. Then*

$$\xi(t) = \begin{cases} x_0 e^{\frac{t^2 - t_0^2}{2\varepsilon}} + \pounds \frac{e^{-\frac{A}{\varepsilon}}}{\sqrt{\varepsilon} t}, & t < \pounds\sqrt{\varepsilon} \\ x_0 e^{\frac{t^2 - t_0^2}{2\varepsilon}} + \pounds e^{\frac{t^2 - 2A}{2\varepsilon}}, & t \geq \pounds\sqrt{\varepsilon}. \end{cases} \tag{9.38}$$

To prove the approximations of (9.38), we solve the linear equation (9.37) by the usual Lagrange method, and derive asymptotic approximations of integrals and other terms figuring in the solution.

Proposition 9.4.2. *Let $A > 0$, $t_0 < 0$ be appreciable and $x_0 \in \mathbb{R}$. Let x be the solution of* (9.37) *with initial condition $x(0) = x_0$. Then*

$$x(t) = x_0 e^{\frac{t^2 - t_0^2}{2\varepsilon}} + \frac{e^{\frac{t^2 - 2A}{2\varepsilon}}}{\sqrt{\varepsilon}} \int_{t_0}^{t} f(s) e^{-\frac{s^2}{2\varepsilon}} ds. \tag{9.39}$$

Proof. Applying the Lagrange method, we derive that the solutions starting at $t = t_0$ take the form

$$x(t) = \frac{e^{\frac{t^2-2A}{2\varepsilon}}}{\sqrt{\varepsilon}} \int_{t_0}^{t} f(s)e^{-\frac{s^2}{2\varepsilon}} ds + C,$$

where C is a constant. By adding the initial condition $x(t_0) = x_0$, we derive (9.39). □

Remark 9.4.3. To prove Theorem 9.4.1 and describe the canard behaviour of entering the halo of the slow curve when it is attractive, and leaving it appreciably beyond the time it turns repulsive, for simplicity we put $t_0 = -1$ and $x_0 = x(-1) = 1$; the general case is very similar. Under these conditions (9.39) becomes

$$x(t) = e^{\frac{t^2-1}{2\varepsilon}} + \frac{e^{\frac{t^2-2A}{2\varepsilon}}}{\sqrt{\varepsilon}} \int_{-1}^{t} f(s)e^{-\frac{s^2}{2\varepsilon}} ds. \tag{9.40}$$

Definition 9.4.3. *In the context of* (9.40) *we define functions* γ, δ_f, I_f *and* N *by*

$\gamma(t) := e^{\frac{t^2-1}{2\varepsilon}}$.

$\delta_f(t) := \frac{e^{\frac{t^2-2A}{2\varepsilon}}}{\sqrt{\varepsilon}} \int_{-1}^{t} f(s)e^{-\frac{s^2}{2\varepsilon}} ds.$

$I_f(t) := \int_{-1}^{t} f(s)e^{-\frac{s^2}{2\varepsilon}} ds.$

$N(t) := \int_{-1}^{t} e^{-\frac{s^2}{2\varepsilon}} ds.$

We will also write

$I(t) := \int_{-1}^{t} \pounds e^{-\frac{s^2}{2\varepsilon}} ds.$

$\delta(t) := \frac{e^{\frac{t^2-2A}{2\varepsilon}}}{\sqrt{\varepsilon}} I(t),$

where we assume that the external integral $I(t)$ is an inner-integral in the sense of Chapter 7 (Definition 7.2.2).

Observe that for $f \equiv 0$, we obtain $x(t) = \gamma(t)$, which is the solution of the homogeneous equation $\varepsilon \frac{dx}{dt} = tx$.

The function N of Definition 9.4.3 is related to the standard normal distribution $\mathcal{N}$ (see Exercise 1.4.6). Being an inner-integral, $I(t) = \bigcup_{f \in F} I_f(t)$. However in Lemma 9.4.5 below we show that one simply has

$$I(t) = \bigcup_{L \in \pounds} I_L(t) = \pounds N(t).$$

To derive (9.40), we need to determine the external integral I. This will follow from an asymptotic approximation of the integral N on successive domains, as given in the next lemma.

Lemma 9.4.4. *One has*

$$N(t) \in \begin{cases} -\frac{\varepsilon}{t}e^{-\frac{t^2}{2\varepsilon}}\left(1-e^{\frac{t+1}{\varepsilon}}+\oslash\right), & t \in -1+\pounds\varepsilon \\ -\frac{\varepsilon}{t}e^{-\frac{t^2}{2\varepsilon}}(1+\oslash), & -1+\pounds\varepsilon < t < \pounds\sqrt{\varepsilon} \\ \sqrt{2\pi\varepsilon}\left(\mathcal{N}\left(\frac{t}{\sqrt{\varepsilon}}\right)+\oslash\right), & t \in \pounds\sqrt{\varepsilon} \\ \sqrt{2\pi\varepsilon}(1+\oslash), & t > \pounds\sqrt{\varepsilon}. \end{cases}$$

Proof. For the first and second formulas, we put $s = t+\varepsilon\frac{u}{t}$. Then

$$N(t) = \int_{-1}^{t} e^{-\frac{s^2}{2\varepsilon}}ds = -\frac{\varepsilon}{t}e^{-\frac{t^2}{2\varepsilon}}\int_0^{-\frac{t+t^2}{\varepsilon}} e^{-u-\frac{\varepsilon u^2}{2t^2}}du.$$

If $t \in -1+\pounds^{+}\varepsilon$, we have $-\frac{t+t^2}{\varepsilon} = -\frac{t(t+1)}{\varepsilon} \simeq \frac{t+1}{\varepsilon}$. Then it follows from the Dominated Approximation lemma (Lemma 1.5.8) that

$$\int_0^{-\frac{t+t^2}{\varepsilon}} e^{-u-\frac{\varepsilon u^2}{2t^2}}du \simeq \int_0^{\frac{t+1}{\varepsilon}} e^{-u}du \simeq 1-e^{\frac{t+1}{\varepsilon}}.$$

This implies the first formula.

As for the second formula, observe that

$$-\frac{t+t^2}{\varepsilon} \in \pounds \Leftrightarrow -t(t+1) \in \pounds\varepsilon \Leftrightarrow t \in -1+\pounds\varepsilon \vee t \in \pounds\varepsilon.$$

So $-\frac{t+t^2}{\varepsilon} > 0$ is unlimited for $-1+\pounds\varepsilon < t < \pounds\sqrt{\varepsilon}$. Also $-\frac{\varepsilon u^2}{2t^2} < 0$ for all u, while $-\frac{\varepsilon u^2}{2t^2} \simeq 0$ for limited u. Then, applying the external Rule of Chasles of Theorem 7.1.1

$$\int_0^{-\frac{t+t^2}{\varepsilon}} e^{-u-\frac{\varepsilon u^2}{2t^2}}du \in \int_0^{\pounds} e^{(-u+\oslash)}du + \int_{\not\infty}^{-\frac{t+t^2}{\varepsilon}} e^{-u-\frac{\varepsilon u^2}{2t^2}}du.$$

As for the first integral,

$$\begin{aligned}\int_0^{\pounds} e^{(-u+\oslash)}du &= \int_0^{\pounds}(1+\oslash)e^{-u}du = (1+\oslash)\int_0^{\pounds} e^{-u}du \\ &= (1+\oslash)(1-e^{-\not\infty}) = (1+\oslash)(1+\oslash) = 1+\oslash.\end{aligned}$$

For the second integral we have

$$\int_{\not\infty}^{-\frac{t+t^2}{\varepsilon}} e^{-u-\frac{\varepsilon u^2}{2t^2}}du \leq \int_{\not\infty}^{-\frac{t+t^2}{\varepsilon}} e^{-u}du \in e^{-\not\infty} - e^{-\frac{t+t^2}{\varepsilon}} \subseteq \oslash.$$

Hence

$$\int_0^{-\frac{t+t^2}{\varepsilon}} e^{-u-\frac{\varepsilon u^2}{2t^2}}du \in 1+\oslash+\oslash = 1+\oslash,$$

which implies the second formula.

To prove the third and fourth formulas, we put $s = \sqrt{\varepsilon}v$. Then

$$N(t) = \int_{-1}^{t} e^{-\frac{s^2}{2\varepsilon}} ds = \sqrt{\varepsilon} \int_{-\frac{1}{\sqrt{\varepsilon}}}^{\frac{t}{\sqrt{\varepsilon}}} e^{-\frac{v^2}{2}} dv.$$

For $t \in \pounds\sqrt{\varepsilon}$ we have the approximation

$$\int_{-\frac{1}{\sqrt{\varepsilon}}}^{\frac{t}{\sqrt{\varepsilon}}} e^{-\frac{v^2}{2}} dv \simeq \int_{-\infty}^{\frac{t}{\sqrt{\varepsilon}}} e^{-\frac{v^2}{2}} dv = \sqrt{2\pi}\mathcal{N}\left(\frac{t}{\sqrt{\varepsilon}}\right);$$

this implies the third formula. For $t > \pounds\sqrt{\varepsilon}$ we obtain

$$\int_{-\frac{1}{\sqrt{\varepsilon}}}^{\frac{t}{\sqrt{\varepsilon}}} e^{-\frac{v^2}{2}} dv \simeq \int_{-\infty}^{\infty} e^{-\frac{v^2}{2}} dv = \sqrt{2\pi}.$$

Hence also the fourth formula is proved. □

The following lemma expresses the external integral I in terms of N. It will lead to the asymptotic approximation of Lemma 9.4.6, and is also the principal tool in justifying that the set of solutions of the external Riccati-Hermite equation is already determined by constant representative functions.

Lemma 9.4.5. *For each $t \geq -1$ one has*

$$I(t) = \bigcup_{L \in \pounds} \int_{-1}^{t} Le^{-\frac{s^2}{2\varepsilon}} ds = \pounds N(t).$$

Proof. Each function f occurring in (9.37) is bounded in absolute value by some positive limited number L. Then for all $t \geq -1$ we have $|I_f(t)| \leq LN(t)$. Also for $f \equiv L$, we have $I_f(t) = L$. Hence $I(t) = \pounds N(t)$. □

Lemma 9.4.6. *One has*

$$I(t) = \begin{cases} \pounds \frac{\varepsilon}{t} e^{-\frac{t^2}{2\varepsilon}}, & -1 \leq t \leq \pounds\sqrt{\varepsilon} \\ \pounds\sqrt{\varepsilon}, & t > \pounds\sqrt{\varepsilon}. \end{cases}$$

Proof. The result is a consequence of Lemmas 9.4.5 and 9.4.4. □

Proposition 9.4.7. *Let $A > 0$ be appreciable. Let $t_0 < 0$ be appreciable. Let ξ be the solution of* (9.36) *with limited initial condition $\xi(-1) = 1$. Then*

$$\xi(t) = \begin{cases} e^{\frac{t^2-1}{2\varepsilon}} + \pounds \frac{e^{-\frac{A}{\varepsilon}}}{\sqrt{\varepsilon}t}, & t \leq \pounds\sqrt{\varepsilon} \\ e^{\frac{t^2-1}{2\varepsilon}} + \pounds e^{\frac{t^2-2A}{2\varepsilon}}, & t > \pounds\sqrt{\varepsilon}. \end{cases} \tag{9.41}$$

Proof. It follows from (9.35) and (9.40) that for all $t \geq -1$

$$\xi(t) = e^{\frac{t^2-1}{2\varepsilon}} + \frac{e^{\frac{t^2-2A}{2\varepsilon}}}{\sqrt{\varepsilon}} I(t).$$

Then by Lemma 9.4.6, for $t \leq £\sqrt{\varepsilon}$ we have

$$\xi(t) = e^{\frac{t^2-1}{2\varepsilon}} + \frac{e^{\frac{t^2-2A}{2\varepsilon}}}{\sqrt{\varepsilon}} \cdot £\frac{\varepsilon}{t} e^{-\frac{t^2}{2\varepsilon}} = e^{\frac{t^2-1}{2\varepsilon}} + £\frac{e^{-\frac{A}{\varepsilon}}}{\sqrt{\varepsilon}t},$$

and for $t > £\sqrt{\varepsilon}$ it holds that

$$\xi(t) = e^{\frac{t^2-1}{2\varepsilon}} + \frac{e^{\frac{t^2-2A}{2\varepsilon}}}{\sqrt{\varepsilon}} \cdot £\sqrt{\varepsilon} = e^{\frac{t^2-1}{2\varepsilon}} + £e^{\frac{t^2-2A}{2\varepsilon}}. \qquad \square$$

As already said in Remark 9.4.3, the proof of Theorem 9.4.1 is based on similar calculations and approximations.

9.4.4 Description of the canard behaviour

For simplicity we take again $t_0 = -1$ and $x_0 = -1$. We write the solution ξ of (9.36) with initial condition $\xi(-1) = 1$ in the form of $\xi = \gamma + \delta$, with γ and δ given by Definition 9.4.3, and use formula (9.41).

We consider first the term $\gamma(t) = e^{\frac{t^2-1}{2\varepsilon}}$. The function γ is symmetrical in time with respect to 0. It enters the halo of the slow curve at times infinitely close to -1, and leaves the slow curve at times infinitely close to 1.

To be more precise, let $t = -1 + \varepsilon u$ with u limited. Then

$$\gamma(-1 + \varepsilon u) = e^{-u + \frac{u^2}{2\varepsilon}} \simeq e^{-u}.$$

So γ converges exponentially to the halo of 0 on the ε-galaxy of -1, and by Theorem 9.3.3 it reaches the halo of 0 immediately afterwards; in fact it has a downward jump of thickness $£\varepsilon$. It continues decreasing, and $\gamma(t) = e^{\frac{t^2-1}{2\varepsilon}}$ becomes exponentially small if $t-1$ becomes appreciable. For $t \geq 0$ we observe the inverse behaviour. The function γ leaves the halo of 0, when $t = 1 + \varepsilon u$ with u limited, with the exponential approximation

$$\gamma(1 + \varepsilon u) = e^{u + \frac{u^2}{2\varepsilon}} \simeq e^{u}. \tag{9.42}$$

So it has here an upward jump of thickness $£\varepsilon$.

We now turn to the term $\delta(t)$. Observe that $\delta(t) = £\frac{e^{-\frac{A}{\varepsilon}}}{\sqrt{\varepsilon}t}$ is exponentially small for $t \leq £\sqrt{\varepsilon}$, so ξ behaves like γ for $t \leq £\sqrt{\varepsilon}$, as described above. For $t > £\sqrt{\varepsilon}$ it holds that $\delta(t) = £e^{\frac{t^2-2A}{2\varepsilon}}$. Hence δ is increasing, but remains exponentially small for $t \underset{\not\approx}{<} \sqrt{2A}$.

We distinguish three cases: (i) $A > \frac{1}{2} + £\varepsilon$, (ii) $A < \frac{1}{2} + £\varepsilon$ and (iii) $A \in \frac{1}{2} + £\varepsilon$.

(i) For $£\sqrt{\varepsilon} < t \leq 1 + £\varepsilon$ one has $\delta(t) \subseteq \oslash$. Hence $\xi(t) \subseteq \gamma(t) + \oslash = e^{\frac{t^2-1}{2\varepsilon}} + \oslash$ for these values of t. Like γ, the canard ξ leaves the slow curve at $1 + £\varepsilon$, with thickness of jump $£\varepsilon$.

(ii) For $t > £\sqrt{\varepsilon}$ it holds that $e^{\frac{t^2-1}{2\varepsilon}} \in \oslash e^{\frac{t^2-2A}{2\varepsilon}}$, and we have

$$\xi(t) = \delta(t) = £e^{\frac{t^2-2A}{2\varepsilon}}.$$

For $t = \sqrt{2A} + \varepsilon v$ and v limited we have

$$e^{\frac{t^2-2A}{2\varepsilon}} = e^{\sqrt{2A}v + \frac{\varepsilon v^2}{2}} \simeq e^{\sqrt{2A}v}.$$

So

$$\xi(\sqrt{2A} + \varepsilon v) = £e^{\sqrt{2A}v} = £, \tag{9.43}$$

and the canard will take appreciable values already at $t = \sqrt{2A} + \varepsilon £$, which is earlier than at $t = 1$.

(iii) For $t = 1 + \varepsilon u$ with u limited we see that the canard becomes appreciable again for $t \in 1 + \varepsilon £$, for one has, using (9.42),

$$\xi(1 + \varepsilon u) = \gamma(1 + \varepsilon u) + \delta(1 + \varepsilon u) = e^u + £e^{2(u-a)} = £. \tag{9.44}$$

We make two further remarks with respect to the cases (ii) and (iii).

Firstly, formulas (9.43) and (9.44) show that not all values of the canard are appreciable. Indeed, there are representatives x of ξ which do not leave the slow curve at $t = \sqrt{2A} + \varepsilon £$. Constant representatives are determined by their initial condition x_0, and when they do not jump at $t = \sqrt{2A} + \varepsilon £$ one has $x_0 \simeq 0$. This is an exceptional case, the set of infinitesimals being much smaller than the set of appreciables. One may use the *Loeb measure* $\mathcal{L}$ [82], corresponding to the Lebesgue measure, to express this clearly. Indeed, the Loeb measure of many external sets is well defined, and $\mathcal{L}(\oslash) = 0$, while the Loeb measure of @ is total.

Secondly, formulas (9.43) and (9.44) do not discriminate positive and negative values, so we cannot specify whether the canards jump upwards or downwards. For constant representative functions we clearly have an upward jump for $x_0 \gtrsim 0$ and a downward jump for $x_0 \lesssim 0$. We will see in the next subsection that for non-constant representatives only the behaviour infinitely close to the singular point counts.

9.4.5 Influence of the singular point on the localization of canards

We present two theorems on representatives f of the flexible function defining equation (9.36), which indicate that the time period during which the canards jump depends on the behaviour of f close to the singular point $(0, 0)$.

Theorem 9.4.8 concerns S-continuous representatives and Theorem 9.4.10, which is more general, considers continuously differentiable functions which may have more rapid fluctuations, but admit a local mean close to $t = 0$ (which is nearly equal to $f(0)$ if f is S-continuous). We will see that not the initial condition but the behaviour of the function close to 0 is decisive for the direction in which the canard is leaving the slow curve.

Theorem 9.4.8. *Consider equation* (9.37), *where $A > 0$ is appreciable and f is limited and S-continuous for $t \geq t_0$. Let $A > 0$ be appreciable. Let $\overline{x}$ be the solution with initial condition $\overline{x}(t_0) = x_0$, where $t_0 < 0$ is appreciable and x_0 is limited. Then*

1. $\overline{x}(t) = x_0 e^{\frac{t^2 - t_0^2}{2\varepsilon}} + \left(1 - e^{\frac{t - t_0}{\varepsilon}} + \oslash\right) f(t) \frac{e^{-\frac{A}{\varepsilon}}}{\sqrt{\varepsilon} t}$, *if* $t \in -t_0 + \pounds\varepsilon$.

2. $\overline{x}(t) = x_0 e^{\frac{t^2 - t_0^2}{2\varepsilon}} + (1 + \oslash) f(t) \frac{e^{-\frac{A}{\varepsilon}}}{\sqrt{\varepsilon} t}$, *if* $-t_0 + \pounds\varepsilon < t < \pounds\sqrt{\varepsilon}$.

3. $\overline{x}(t) = x_0 e^{\frac{t^2 - t_0^2}{2\varepsilon}} + \sqrt{2\pi}\mathcal{N}\left(\frac{t}{\sqrt{\varepsilon}}\right)(f(0) + \oslash) e^{-\frac{t^2 - 2A}{2\varepsilon}}$, *if* $t \in \pounds\sqrt{\varepsilon}$.

4. $\overline{x}(t) = x_0 e^{\frac{t^2 - t_0^2}{2\varepsilon}} + \sqrt{2\pi}(f(0) + \oslash) e^{\frac{-t^2 - 2A}{2\varepsilon}}$, *if* $t > \pounds\sqrt{\varepsilon}$.

To describe the behaviour of the solutions where the slow curve is repulsive, in Part 3 of Theorem 9.4.8, consider the case $A < \frac{t_0^2}{2} + \pounds\varepsilon$, where the second term of Part 4 dominates the first term. As far as S-continuous functions f are concerned, almost all leave the halo of the slow curve at the same time-period $\sqrt{2A + \varepsilon\pounds}$, which corresponds to $f(0)$ being appreciable. More precisely, given, say, the Lebesgue measure on the values $f(0)$, the Loeb measure of the exceptional case of leaving later, which corresponds to $f(0)$ being infinitesimal, is zero. Note that the value of the function at 0 decides whether they leave upwards (for $f(0) \gtrsim 0$) or downwards (for $f(0) \lesssim 0$), the chances being equal.

Again for reasons of simplicity of notation we prove Theorem 9.4.8 for the case where $t_0 = -1$ and $x_0 = 1$. Then we need to give an approximation of formula (9.40), which depends on the value of I_f as given by Definition 9.4.3.

Lemma 9.4.9. *If f is S-continuous, then*

$$I_f(t) \in \begin{cases} -\frac{\varepsilon}{t} e^{-\frac{t^2}{2\varepsilon}} \left(1 - e^{\frac{t+1}{\varepsilon}}\right)(f(t) + \oslash), & t \in -1 + \pounds\varepsilon \\ -\frac{\varepsilon}{t} e^{-\frac{t^2}{2\varepsilon}} (f(t) + \oslash), & -1 + \pounds\varepsilon < t < \pounds\sqrt{\varepsilon} \\ \sqrt{2\pi\varepsilon}\mathcal{N}\left(\frac{t}{\sqrt{\varepsilon}}\right)(f(0) + \oslash), & t \in \pounds\sqrt{\varepsilon} \\ \sqrt{2\pi\varepsilon}(f(0) + \oslash), & t > \pounds\sqrt{\varepsilon}. \end{cases}$$

Proof. Using the S-continuity of f, for $t \in -1 + \pounds\varepsilon$ we obtain

$$\begin{aligned} I_f(t) &\in \int_{-1}^{t} (f(-1) + \oslash) e^{-\frac{s^2}{2\varepsilon}} ds \\ &= f(-1)N(t) + \oslash N(t) \\ &= (f(-1) + \oslash)N(t) \\ &= -\frac{\varepsilon}{t} e^{-\frac{t^2}{2\varepsilon}} \left(1 - e^{\frac{t+1}{\varepsilon}}\right)(f(t) + \oslash). \end{aligned}$$

This proves the first formula.

For $-1 + \pounds\varepsilon < t < \pounds\sqrt{\varepsilon}$ we use the change of variable $s = t + \frac{\varepsilon u}{t}$ of Lemma 9.4.4. Then

$$I_f(t) = f(t)N(t) - \frac{\varepsilon}{t} e^{-\frac{t^2}{2\varepsilon}} \int_0^{-\frac{t+t^2}{\varepsilon}} \left(f\left(t + \frac{\varepsilon u}{t}\right) - f(t)\right) e^{\left(-u - \frac{\varepsilon u^2}{2t^2}\right)} du.$$

The second term is infinitesimal. Indeed, applying Theorem 7.1.1 we obtain

$$\begin{aligned} &\int_0^{-\frac{t+t^2}{\varepsilon}} \left(f\left(t + \frac{\varepsilon u}{t}\right) - f(t)\right) e^{\left(-u - \frac{\varepsilon u^2}{2t^2}\right)} du \\ &= \int_0^{\pounds} \left(f\left(t+\frac{\varepsilon u}{t}\right) - f(t)\right) e^{\left(-u - \frac{\varepsilon u^2}{2t^2}\right)} du + \int_{\not\infty}^{-\frac{t+t^2}{\varepsilon}} \left(f\left(t+\frac{\varepsilon u}{t}\right) - f(t)\right) e^{\left(-u - \frac{\varepsilon u^2}{2t^2}\right)} du. \end{aligned}$$

Using the S-continuity of f, we find

$$\int_0^{\pounds} \left(f\left(t + \frac{\varepsilon u}{t}\right) - f(t)\right) e^{\left(-u - \frac{\varepsilon u^2}{2t^2}\right)} du \subseteq \int_0^{\pounds} \oslash e^{(-u+\oslash)} du = \oslash,$$

while

$$\begin{aligned} &\int_{\not\infty}^{-\frac{t+t^2}{\varepsilon}} \left(f\left(t + \frac{\varepsilon u}{t}\right) - f(t)\right) e^{\left(-u - \frac{\varepsilon u^2}{2t^2}\right)} du \\ &\qquad \subseteq \int_{-\not\infty}^{\not\infty} \pounds e^{(-u+\oslash)} du = \pounds \cdot \oslash = \oslash. \end{aligned}$$

This proves the second formula.

For the two remaining formulas we use the fact that the mass of the standard normal distribution is $\pounds$. For $t \in \pounds\sqrt{\varepsilon}$ we have the approximation

$$\begin{aligned} \int_{-\frac{1}{\sqrt{\varepsilon}}}^{\frac{t}{\sqrt{\varepsilon}}} f(\sqrt{\varepsilon} v) e^{-\frac{v^2}{2}} dv &\in \int_{-\not\infty}^{\frac{t}{\sqrt{\varepsilon}}} f(\sqrt{\varepsilon} v) e^{-\frac{v^2}{2}} dv + \oslash \\ &= \int_{-\not\infty}^{\frac{t}{\sqrt{\varepsilon}}} f(0) e^{-\frac{v^2}{2}} dv + \oslash + \oslash \\ &= \sqrt{2\pi}\mathcal{N}\left(\frac{t}{\sqrt{\varepsilon}}\right)(f(0) + \oslash). \end{aligned}$$

This proves the third formula.

For $t > \pounds\sqrt{\varepsilon}$ we obtain similarly

$$\int_{-\frac{1}{\sqrt{\varepsilon}}}^{\frac{t}{\sqrt{\varepsilon}}} e^{-\frac{v^2}{2}}\,dv \in \int_{-\not\infty}^{\not\infty} f(\sqrt{\varepsilon}v)e^{-\frac{v^2}{2}}\,dv + \oslash$$
$$= \int_{-\not\infty}^{\not\infty} f(0)e^{-\frac{v^2}{2}}\,dv + \oslash + \oslash$$
$$= \sqrt{2\pi}(f(0) + \oslash).$$

So the last formula is proved as well. □

Theorem 9.4.8 follows directly from (9.40) and Lemma 9.4.9.

Finally we consider the case where f is not necessarily S-continuous. For instance, it could be a rapidly oscillating function, like $t \mapsto \sin(\omega t)$, with ω unlimited, or f could be a piece-wise linear function, obtained from the trajectory of the nonstandard Wiener walk of [94] by joining successive points; such a function is not C^1; still it is Lipschitz and allows for uniqueness of solution in the differential equation.

We do not give an exhaustive treatment, and consider only properties relevant for the canard phenomena. These correspond to the last formula of Theorem 9.4.8. Then for the canard solution to leave upwards or downwards, instead of the (approximate) value of f at or close to 0, now the local mean of f at 0 with respect to the normal distribution is decisive. To state and prove the second theorem we use the local averages of Chapter 7.

Theorem 9.4.10. *Consider (9.37), where $A > 0$ is appreciable and f is limited for $t \geq t_0$ and averageable on $\pounds\sqrt{\varepsilon}$ with respect to the normal density $x \mapsto \frac{1}{\sqrt{2\pi\varepsilon}}e^{-\frac{x^2}{2\varepsilon}}$. Let $\overline{x}$ be the solution with initial condition $\overline{x}(t_0) = x_0$, where $t_0 < 0$ is appreciable and x_0 is limited. Then*

$$\overline{x}(t) \in x_0 e^{\frac{t^2 - t_0^2}{2\varepsilon}} + \sqrt{2\pi} M_{\pounds\sqrt{\varepsilon}}(f) e^{-\frac{t^2 - 2A}{2\varepsilon}}. \tag{9.45}$$

Proof. By Part 4 of Theorem 9.4.8

$$\overline{x}(t) = x_0 e^{\frac{t^2 - t_0^2}{2\varepsilon}} + \frac{e^{\frac{t^2 - 2A}{2\varepsilon}}}{\sqrt{\varepsilon}} \int_{t_0}^{t} f(s) e^{-\frac{s^2}{2\varepsilon}}\,ds.$$

Now $\frac{t_0}{\sqrt{\varepsilon}} < \pounds < \frac{t}{\sqrt{\varepsilon}}$, so the mass of the normal distribution with density $s \mapsto \frac{1}{\sqrt{2\pi\varepsilon}}e^{-\frac{s^2}{2\varepsilon}}$ is $\frac{\pounds}{\sqrt{\varepsilon}}$. Then

$$\frac{1}{\sqrt{\varepsilon}} \int_{t_0}^{t} f(s) e^{-\frac{s^2}{2\varepsilon}}\,ds \in \sqrt{2\pi} \int_{\pounds\sqrt{\varepsilon}} f(s) \frac{e^{-\frac{s^2}{2\varepsilon}}}{\sqrt{2\pi\varepsilon}}\,ds = \sqrt{2\pi} M_{\pounds\sqrt{\varepsilon}}(f),$$

which entails (9.45). □

Consider equation (9.37), with the second term of Part 4 of Theorem 9.4.8 dominating the first term, i.e. with $\oslash < A < \frac{t_0^2}{2} + \pounds\varepsilon$. Assume f is limited for $t \geq t_0$ and averageable on $\pounds\sqrt{\varepsilon}$ with respect to the normal density $x \mapsto \frac{1}{\sqrt{2\pi\varepsilon}} e^{-\frac{x^2}{2\varepsilon}}$. By Theorem 9.4.10, if the average of f on $\pounds\sqrt{\varepsilon}$ with respect to the mentioned density is non-infinitesimal the canard solution defined by f indeed leaves the halo of the slow curve at the times $t \in A + \pounds\varepsilon$, and its sign decides if the canard leaves upwards or downwards.

Chapter 10

Applications in other fields

10.1	The Sorites paradox in philosophy	244
	10.1.1 Forms of the paradox	245
	10.1.2 Response proposals	247
	10.1.3 External numbers as a model	255
10.2	External recurrence relations and near stability	258
10.3	On the size of fluctuations of the financial market	264
10.4	Further applications of external numbers	269
	10.4.1 Near-optimization with uncertainties	269
	10.4.2 On statistical estimation of uncertainties	270

The external numbers are intended to formalize orders of magnitude and vagueness of the Sorites type. This chapter deals with attempts to model mathematically phenomena with intrinsic indeterminateness of the mentioned type.

The chapter starts with the Sorites paradox itself. We investigate to which extent nonstandard analysis and in particular the neutrices and external numbers capture the Sorites property. We also compare with some existing approaches.

In Section 10.2 a model for near stability is proposed for recurrence relations. In contrast to ordinary stability, restricted fluctuations are allowed; we may think of them as models for steering and other types of control where ultimate precision cannot be attained (see also [80]).

Section 10.3 deals with orders of magnitude in economy. The use of scaling in economy is fundamental, as is reflected by the distinction between microeconomics and macroeconomics. Already Robinson and Brown [22] formulated a nonstandard model in which macroscopic behaviour results from an unlimited number of infinitesimal, microscopic influences (see also [107]). Here we try to relate microscopic and macroscopic price-fluctuations. It follows from stochastic analysis (see e.g. [4, 94, 73]) that perceptible price-fluctuations over perceptible time-intervals can only exist when microscopic trading periods δt induce microscopic price-fluctuations of bigger order, *in casu* $\sqrt{\delta t}$. Here we aim to explain this at the micro-level.

In the final section we describe briefly some attempts of mathematical modelling with external numbers by other authors, in particular Nam Tran Van (linear optimization), and J. Bosgiraud (statistics). These problems typically deal with more than one variable.

Optimization under uncertainties is modelled in Subsection 10.4.1. A notion of near-optimum was introduced in [147]. By means of an example, we show how the parameter method of Section 8.4 leads to the localization of near-optima in linear programming with external numbers.

The objective of statistics is to measure uncertainties related to a hypothesis, but in the final Subsection 10.4.2 several grades of uncertainties are allowed in the hypothesis itself. We describe shortly this approach, which is due to J. Bosgiraud [15, 14, 16, 17, 18], and includes multivariate statistics.

10.1 The Sorites paradox in philosophy

In this section we give an application of the external numbers to philosophy. More precisely, we present a way to model the Sorites paradox using nonstandard analysis, via the axiom scheme of External Induction and propose the external numbers as models for the kind of vagueness expressed by the Sorites paradox. The content of this section is adapted from [39] (see also [40]).

Paradoxes arise when several orders of magnitude are considered. These paradoxes are part of the larger phenomenon known as *vagueness*. Vague predicates share (at least) the following features.

1. Admit borderline cases.

2. Lack sharp boundaries.

3. Are susceptible to Sorites paradoxes.

For *borderline* cases it is not clear whether or not the predicate applies, independently of how much one knows about it. For instance, most basketball players are clearly tall and most jockeys are clearly not tall. But, even if one knows the height of a person with great precision, in many cases it is rather unclear if the person should be classified as tall or not. Furthermore, the set of all tall people and the set of people that are not tall lack sharp boundaries for there is no clear distinction between them.

This sort of vagueness is related to a collection of paradoxes called the *Sorites* paradoxes. These paradoxes were first formulated by the Ancient Greek philosopher Eubulides of Miletus in the IVth century BCE and can be stated in the following way: a single grain of wheat cannot be considered as a heap. Neither can two grains of wheat. One must admit the presence of a heap sooner or later, so where to draw the line? In fact, the name Sorites derives from the Greek word *sorós* which means heap. However, one can reconstruct the paradox by replacing the term 'heap' by other vague concepts such as 'tall', 'beautiful', 'bald', 'heavy', 'cold', 'rich',...

The argument consists of a predicate S, called the *soritical predicate* and a subject expression in the sequence in regard to which S is soritical which consists of a finite sequence (a_n). According to Barnes [5] a predicate S must satisfy three constraints in order to be considered soritical:

1. Appear to be valid for a_1, the first term of the sequence.
2. Appear to be false for a_i, the last term of the sequence.
3. Each adjacent pair of terms of the sequence, a_n and a_{n+1} must be sufficiently similar as to appear indiscriminable with respect to S.

This means that the predicate S needs to be sufficiently vague in order to allow for small changes. Clearly, small changes do not determine the difference between a set of individual grains and a heap, between a bald man and a hairy one, between a rich person and a poor one, etc. However, and in spite of the vagueness involved, it also needs to have a certain area on which S is clearly true and an area on which S is clearly false, because big changes make a difference. It would seem that one grain, one hair, etc. would be insignificant to make any difference to the application of the predicate, yet after a while, the accumulation of the apparently insignificant changes does not seem insignificant anymore.

> The difference of one grain would seem to be too small to make any difference to the application of the predicate; it is a difference so negligible as to make no apparent difference to the truth-values of the respective antecedents and consequents.
>
> Yet the conclusion seems false. [65]

Many approaches have been proposed in order to deal with the phenomenon of vagueness and with the Sorites type paradoxes which arise when vague predicates are used. We review some of these approaches in Section 10.1.2.

10.1.1 Forms of the paradox

The Sorites paradox can be stated in various ways. This implies that one cannot hope to solve the paradox by pointing out a fault particular to any one of those. One should instead try and reveal a common fault to all possible forms that the paradox can take. We do so by considering the (standard) mathematical induction and conditional schemata. These are revisited, in Section 10.1.3, after a nonstandard point of view is adopted.

Inductive form

Mathematical Induction is generally used (within standard mathematics) to prove that a certain mathematical statement involving a natural number n holds for all possible values of n. This is done in two steps. First, one proves that there is a first element for which the statement holds (the *basis*). On the second step (*inductive step*) one shows that if the statement holds for some n then it also holds for $n+1$. Then the Induction Principle allows us to conclude that the statement is valid for all natural numbers n. Let S represent a soritical predicate, for example 'is not a heap', and let a_n represent the n-th element in the soritical sequence. In the example above it would be the predicate representing that we have 'n grains of wheat'.

The Sorites paradox can now be represented in the following way:

$$\begin{cases} (Sa_1 \wedge \forall n\,(Sa_n \Rightarrow Sa_{n+1})) \Rightarrow \forall n Sa_n \\ \exists \omega\,(\neg Sa_\omega)\,. \end{cases} \tag{10.1}$$

So, if one admits that,

1. A single grain of wheat is not a heap,
2. If a collection of n grains of wheat is not a heap then a collection of $n+1$ grains of wheat is also not a heap,

by Induction one concludes that the heap will never appear. Since at some point the heap is obviously there one might come to the conclusion that there is something wrong with Induction or with its application to vague predicates.

Conditional form

The conditional form of the Sorites paradox is the most common form throughout the literature. Using the notation of the previous section it can be formalized in the following way:

$$\begin{cases} Sa_1 \\ Sa_1 \Rightarrow Sa_2 \\ Sa_2 \Rightarrow Sa_3 \\ \cdots \\ Sa_i \Rightarrow Sa_{i+1} \\ \exists j\,(\neg Sa_j)\,. \end{cases} \tag{10.2}$$

Assuming Sa_1, $Sa_1 \Rightarrow Sa_2$, $Sa_2 \Rightarrow Sa_3$, ..., $Sa_i \Rightarrow Sa_{i+1}$, by *modus ponens,* the conclusion is Sa_i, where i can be arbitrarily large. This is a fairly simple reasoning where the premises are: a single grain of wheat does not make a heap; if one grain of wheat does not make a heap then two grains of wheat do not form a heap either; if two grains of wheat do not make a heap then three grains of wheat do not form a heap either... if i grains of wheat do not make a heap then $i+1$ grains of wheat do not form a heap either. The conclusion is that a set of an arbitrarily large number of grains i does not make a heap. However if one observes that there is a set of j grains that form a heap it generates a paradox.[1]

[1]Classical logic reasoning was used here and one could think that the paradox may be circumvented if one would use, say, intuitionistic logic instead. In fact, there have been

10.1.2 Response proposals

There are several attempts to solve the Sorites paradox. According to Keefe [72, p. 19-20], these are divided into four types. A first type of response would be to deny the validity of the argument, refusing to grant that the conclusion follows from the premises. Alternatively one can question the strict truth of the inductive premise (or of one of the conditionals). A third possibility is to accept the validity of the argument and the truth of its inductive premise (or of all the conditional premises) but contest or refute the truth of the conclusion. Finally one can grant that there are good reasons to consider both the argument form as valid, accepting the premises, and denying the conclusion hence proving that the predicate is incoherent.

We will review some of the responses to the paradox. For a wider account on this matter we refer to [72, 120, 117, 150]. We would like to emphasize that the theories presented below correspond to a wide variety of related points of view. This means that there are many versions of the theories presented and many of the more subtle aspects are omitted. So, we tend to give only the general lines, common to the various versions of the theory in question. The theories presented are used to analyse various types of vagueness. As we will see below several authors argued that, in spite of their merits, they do not give a definite response to the Sorites paradox.

Ideal languages

Natural languages distinguish between intension and extension of terms. The *intension* is the internal content of a term or concept while the *extension* is the range of applicability of a term by naming the particular objects that the term denotes. The two predicates 'is a creature with a heart' and 'is a creature with a kidney' (see [106]) have the same extension because the set of creatures with hearts and the set of creatures with kidneys are the same. However, having a heart and having a kidney are very different things, so one concludes that

> terms can name the same thing but differ in meaning. [106]

Hence, the distinction between intension and extension leads necessarily to vagueness, ambiguity and indeterminacy of meaning for words and phrases. Therefore, if one requires clarity and precision of language then one is forced to conclude that natural languages are not suitable. An ideal language would leave out all such factors in order to eliminate any vagueness.

---

attempts to use intuitionistic logic to deal with the paradox (see [105] and its defence [116]). However, according to Keefe,

> [...] The bulk of the criticisms point to the conclusion that there is no sustainable account of vagueness that emerges from rejecting classical logic in favour of intuitionistic logic [72, p. 22].

For the criticism referred by Keefe the reader may consult [108, 25].

The defenders of this response, among them Frege [51], Russell [114] and Wittgenstein [152], tend to consider vagueness as a non-eliminable feature of natural language. In their point of view, the way to avoid vagueness is by creating and using ideal languages instead. This would mean that arguments of the Sorites type are not valid since they contain vague expressions.

As stated by Russell,

> The fact is that all words are attributable without doubt over a certain area, but become questionable within a penumbra, outside which they are again certainly not attributable. Someone might seek to obtain precision in the use of words by saying that no word is to be applied in the penumbra, but unfortunately the penumbra is itself not accurately definable, and all the vagueness which apply to the primary use of words apply also when we try to fix a limit to their indubitable applicability. [114]

So, this response implies that it is the philosopher's job to discover a logically ideal language. Russell also believed that vagueness is a matter of degree, and accuracy would be an "ideal limit".

Ideal languages as a response to the Sorites paradox seem to have unsatisfying features for, according to Keefe [72, p. 20], they seem to require giving up fundamental rules of inference such as *modus ponens* and Mathematical Induction. Furthermore, by eliminating vague predicates from the language one is not solving the paradox but simply avoiding it by "sweeping it under the rug". In fact, philosophers nowadays tend to see vagueness as an important part of natural language that should not be separated from it.

The Epistemic theory

The Epistemic theory is based on the idea that the precise boundaries to knowledge itself cannot be known. Vagueness is therefore a particular type of ignorance.

The fact that this theory is built in the classical logic framework implies that there are precise bounds for the extensions of vague predicates even if one does not know where they are located. For instance, the defenders of the Epistemic theory claim that there is in fact a last grain of wheat in the sequence before the heap turns up, even if one is not (nor ever will be) able to identify it definitively. In fact, Williamson [150] has shown that if there is a precise boundary for penumbral cases one cannot know where it is. So, soritical predicates are indeterminate in extension but not semantically. This position has been notably defended by Williamson and Sorensen [150, 151, 122, 123].

The first and major objection to the Epistemic theory is its counter-intuitive nature. The meaning of a word is (usually) determined by its use.[2]

[2]This is Wittgenstein's *meaning as use*. According to this author "if we had to name anything which is the life of the sign, we should have to say that it is its use" [153] and "the meaning of a word is its use in the language" [152].

For instance, by the word 'guitar' one means an actual guitar because one uses that word to mean an actual guitar (even if one does not know how to play). Now, one does not usually use the word 'heap' as if a single grain of wheat could make a difference. Neither, more generally, does one use any vague term as if it were not tolerant to small changes. One does not use vague terms as if they had precise borders. In this sense, Smith [120] claims that the epistemicist is forced to deny a link between meaning and use.

Another point that deserves criticism is that nothing is said about how predicates get the precise extensions that they do. It is claimed that there is in fact a last grain of wheat in the sequence before the heap turns up. So there should be attempts to find which one it is [72]. Ignorance is no excuse for the lack of attempts to find the precise boundaries of vague concepts so there should be at least some reasons to believe about where these boundaries are.

Supervaluationism

According to Fine, vagueness is a semantic notion not to be confused with ambiguity nor undecidability:

> Very roughly, vagueness is deficiency of meaning. As such, it is to be distinguished from generality, undecidability, and ambiguity. These latter are, if you like, lack of content, possible knowledge, and univocal meaning, respectively. [48]

Supervaluationism proposes to solve the problem of vagueness by modifying classical semantics, using Van Fraassen's supervaluations [145]. According to the latter, a *supervaluation over a model* is a function that assigns the value T (respectively F) exactly to those statements assigned the value T (respectively F) by all the classical valuations over that model. This implies that supervaluations must have truth-value gaps.

In classical logic the connectives have truth values in a functional way. The classical logic connectives are all *truth-functional*; i.e. the truth value of any compound statement obtained by applying that connective is a function of the individual truth values of the constituent statements that form the compound.[3] Supervaluationists abandon the concept of truth-functionality.

Fine applies the distinction between extension and intension [106] to vagueness:

> Extensional vagueness is deficiency of extension, intensional vagueness deficiency of intension. Moreover, if intension is the possibility of extension, then intensional vagueness is the possibility of extensional vagueness. [48]

According to supervaluationism, a vague predicate does not need to have a unique, sharply bounded, truth function. Vague predicates have things to

[3]This is immediately visible if one computes the logical value of a given sentence using truth tables or a proof calculus like natural deduction [103].

which they definitely apply (*positive extension*), things to which they definitely do not (*negative extension*) and a penumbra (*penumbral connections*). The penumbra involves cases which seem to be neither true nor false[4] (*borderline cases*). These penumbral connections are instances of truth-value gaps. Truth-value gaps are related to extensional vagueness. However,

> Despite the connection, extensional vagueness should not be defined in terms of truth-value gaps. This is because gaps can have other sources, such as failure of reference or presupposition. [48]

Supervaluationists claim, roughly speaking, that a vague sentence is true if and only if it is true for all ways of making it completely precise [48, 72], called *precisifications.* There are then many interpretations or precisifications. Each one of these precisifications has no penumbra because it behaves according to classical bivalence. The assignment of truth value for all such precisifications is a supervaluation.

A sentence which is true in all precisifications is called supertrue and a sentence which is false in all precisifications is called superfalse. A sentence which is true for some precisifications and false for others is neither true nor false[5]. This means in particular that tautologies from classical logic are supertrue.

According to Keefe [72], *truth is supertruth,* meaning that a sentence is true if and only if it is true on all admissible precisifications. A precisification is acceptable only if the extensions of the concepts do not overlap. The truth of a compound sentence is determined by its truth for every precisification.

Fodor and Lepore are particularly critical of the supervaluationist approach to vagueness. They argue that

> [...] there is something fundamentally wrong with using supervaluation techniques either for preserving classical logic or for providing a semantics for linguistic expressions ordinarily thought to produce truth-value gaps. [49]

However the fault that they point out is not of a logical nature but only *philosophical.* Indeed, according to Fodor and Lepore, the main flaws of supervaluationism are the violation of intuitive semantic principles concerning disjunctions and existential quantification, the abandonment of classical rules of inference and the violation of core principles concerning the concept of truth.

Let S represent the predicate 'is not a heap'. Since all tautologies are supertrue, $\neg(\forall n(Sa_n \Rightarrow Sa_{n+1}))$ is equivalent to $\exists n(Sa_n \wedge \neg Sa_{n+1})$ which,

---

[4]Fine warns about the general confusion of under- and over-determinacy.

> A vague sentence can be made more precise; and this operation should preserve truth-value. But a vague sentence can be made to be either true or false, and therefore the original sentence can be neither. [48]

[5]In fact, not all supervaluationists accept this last point of view.

semantically speaking, seems to postulate the existence of a sharp boundary and looks for that matter like a step back towards the Epistemic theory. Also $(S \vee \neg S)$ is supertrue. So, for all precisifications one of the statements is true. However, the statement S is borderline and therefore neither true nor false.

Keefe [72] argues that it is possible to surpass these difficulties at the price of adding a new operator to the language: the 'definitely' operator D. This operator is however not closed under certain operations such as contraposition and conditional introduction. So alternatives to the classical closure principles are proposed. However, this implies that the logic used is no longer classical.

Another argument against supervaluationism is that little information is given on what makes a precisification acceptable other than saying that precisifications must respect penumbral connections and therefore their admissibility is a vague matter. Also, supervaluationism states that precisifications behave in a classical way and have no penumbra. However, each precisification may divide the positive and negative extensions in different places.

For more on objections to supervaluationism and attempts to respond to those objections the reader is referred to [28].

Many-valued logics

Many-valued logic is a general term that refers to any logic which has more than two truth-values [58]. In these logics the principle of truth-functionality is accepted and so a sentence remains unaffected when one of its components is replaced by another with the same truth value. Many-valued logics became accepted as an independent part of logic with the works of Łukasiewicz [83] and Post [102] in the 1920's. Since then several other many-valued logics emerged (e.g. [55, 59]). We discuss an application of Kleene's three-valued logic and applications of fuzzy logics because these seem to be the most relevant in what concerns the phenomenon of vagueness.

Kleene's logic

Perhaps one of the simplest and best-known examples of a many-valued logic is Kleene's three-valued logic [74]. Kleene thought of the third truth value as undefined or underdetermined[6]. So one has three truth-values: 1 (true), 0 (false) and $\frac{1}{2}$ (undefined or unknown). One has truth-tables for which the connectives are *regular*; i.e. in terms of ordering, undefined is placed below both true and false. This means that the behaviour of the third truth value should be compatible with any increase in information. In Kleene's logic we have the following truth-table for negation.

p	$\neg p$
1	0
0	1
$\frac{1}{2}$	$\frac{1}{2}$

[6]Priest gave in [104] an alternative three-valued logic conceiving the third truth-value as overdetermined, interpreting the symbol $\frac{1}{2}$ as being both true and false.

As for the remaining *strong* connectives the truth-table is the following.

p	q	$p \vee q$	$p \wedge q$	$p \Rightarrow q$	$p \Leftrightarrow q$
1	1	1	1	1	1
1	0	1	0	0	0
1	$\frac{1}{2}$	1	$\frac{1}{2}$	$\frac{1}{2}$	$\frac{1}{2}$
0	1	1	0	1	0
0	0	0	0	1	1
0	$\frac{1}{2}$	$\frac{1}{2}$	0	1	$\frac{1}{2}$
$\frac{1}{2}$	1	1	$\frac{1}{2}$	1	$\frac{1}{2}$
$\frac{1}{2}$	0	$\frac{1}{2}$	0	$\frac{1}{2}$	$\frac{1}{2}$
$\frac{1}{2}$	$\frac{1}{2}$	$\frac{1}{2}$	$\frac{1}{2}$	$\frac{1}{2}$	$\frac{1}{2}$

These tables are uniquely determined as the strongest possible regular extensions of the classical two-valued tables. Quantifiers can be defined in the following way: $\exists x\, P(x)$ is true if $P(x)$ is true for some value of x and it is false if $P(x)$ is false for all values and indefinite otherwise; $\forall x\, P(x)$ is true if $P(x)$ is true for all values of x and false if $P(x)$ is false for some value and indefinite otherwise. In [127], Tye applies Kleene's three-valued logic to the Sorites paradox. However, the objections made to the bipartite division can also be used to refute a tripartite division. In fact, Tye himself claims that "vagueness cannot be reconciled with any precise dividing lines" because "there is no determinate fact of the matter about where truth-value changes occur". That is to say that there is no way to assign precise truth-values to vague terms. As a solution, Tye proposes to use a vague metalanguage, claiming that there are sets which are genuinely vague items. For instance the set of tall men has borderline members (men who are neither clearly members nor clearly non-members of the set).

Kleene's three-valued logic has the undesirable feature of having no tautologies, because the two-valued tautologies can take the value $\frac{1}{2}$ in the three-valued case. As an example consider the law of excluded middle $p \vee \neg p$. In Kleene's three-valued logic the truth table is the following:

p	$\neg p$	$p \vee \neg p$
1	0	1
0	1	1
$\frac{1}{2}$	$\frac{1}{2}$	$\frac{1}{2}$

Tye tries to avoid this flaw by saying that a statement is a *quasi-tautology* if it has no false substitution instances. So two-valued tautologies become three-valued quasi-tautologies.

Kleene's three-valued logic is still a precise formalization and having no tautologies seems a price too high to pay in order to be able to deal in the above sense with vagueness. Also, according to Keefe,

> [...] the appeal to quasi-tautologies adds nothing: if earning this title is enough for his [Tye's] purposes, then the fact that $p \vee \neg p$ also earns it should be of concern. Moreover, what matters for validity does not relate to quasi-tautologies, and assertion depends on sentences being true not being either true or indefinite, so the role for the notion seems to be merely one of appeasement. [72, p. 111]

Fuzzy logics

Fuzzy logics propose a graded notion of inference. Truth-values range in degree between 0 and 1 in order to capture different degrees of truth. In this way, the value 0 is attributed to sentences which are completely false and the value 1 to sentences which are completely true. The remaining sentences are truer than the false sentences, but not as true as the true ones so they have intermediate logical values according to "how true" they are. Truth is therefore a matter of degree [11].

Fuzzy logic is related to Zadeh's work on fuzzy sets [154]. A *fuzzy subset* A of a set X is characterized by a membership function $f_A(x)$ with values in the interval $[0, 1]$. So, a fuzzy set A is a class of objects that allow a continuum of grades of membership. The membership degree is then the degree to which the sentence 'x is a member of A' is true. So, one can interpret the membership degrees of fuzzy sets as truth degrees of the membership predicate in a suitable many-valued logic.

Theories of vagueness which recourse to fuzzy logics are advocated most notably by Machina and Smith [87, 120]. According to these authors the notion of heap is a vague one and it may hold true of given objects only to some (truth) degree. The premises should be considered partially true to a degree quite near to the maximal degree 1. This inference has to involve truth degrees for the premises and has to provide a truth degree for the conclusion in a way that in each step the truth degree becomes smaller. The sentence 'n grains of sand do not make a heap' tends towards being false for an increasing number of grains.

The problem of saying whether the sentence 'a set of n grains makes a heap' is true or not is essentially the same as to say that that sentence is true with a certain (precise) fixed degree. This false precision is perhaps the main objection to the application of many-valued logics to the Sorites paradox. According to Keefe,

> the degree theorist's assignments impose precision in a form that is just as unacceptable as a classical true/false assignment. In so

> far as a degree theory avoids determinacy over whether a is F, the objection here is that it does so by enforcing determinacy over the degree to which a is F. All predications of "is red" will receive a unique, exact value, but it seems inappropriate to associate our vague predicate "red" with any particular exact function from objects to degrees of truth. For a start, what could determine which is the correct function, settling that my coat is red to degree 0.322 rather than 0.321? [72, p. 113]

Also, Urquhart states that

> One immediate objection which presents itself to [fuzzy logic's] line of approach is the extremely artificial nature of the attaching of precise numerical values to sentences like '73 is a large number' or 'Picasso's Guernica is beautiful'. In fact, it seems plausible to say that the nature of vague predicates precludes attaching precise numerical values just as much as it precludes attaching precise classical truth values. [129]

Smith [120] tries to solve this problem, suggesting several possible solutions and concluding that the best answer is to mix fuzzy logic with a theory called plurivaluationism[7] called *fuzzy plurivaluationism.* So, Smith accepts the semantic realism implied by the Epistemic view, but denies that vague predicates have to refer to a single bivalent model.

The fuzzy logic approach has been quite successful in solving, for example, vagueness related to traffic and transportation processes (see for example [125, 11, 156] for other examples of applications of fuzzy set theory). However the fuzzy logic approach is also not without fault as a model of imprecision, because it ultimately recourses to precise intervals to model imprecise situations.

Contextualism

Contextualism defends that interpretations change over time and according to context. Such shifts of context may occur instantaneously. For instance, at the beginning of a conversation the context is empty. Then, as the conversation goes along, these notions are sharpened in such a way that borderline cases (undecided so far) get assigned to either the extension or the anti-extension of the vague predicates in question. In fact, borderline sentences can express something true in one context and something false in another, so they are context-sensitive. In this way one can disagree about the truth-values of the propositions expressed by borderline sentences, even in situations where all the relevant information is available. This view is most prominently elaborated by Shapiro and DeRose [117, 118, 32].

[7] Supervaluationism involves only one intended (non-classical) model relevant to questions concerning meaning and truth, while plurivaluationism allows that there may be multiple (classical) models.

Besides context-sensitivity Shapiro defines as central the concepts of *judgement dependence*, *open texture*, and the *principle of tolerance*. Judgement dependence means that both the extensions and anti-extensions for the borderline cases are solely determined by the decisions of competent speakers. These decisions are put in (and can be removed from) the *conversational record*. Open texture means that for a vague predicate S there exists an object a such that a competent speaker can decide whether Sa holds or not without her competency being compromised. The principle of tolerance is defined as follows. Suppose that two objects a, b differ only marginally in the relevant respect on which a vague predicate S is tolerant. Then if one competently judges Sa to hold, then Sb also holds.

One objection to Contextualism is that the problems with vague expressions seem to remain whether context-sensitivity is taken into account or not. By taking the context into account one can reduce vagueness but not eliminate it completely. Indeed, sets with vague boundaries are invariant to some translations. Take for instance the word 'ugly'. Even if a particular context is given (and even if one knows a great deal about another one's ugliness) there is still no reason to suppose that there is a sharp boundary between what 'ugly' applies to and what it does not.[8]

Furthermore, Smith [120] argues that Contextualism should not be seen as a theory of vagueness in its own right since it is is compatible with all other mentioned theories.

10.1.3 External numbers as a model

We want to emphasize that we are not trying to give a theory for vagueness in general but model only a specific type of vagueness i.e. of the type Sorites.

As mentioned above, soritical arguments are tolerant to small changes but not tolerant to large changes in relevant aspects. In fact, using nonstandard analysis it is possible to define rigorously what is meant with terms such as 'small changes' or 'large changes'. The fact that large changes come as the result of the accumulation of small changes is no surprise because it is a very basic and well-known fact from nonstandard analysis that an infinitely large sum of infinitesimals may very well become appreciable or even infinitely large (see Exercise 1.1.5). In this way, one can make a rigorous claim that a heap and a set of individual grains of wheat are indeed not of the same order of magnitude.

The "economical" version of Nonstandard Analysis ENA presented in Chapter 1 is enough for the current purpose.

If one replaces Mathematical Induction by External Induction, the reasoning (10.1) becomes:

---

[8]The reader interested in the criticism to Shapiro's ideas may consult the review of [117] made by Matti Eklund, available online at `http://ndpr.nd.edu/news/vagueness-in-context/`.

$$\begin{cases} (Sa_1 \wedge \forall^{\mathrm{st}} n\,(Sa_n \Rightarrow Sa_{n+1})) \Rightarrow \forall^{\mathrm{st}} n(Sa_n) \\ \exists \omega\,(\neg Sa_\omega)\,. \end{cases} \tag{10.3}$$

So, if one admits that,

1. A single grain of wheat is not a heap,
2. If n is a standard number and if a set of n grains of wheat is not a heap then a set of $n+1$ grains of wheat is also not a heap,

one concludes that in the presence of a standard number of grains of wheat one does not have a heap. Numbers indicating a heap are of a different nature: the heap only arises when one has a nonstandard number $\omega \simeq +\infty$ of grains of wheat.

The conditional form (10.2), using nonstandard analysis, becomes the following.

Let i be a standard natural number. If

$$\begin{cases} Sa_1 \\ Sa_1 \Rightarrow Sa_2 \\ Sa_2 \Rightarrow Sa_3 \\ \ldots \\ Sa_i \Rightarrow Sa_{i+1} \end{cases} \tag{10.4}$$

then, by *modus ponens,* the conclusion is Sa_i, for i an arbitrarily large but standard number. In nonstandard analysis this is modelled by allowing *modus ponens* but only a standard number of times, so it does not contradict $\neg Sa_\omega$, for some nonstandard ω. Working with the naive numbers of Reeb's slogan "the naive numbers do not fill up $\mathbb{N}$" (1.2) we have a second way to avoid the paradox. We recall that the "naive" natural numbers are the ones that can be obtained from zero by the successive addition of one. So, analysing (10.2) gives us Sa_i for i a naive natural number. There is no contradiction with the existence of a non-naive number j with the property $\neg Sa_j$.

We are in fact claiming that the formalization of the predicate 'is not a heap' should be considered to be an external predicate, where not being a heap means to possess a standard number of grains.

The forms (10.3),(10.4) suggest that the set of individual grains may be modelled by the external set of limited numbers (positive part of a neutrix) and the set of grains that form a heap may be modelled by the external set of the infinitely large numbers. Indeed 'precise' objects possess sharp bounds and can be modelled by standard sets while 'vague' objects have no clear bounds and should be for this matter modelled by nonstandard sets which are given by external properties.

Soritical arguments share with external numbers the fact of being tolerant to small changes but not tolerant to large changes in relevant aspects. In fact, with external numbers, using the different orders of magnitude, it is possible to define rigorously what one means with terms such as 'small changes' or 'large changes'. The fact that large changes come as the result of the accumulation of small changes is a very natural consequence of the theory.

We stress the point that our approach is semantical and that we have made a model. A million grains of wheat should form a heap and yet that is clearly a standard number of grains.

With neutrices and external numbers built upon them it is possible to incorporate several transitions of the Sorites type and to calculate effectively with them. Indeed it is shown in the present book that, firstly, there is an infinity of neutrices which do not have precise upper bounds, for they have neither infimum nor supremum and are tolerant to (not too large) changes. Secondly, since external numbers satisfy strong algebraic laws, similar to the ones of the real numbers, calculations are quite simple to carry on.

Let us compare this approach with some of the theories presented above.

Epistemicists believe in the existence of sharp bounds for vague concepts, claiming that ignorance is somehow inevitable. The current proposal takes the opposite direction. Indeed, the tolerance of vague terms, such as 'heap', to small changes indicates that such terms do not have a sharp, definite bound. By using neutrices to model such terms it is possible to avoid the paradox and explain the tolerance to small changes.

According to Keefe [72], degree theories fail to provide an acceptable account of vagueness and are forced to make an implausible commitment to a unique numerical assignment for each sentence. Smith [120] argues that an adequate account of vagueness must involve degrees of truth and that the main objections to this theory may be overcome. His fuzzy plurivaluationism theory seems overcomplicated with respect to the present approach to the Sorites paradox. Also we believe that the problem with the fuzzy logic approach is the fact that ultimately precise numbers are used to model imprecise predicates.

A final remark concerns the strength of the nonstandard axioms, which may introduce undesirable consequences of external modelling. As such, within a nonstandard theory, the proposed solution of the Sorites paradox

$$\begin{cases} \mathrm{st}(0) \\ \forall n(\mathrm{st}(n) \Rightarrow \mathrm{st}(n+1)) \\ \exists \omega(\neg \mathrm{st}(\omega)) \end{cases} \tag{10.5}$$

implies, by the group property of the standard numbers, invariance by doubling, i.e.

$$\forall n\,(\mathrm{st}\,(n) \Rightarrow \mathrm{st}\,(2n))\,.$$

One easily imagines a soritical context where this is inappropriate. However,

$$\exists n\,(\mathrm{st}\,(n) \wedge \neg \mathrm{st}\,(2n)) \tag{10.6}$$

is consistent with (10.5). To see this, take a new predicate **st** and a nonstandard number ω, and set $\mathbf{st}(k)$ for all $k \in \mathbb{N}$ which are less than or equal to $\omega + n$ for some $\mathrm{st}(n)$. Then $\neg\mathbf{st}(2\omega)$.

In this context $\{(10.5), (10.6)\}$ is an acceptable axiom system indeed, though of course at the price of losing some calculation properties.

Exercises

Exercise 10.1.1. *Show that the following axiom systems are consistent with* ZFC.

1. $$\begin{cases} \mathrm{st}(0) \\ \forall n(\mathrm{st}(n) \Rightarrow \mathrm{st}(n+1)) \\ \forall m,n((\mathrm{st}(m) \wedge \mathrm{st}(n)) \Rightarrow \mathrm{st}(m+n)) \\ \exists m,n((\mathrm{st}(m) \wedge \mathrm{st}(n)) \wedge \neg\mathrm{st}(mn)). \end{cases}$$

2. $$\begin{cases} \mathrm{st}(0) \\ \forall n(\mathrm{st}(n) \Rightarrow \mathrm{st}(n+1)) \\ \forall m,n((\mathrm{st}(m) \wedge \mathrm{st}(n)) \Rightarrow \mathrm{st}(m+n)) \\ \forall m,n((\mathrm{st}(m) \wedge \mathrm{st}(n)) \Rightarrow \mathrm{st}(mn)) \\ \exists m,n((\mathrm{st}(m) \wedge \mathrm{st}(n)) \wedge \neg\mathrm{st}(m^n)). \end{cases}$$

10.2 External recurrence relations and near stability

Stability and asymptotic stability are abstractions, and are not easily attained in practice. Also, there are natural situations of incomplete stability. For instance in steering, due to the permanent attention to a stream of new observations, one is frequently making small corrections to maintain the desired trajectory. Also the outcome of a production process may undergo small fluctuations to small impurities in ingredients, machine irregularities or adjustments by hand.

We propose notions of stability and asymptotic stability up to a neutrix, that can help to model such situations, which are stable on a macroscopic scale, while on a microscopic scale are subject to small uncertainties or instabilities. We indicate a class of recurrence relations and differential equations for which our notions are relevant. If the neutrix in question is $\oslash$, our notions come close to the nonstandard characterizations of near-stability by Lobry, Sari and Touhami [81]. For the classical notions we refer to [3].

We start by defining recurrence relations giving rise to flexible sequences. This will be done in Definition 10.2.1, through internal representatives. Some care is needed, as is illustrated by the following example.

Consider the recurrent relation

$$\begin{cases} \alpha_0 = 1 \\ \alpha_n = \oslash \alpha_{n-1}. \end{cases} \tag{10.7}$$

It is tempting to solve the equation by the sequence $\alpha_n = \oslash^n$ for $n \in \mathbb{N}$. However we encounter several problems. Firstly, if we put $\alpha_0 = 1$ and

$$\alpha_n = \{x_n : x_n = \varepsilon x_{n-1} \wedge \varepsilon \simeq 0\},$$

we will not get a flexible sequence (of external numbers), because for standard $n \geq 1$

$$\alpha_n = \begin{cases} \oslash, & n = 2k - 1 \\ \oslash^+, & n = 2k. \end{cases} \tag{10.8}$$

Also, we would obtain only part of the solution. Indeed, External Induction applied to the rules of multiplication of external numbers yields for standard $n \in \mathbb{N}$, $n \geq 1$

$$\alpha_n = \oslash, \tag{10.9}$$

for $\alpha_1 = \oslash$, and if $\alpha_n = \oslash$, we have $\alpha_{n+1} = \oslash \alpha_n = \oslash \cdot \oslash = \oslash$. So $\alpha_n = \oslash$ also for even standard $n \in \mathbb{N}$.

In addition, we cannot as such apply the classical Induction Principle to the external recurrence relation (10.9) to extend the local sequence $(\alpha_n)_{\mathrm{st}(n) \in \mathbb{N}}$ to a global sequence. Again because the classical Induction Principle is not valid for external formulas, for unlimited $\omega \in \mathbb{N}$ the expression $\oslash^\omega$ is not well-defined. Due to the presence of negative numbers, also the definition of $\oslash^\omega$ through exponential and logarithmic functions is not straightforward. In [75] it is defined by

$$\oslash^\omega = \exp_s(\omega \log_s \oslash),$$

where the *symmetrical logarithm* $\log_s$ of a neutrix $N > 0$ is given by

$$\log_s N := \log(N^+ \backslash \{0\}),$$

and the *symmetrical exponential* of a symmetric set S by

$$\exp_s S := \bigcup_{t \in e^S} [-t, t].$$

In this way $\oslash^\omega = \exp_s(-\omega \not\infty) = £e^{-\omega \not\infty}$. Observe that $£e^{-n \not\infty} = \oslash$ for standard $n \in \mathbb{N}$, so the flexible sequence (α_n) given by

$$\alpha_n = £e^{-n \not\infty} \tag{10.10}$$

satisfies (10.7). However, it is not the only solution, for if we choose $\omega \in \mathbb{N}$, a second solution is given by the sequence (β_n) defined by

$$\beta_n = \begin{cases} \oslash, & n \text{ is limited} \\ £e^{-(\omega+n)\not\infty}, & n \text{ is unlimited.} \end{cases} \tag{10.11}$$

We will now consider a general class of external recursive relations, for which we define admissible solutions through sequences of internal representatives. In this interpretation the solution of (10.7) is given by (10.10), and not by (10.11). We use the interpretation of flexible functions with parameters of Definition 4.4.3.

Definition 10.2.1. *Let $k \in \mathbb{N}$ be standard and $f : \mathbb{R}^{k+1} \to \mathbb{R}$ be internal. Let $f : \mathbb{R}^{k+2} \to \mathbb{R}$ be internal and $\alpha_1, ..., \alpha_k$ be external numbers. A* flexible recurrence relation *is a recurrent relation of the form*

$$u_{n+1} = f(n, u_n, \alpha_1, ..., \alpha_k). \tag{10.12}$$

The solution of (10.12) is defined to be the sequence of external sets (u_n), where $u_0 := \alpha_0$ is an external number, and for every $n \in \mathbb{N}$

$$\begin{aligned} u_{n+1} := \{t_{n+1} : \exists a_1, ..., a_k \forall i \in \{1, ..., k\}(a_i : \mathbb{N} \to \alpha_i, a_i \text{ internal} \wedge \\ t_{n+1} = f(n, u_n, a_1(n), ..., a_k(n)))\}. \end{aligned} \tag{10.13}$$

We call the solution admissible *if u_n is an external number, for all $n \in \mathbb{N}$.*

Proposition 10.2.1 determines an admissible solution to the equation (10.7) in the sense of Definition 10.2.1. It was already seen in (10.8) that (10.7) cannot be solved by taking constant representative functions a_i. We will use a symmetry argument.

Proposition 10.2.1. *The solution of (10.7) is given by $\alpha_n = \pounds e^{-n\not\infty}$ for all $n \in \mathbb{N}$.*

Proof. Let $n \in \mathbb{N}$. Let (α_n) be the solution of (10.7) in the sense of Definition 10.2.1. We write $\overline{\oslash}^n := \{\varepsilon^n : \varepsilon \in \oslash \wedge \varepsilon \geq 0\}$. We show first that $\alpha_n^+ = \overline{\oslash}^n$. Both sets contain 0. As for the non-zero elements, let $y \in \alpha_n, y > 0$. Then there exists an internal sequence of positive infinitesimals $(\varepsilon_k)_{1\leq k\leq n}$ such that $\prod_{1\leq k\leq n} \varepsilon_k = y$. Let $\varepsilon_- := \min\{|\varepsilon_k| : 1 \leq k \leq n\}$ and $\varepsilon_+ := \max\{|\varepsilon_k| : 1 \leq k \leq n\}$. Then $\varepsilon_-, \varepsilon_+ \simeq 0$, so $\varepsilon_-^n, \varepsilon_+^n \in \alpha_n^+$. By continuity there exists ε with $\varepsilon_- \leq \varepsilon \leq \varepsilon_+$ such that $\varepsilon^n = y$, clearly $\varepsilon \simeq 0, \varepsilon > 0$. Hence $\alpha_n^+ \subseteq \overline{\oslash}^n$. The other inclusion is obvious. Hence $\alpha_n^+ = \overline{\oslash}^n$.

Secondly, we prove that $\overline{\oslash}^n = \pounds^+ e^{-n\not\infty}$. Clearly 0 is contained in both sets. Let $\delta \simeq 0, \delta > 0$. Then $\log(\delta) \in -\not\infty$. Hence

$$\delta^n \in \pounds^+\delta^n = \pounds^+ e^{n\log(\delta)} \subseteq \pounds^+ e^{-n\not\infty},$$

implying that $\overline{\oslash}^n \subseteq \pounds^+ e^{-n\not\infty}$. The other inclusion is a consequence of the fact that a number of the form $e^{-\omega n}$ with ω unlimited is equal to ε^n, with $\varepsilon := e^{-\omega}$ infinitesimal. We conclude that $\alpha_n^+ = \overline{\oslash}^n = \pounds^+ e^{-n\not\infty}$. Then $\alpha_n = \pounds e^{-n\not\infty}$ by symmetry. □

Below we adapt some classical notions of stability like Lyapunov stability and asymptotic stability to our setting.

Definition 10.2.2. *Let N be a neutrix. A solution (u_n) of the flexible recurrence relation* (10.12) *is called N-*stable *if $u_n \in \mathbb{N}$ for all $n \in \mathbb{N}$, and for every solution (v_n) such that $v_0 - u_0 \subseteq N$ it holds that $v_n - u_n \subseteq N$ for all $n \in \mathbb{N}$.*

Definition 10.2.3. *Let N be a neutrix. A solution (u_n) of the flexible recurrence relation* (10.12) *is called N-*asymptotically stable *if $u_n \in \mathbb{N}$ for all $n \in \mathbb{N}$, and there exists $\varepsilon > N$ such that every solution (v_n) with $|v_0 - u_0| < \varepsilon$ satisfies $\lim_{n\to\infty} v_n - u_n = N$. It is called* strongly N-asymptotically stable *if for all such solutions (v_n) the difference $(v_n - u_n)$ has N as a strong limit.*

Definitions 10.2.2 and 10.2.3 are natural extensions of $\oslash$-stability and $\oslash$-asymptotic stability. Written in terms of near-equalities, and applied to equilibrium solutions, they correspond to notions of nonstandard (asymptotic) stability as given by Lobry, Sari and Touhami in [81]. They showed equivalence with classical notions. We rewrite their proof in terms of $\oslash$-stability and $\oslash$-asymptotic stability.

Proposition 10.2.2. *Consider a standard recurrence relation. Then a stable standard solution (u_n) is $\oslash$-stable and an asymptotically stable standard solution (u_n) is $\oslash$-asymptotically stable.*

Proof. Without restriction of generality we may assume that (u_n) is identically zero. Also, by Definition 10.2.1 we only need to consider internal solutions. As regards stability, let (v_n) be an internal solution such that $v_0 \simeq 0$. Suppose that there exist a standard $\varepsilon > 0$ and $n \in N$ such that $|v_n| > \varepsilon$. Now there exists $\delta > 0$ such that for all solutions v with $|v_0| < \delta$ it holds that $|v_n| < \varepsilon$ for all $n \in \mathbb{N}$. By Transfer $\delta > 0$ may be considered standard, a contradiction. Hence $v_n \simeq 0$ for all $n \in \mathbb{N}$, meaning that 0 is $\oslash$-stable.

As regards asymptotic stability, there exists $\eta > 0$ such that for every internal solution v with $|v_0| < \eta$ it holds that $\lim_{n\to\infty} v_n = 0$. By the Transfer Principle $\eta > 0$ may be considered standard, i.e. $\eta > \oslash$. Because $\lim_{n\to\infty} v_n = 0$ for all solutions v with $|v_0| < \eta$, clearly $\lim_{n\to\infty} v_n = \oslash$. Hence 0 is $\oslash$-asymptotically stable. □

As a consequence of Theorem 5.1.11 solutions nearby an N-asymptotically stable sequence satisfy a strong convergence property, as given by the next proposition.

Proposition 10.2.3. *Let $N \neq 0$ be a neutrix. If a solution (u_n) of the flexible recurrence relation* (10.12) *is N-asymptotically stable, then it is strongly N-asymptotically stable.*

Proof. There exists $\varepsilon > N$ such that for every solution (v_n) such that $|v_0 - u_0| < \varepsilon$ it holds that $(v_n - u_n)$ converges to N. Let (t_n) be an internal representative of $(v_n - u_n)$. Then (t_n) converges to N. By Theorem 5.1.11 it converges strongly to N. Hence $(v_n - u_n)$ strongly converges to N. Hence (u_n) is strongly N-asymptotically stable. □

The following class of nearly homogeneous linear difference equations exhibit asymptotically stable solutions with respect to a neutrix, which can be determined.

Proposition 10.2.4. *Let $N \neq 0$ be a neutrix and $\alpha \in \mathbb{E}, |\alpha| < 1$. Consider the recurrence relation*

$$u_{n+1} = \alpha u_n + N. \tag{10.14}$$

Then 0 is a strongly $\left(\frac{N}{1-|\alpha|}\right)$-asymptotically stable solution.

Proof. Let (u_n) be a solution and (t_n) be an internal representative of (u_n). Then there are internal sequences (a_n) of elements of α and (b_n) of elements of N such that $t_{n+1} = a_n t_n + b_n$ for all $n \in \mathbb{N}$. There exists $q \in |\alpha|$ and $c \in N$, $c > 0$ such that $|a_n| < q$ and $|b_n| < c$ for all $n \in \mathbb{N}$. Then $|t_{n+1}| \leq q|t_n| + c$ for all $n \in \mathbb{N}$, hence also $|t_n| \leq \left(|t_0| + \frac{c}{1-q}\right) q^n + \frac{c}{1-q}$. Hence 0 is a $\left(\frac{N}{1-|\alpha|}\right)$-asymptotically stable solution and by Proposition 10.2.3 it is strongly $\left(\frac{N}{1-|\alpha|}\right)$-asymptotically stable. □

In contrast to asymptotically stable solutions, a *drain* [133] is a family of solutions of a recurrence relation with "almost parallel" solutions, around a particular slowly varying solution. Theorems 3.1 and 3.4 of [133] give conditions of existence and characterization for drains, and also some examples. By Part 1 of Definition 10.2.4 below the growth of a drain must be less than n^r for all rational $r > 0$. In a sense, drains generalize stability. In fact, we will see that if a drain (u_n) is, say, positive from the beginning, and has a non-zero limit L, it is $\oslash$-stable, but not $\oslash$-asymptotically stable.

Definition 10.2.4. *Consider a standard difference equation $D : x_{n+1} = f(n, x_n)$, where f is of class C^1. A standard solution u is called a* drain *if it is eventually positive or eventually negative, and for all unlimited $\omega \in \mathbb{N}$ the following hold:*

1. *For all appreciable p such that $p\omega \in \mathbb{N}$ it holds that $\frac{u_{pn}}{u_n} \simeq 1$.*
2. *For every solution v such that $\frac{v_\omega}{u_\omega} \simeq 1$ it holds that $\frac{v_n}{u_n} \simeq 1$ for all $n \simeq \infty$.*

Using the S-continuity of f, one shows that all solutions v such that $v_0 \simeq u_0$ satisfy $v_n \simeq u_n$ for all standard $n \in \mathbb{N}$. By permanence this property persists up to some unlimited $\omega \in \mathbb{N}$. Note that L is appreciable; hence all terms u_n, v_n with unlimited n satisfy $u_n \simeq v_n \simeq L$, hence are also appreciable. Then it follows from Part 2 of Definition 10.2.4 that $v_n \simeq u_n$ for all $n \in \mathbb{N}$. Hence the drain is $\oslash$-stable. It is not $\oslash$-asymptotically stable, because if a solution (w_n) satisfies $w_\nu \not\simeq u_\nu$ for some unlimited $\nu \in \mathbb{N}$, there cannot exist $n \in \mathbb{N}$, $n > \nu$ such that $w_n \simeq u_n$.

With the help of the results mentioned above, in [133] it was verified that for standard $a > 1$ the solution $\overline{u}_n = 1 + \frac{(-1)^n}{n^a}$ of

$$u_{n+1} = \left(-1 + \frac{1}{n^a}\right) u_n + 2 - \frac{1}{n^a} + (-1)^n \left(\frac{1}{n^a} - \frac{1}{(n+1)^a} - \frac{1}{n^{2a}}\right)$$

is a drain. Hence $(\overline{u}_n)$ is $\oslash$-stable, but not $\oslash$-asymptotically stable indeed.

We end with a notion of temporal stability. This applies in particular to slow curves of singular perturbations with a singular point and canards.

Definition 10.2.5. *Let N be a neutrix.*

1. *A solution (u_n) of the flexible recurrence relation* (10.12) *is called* temporally N-stable *if there exists $\nu \in \mathbb{N}$ such that $u_n \subseteq N$ for all $n \in \mathbb{N}$, $n \leq \nu$, and if for every solution (v_n) such that $v_0 \subseteq N$ it holds that $v_n \subseteq N$ for all $n \in \mathbb{N}$, $n \leq \nu$.*
2. *A solution (u_n) of the flexible recurrence relation* (10.12) *is called* temporally strongly N-asymptotically stable *if there exist $\varepsilon > N$ and $\nu \in \mathbb{N}$ such that $u_n \subseteq N$ for all $n \in \mathbb{N}$, $n \leq \nu$, and every solution (v_n) with $|v_0| < \varepsilon$, when restricted to $n \leq \nu$, has N as a strong limit.*

We may as well adapt this definition to flexible differential equations, with solutions as given by Definition 9.4.1. *Mutatis mutandis* the zero solutions of the external Riccati-Hermite equation of Section 9.4.2 of Chapter 9 are temporally strongly $\oslash$-asymptotically stable.

Exercises

Exercise 10.2.1. *Assume we drop the condition that solutions of the recurrence relation* (10.14) *should be internally representable. With $\varepsilon \simeq 0$, $\varepsilon > 0$, give a sequence (u_n) satisfying $u_{n+1} = (1 + \varepsilon \pounds)u_n + \oslash$ for all $n \in \mathbb{N}$ which is not internally representable.*

Exercise 10.2.2. *Let $\varepsilon \simeq 0$, $\varepsilon > 0$. Solve*

$$u_{n+1} = \left(\frac{1 + \varepsilon \pounds}{2}\right) u_n + 2 + \varepsilon \oslash$$

for N-asymptotically stable solutions, where N is the smallest possible neutrix.

Exercise 10.2.3. *Consider the logistic equation $\frac{dx}{dt} = x(1-x)$. Show that it has an $\oslash$-asymptotic stable solution which is not $\pounds$-asymptotic stable.*

Exercise 10.2.4. *Consider a singular perturbation as in Theorem 9.3.2 which has a slow curve which is identically zero and attractive during at least an appreciable time. To be precise, let $\varepsilon \simeq 0$, $\varepsilon > 0$, $a, b > 0$ be standard, f defined on $[0, a] \times [-b, b]$ be internal of class $C^1 \cup S^1$ with $f(t, 0) \simeq 0$ and $\frac{\partial f}{\partial y}(t, y) \lessapprox 0$ for all (t, y) with $0 \leq t \leq a$ and $-b \leq y \leq b$. Consider the singular perturbation $\varepsilon \frac{dy}{dt} = f(t, y)$. By Theorem 9.3.2 we know that for a solution λ with $\lambda(0) = b$ there exist $\tau_1 \simeq 0$ such that $\lambda(\tau_1) \simeq 0$ and $\tau_2 \simeq 0$ such that $\lambda(\tau_2) \in \varepsilon \pounds$, where we may assume that $\tau_1 > \varepsilon \pounds$. For t with $\tau_1 \leq t \leq a$, put $\mu(t) = \max_{\tau_1 \leq s \leq t} \lambda(s)$.*

1. *Show that, given a solution* θ *such that* $0 \lnapprox \theta(t) \lnapprox b$ *for some* t *with* $\tau_1 \lnapprox t \leq b$*, there exists* s *with* $0 \lnapprox s < t$ *such that* $\theta(s) > \theta(t)$.

2. *Deduce that if* $|\mu(t)| \gnapprox 0$ *for some* t *such that* $\tau_1 \leq t \leq a$*, we obtain a contradiction.*

3. *Derive that* $\lambda(t) \simeq 0$ *for all* t *such that* $\tau_1 \leq t \leq a$.

4. *Conclude that* λ *is* $\oslash$*-asymptotically stable on* $[\tau_1, a]$ *(by [81] this is considered a low-dimensional case of Tikhonov's Theorem).*

5. *With similar methods, prove that* λ *is* $\varepsilon\pounds$*-asymptotically stable on* $[\tau_2, a]$.

6. *Assume now that the above condition on* $\frac{\partial f}{\partial y}(t, y)$ *remains the same for* $t \lnapprox a$*, but* $\frac{\partial f}{\partial y}(t, a) \simeq 0$*, which corresponds to a (near) singularity at* $(a, 0)$*. Show that there exists* $\tau_3 \simeq a$ *such that* $\lambda(t) \simeq 0$ *up to* τ_3.

7. *Apply the latter to the canard solutions of Exercise 9.3.3.*

10.3 On the size of fluctuations of the financial market

The current models of insecure financial assets (actions, interest rates, currency exchange rates) relate microscopic assumptions and macroscopic observations. In the context of the stock market, we describe here some of the involved factors, and we will see that several orders of magnitude coexist.

Time-intervals of perceptible size usually come with price-fluctuations of perceptible size. The use of the word "perceptible" depends both on the context and the interpretation, but if we accept fluctuations, from say 5% or more, as perceptible, and one year as a perceptible period, the previous observation makes sense. Smaller time-periods like a day tend to come with fluctuations of less than 1%, which can maybe be classified as small, but there are almost instantaneous fluctuations in periods less than a second. Then we lose human control, and we propose to model the length of these intervals by an infinitesimal δt. In Keisler's model of price adjustment [73], it is assumed that δt is of the order of magnitude of the inverse of the number of traders; this number is thought to be unlimited, like in the model relating microeconomy and macroeconomy given by Robinson and Brown [22] (see also [107]).

Though other approaches are proposed, an important class of models comes from stochastic differential equations. If they are standard, using Loeb-measure theory [82, 2, 29] one shows that infinitesimal discretizations of these stochastic processes have infinitely close trajectories and probability distributions. We consider here discrete processes with infinitesimal time-period

$\delta t > 0$, which correspond to the approach by Nelson in [94]. More precisely, they have the form

$$\delta S_t = \mu(t, S_t)\delta t + \sigma(t, S_t)\delta W_t, \tag{10.15}$$

where S_t indicates the stock price at time t, $\delta S_t := S_{t+\delta} - S_t$ is the increment corresponding to the time-step δt, the process W_t is the *Wiener walk* given by

$$\begin{cases} W_0 &= 0 \\ \delta W_t &= \begin{cases} \sqrt{\delta t}, & \text{with probability pr } = \frac{1}{2} \\ -\sqrt{\delta t}, & \text{with probability pr } = \frac{1}{2}, \end{cases} \end{cases} \tag{10.16}$$

μ is the *trend* of the process S_t, and σ is its *conditional variance.* An important special case is, with some abuse of language, given by $\mu(t, S_t) = \mu S_t$ and $\sigma(t, S_t) = \sigma S_t$ when μ and σ are constant, and then we obtain the *Geometrical Brownian Motion*

$$\delta S_t = \mu S_t \delta t + \sigma S_t \delta W_t. \tag{10.17}$$

Observe that both (10.16) and (10.17) incorporate two orders of magnitude: the instantaneous fluctuations of the stochastic processes are of order $\sqrt{\delta t}$, which is infinitely large with respect to the microscopic time-interval δt.

Now we analyse the order of magnitude and the behaviour of the solutions. Firstly, for appreciable t the value W_t is almost surely limited. Indeed, W_t has a binomial distribution, which due to the DeMoivre-Laplace theorem for appreciable t is nearly equal to the normal distribution with density $\frac{e^{-\frac{x^2}{2t}}}{\sqrt{2\pi t}}$ (see [132]). So its mass M_W satisfies $M_W = \pounds\sqrt{t} = \pounds$. In fact, Nelson shows in [94, Chapter 13], that almost all trajectories are S-continuous, hence also limited, at limited times t and, due to the density property, almost surely with appreciable fluctuations. It is to be observed that it follows from Nelson's theorems that if we take a standard power a other than $\frac{1}{2}$ of δt for the steps in (10.16), either almost all trajectories remain infinitesimal for limited times (for $a > \frac{1}{2}$) or almost all trajectories are unlimited for appreciable times (for $0 < a < \frac{1}{2}$). So necessarily $a = \frac{1}{2}$, and to have trajectories which are almost surely appreciable, the steps in (10.16) must have the same order of magnitude as $\sqrt{\delta t}$, and then we may choose $\delta W_t = \pm\sqrt{\delta t}$ for simplicity.

We argue that necessarily μ is limited and σ appreciable. Indeed, with initial condition S_0, for μ limited and σ appreciable trajectories corresponding to limited trajectories of the Wiener walk satisfy the approximation [30, 7, 132]

$$S_t = (1 + \oslash)S_0 \cdot e^{\left(\mu - \frac{\sigma^2}{2}\right)t + \sigma W_t}. \tag{10.18}$$

Then limited and S-continuous trajectories of class S^0 of the Wiener walk correspond to appreciable and S-continuous trajectories of the Geometrical Brownian Motion, so the latter will be almost surely appreciable, S-continuous and with appreciable fluctuations at limited time, corresponding to our observations of the market.

We show that this does not happen for different orders of magnitude for μ and σ. The analysis is simplified if we may still use formula (10.18), which by the Fehrele Principle is valid for not too extreme values of μ and σ. First, observe that if μ is limited and σ is infinitesimal, all trajectories are always almost the same for limited times, which is not reflected by reality. Second, for limited μ and unlimited σ, the trend $e^{\left(\mu-\frac{\sigma^2}{2}\right)t}$ becomes infinitesimal already at some infinitesimal time, and the trend largely dominates the stochastics fluctuations. Indeed, for appreciable t almost surely $\sigma W_t \in \oslash\cdot\left(-\frac{\sigma^2}{2}t\right)$, because $\sigma W_t \subset \pounds\sigma \subset @\sigma^2$. So almost all trajectories take infinitesimal values at appreciable times. Finally, if σ is limited and μ is unlimited, for $\mu > 0$ almost all trajectories take unlimited values at appreciable times, and for $\mu < 0$ almost all trajectories take infinitesimal values at appreciable times.

The argument above is an *a posteriori* justification, the order of magnitude of microscopic fluctuations being derived from macroscopic observations on financial markets. We will present an intrinsic argument of necessity: only if the price-fluctuation is of order $\sqrt{\delta t}$, an individual economic agent will feel an incentive to trade. It is based on an assumption of myopia: the agent is only partially, incompletely aware of its own preferences. This will be based on a notion of near-indifference in real time, to be introduced below.

Let $u(t, S(t))$ be the utility of the stock for the agent at time t. We make the following assumptions.

1. The utility function is twice continuously differentiable.

2. At time t the value $S(t)$ maximizes the agent's utility function. This means that $\frac{\partial u(t,S(t))}{\partial t} = \frac{\partial u(t,S(t))}{\partial S} = 0$; we assume that $\frac{\partial^2 u(t,S(t))}{\partial t^2}$, $\frac{\partial^2 u(t,S(t))}{\partial S^2} < 0$.

3. The indifference curves for u are curves in the $t - S$ plane satisfying equations of the form $u(t, S) = C$.

4. The agent has imperfect knowledge of the portfolio which optimizes the utility function. This is perhaps an unavoidable fuzziness, related to such phenomena as lack of self-knowledge, lack of time to study all idiosyncrasies of the market, psychological barriers to do away with an enterprise which one liked one day, or to invest in supposedly risky sectors, reluctance to mentally accept objective changes, etc. In particular we assume that the agent cannot distinguish between two utilities such that their difference is small with respect to the trading period. This leads to a relation of *near-indifference* $\sim$ given by

$$u(t, S) \sim u(t', S') \Leftrightarrow u(t, S) - u(t', S') \in \oslash\delta t. \tag{10.19}$$

5. The agent is sufficiently alert and close to the market to restore optimality of equilibrium if it is observed that the difference with respect to the optimality portfolio is of order δt, i.e. $u(t,S) - u(t',S') \in @\delta t$.

It is easy to verify (see Exercise 10.3.1) that $\sim$ is an equivalence relation. It induces a partition of the plane in domains of near-indifference, which are external sets.

So the domains of near-indifference are invariant under certain small changes in time and small price-fluctuations. The intrinsic vagueness due to not knowing to full precision the objective utility is expressed by the fact that when the utilities are appreciable, they are stable under translation by vectors from the two-dimensional neutrix $\oslash\sqrt{\delta t} \times \oslash\sqrt{\delta t}$. Indeed, assume that $t - t'$ is infinitesimal. Then a second-order Taylor expansion shows that

$$\begin{aligned}
U(t', S(t')) - U(t, S(t)) &= \frac{\partial u(t,S(t))}{\partial t}(t'-t) + \frac{\partial u(t,S(t))}{\partial S}(S(t') - S(t)) \\
&+ \frac{1+\oslash}{2}\frac{\partial^2 u(t,S(t))}{\partial t^2}(t'-t)^2 + \frac{1+\oslash}{2}\frac{\partial^2 u(t,S(t))}{\partial S^2}(S(t')-S(t))^2 \\
&+ \frac{\partial^2 u(t,S(t))}{\partial t \partial S}(t'-t)(S(t')-S(t)) + \oslash(t'-t)(S(t')-S(t)) \\
&= \frac{1+\oslash}{2}\frac{\partial^2 u(t,S(t))}{\partial t^2}(t'-t)^2 + \frac{1+\oslash}{2}\frac{\partial^2 u(t,S(t))}{\partial S^2}(S(t')-S(t))^2 \\
&+ \left(\frac{\partial^2 u(t,S(t))}{\partial t \partial S} + \oslash\right)(t'-t)(S(t')-S(t)).
\end{aligned}$$

Hence

$$U(t', S(t')) - U(t, S(t)) \in \oslash\delta t \Leftrightarrow \begin{cases} t - t' = \oslash\sqrt{\delta t} \\ S(t) - S(t') = \oslash\sqrt{\delta t}. \end{cases} \tag{10.20}$$

The agent observes that optimality is lost during the trading period δt by a price-movement of order $@\sqrt{\delta t}$. Indeed, by the above

$$\begin{aligned}
&U(t', S(t+\delta t)) - U(t, S(t)) \\
&\quad \in (1+\oslash)\frac{\partial^2 u(t,S(t))}{\partial t^2}(\delta t)^2 + (1+\oslash)\frac{\partial^2 u(t,S(t))}{\partial S^2}(\delta S(t))^2 \\
&\quad + \left(\frac{\partial^2 u(t,S(t))}{\partial t \partial S} + \oslash\right)\delta t \delta S(t) \in \pounds(\delta t)^2 - @\delta t + \pounds(\delta t)^{3/2} = -@\delta t.
\end{aligned}$$

So there is an incentive to buy or sell, depending on the sign of $\frac{\partial u(t+\delta t, S(t+\delta t))}{\partial S}$. We observe that even if the stock-price increases, the utility of the amount of risky stocks in the portfolio may very well decrease, for desires of taking profit or more risk, or even necessity, if the stocks are used to hedge some option.

By the criterion for indifference at optima (10.20) again we can attain only optimality either by following the market, and adjusting appropriately the portfolio, or by setting a price for offer or demand that has a difference of order of magnitude $@\sqrt{\delta t}$ with the current market price; in case of success this price will be the new market price, and the price-fluctuation is of order $@\sqrt{\delta t}$ indeed.

We make some final remarks on the plausibility of the stochastic difference equations (10.15) and (10.17). Assume we adapt the simple model of a binary tree. First, by the Decomposition Theorem, which states that a stochastic process is the sum of a martingale and a deterministic process, the increments of S_t may be given in the form

$$\delta S_t = \mu_t \delta t + \sigma_t \delta W_t, \tag{10.21}$$

where μ_t and σ_t are also stochastic processes; by the above it holds that σ_t is appreciable. In many price models it is assumed that all the knowledge on the stock at time t is discounted in the price (in our model this is perhaps almost true). Then μ_t and σ_t are of the form $\mu(t, S(t))$ and $\sigma(t, S(t))$, which gives rise to (10.15). If we assume that the drift and the volatility at least locally are constant, we obtain the simplification of the Geometric Brownian motion (10.17).

Exercises

Exercise 10.3.1. *Show that the relation $\sim$ of* (10.19) *is an equivalence relation.*

Exercise 10.3.2. *Above we examined the behaviour of trajectories for μ and σ slightly outside the preferred values of μ being limited and σ appreciable. Show in detail that the observations are correct for limited μ and appreciable σ.*

Exercise 10.3.3. *Show that, under the conditions of the model of this section, away from the optimum we are earlier disturbed: the optimum is not only objectively the best allocation for us, but also the more comfortable. Indeed, let U be the utility function as above. Show that if $\frac{\partial U}{\partial t}$ and $\frac{\partial U}{\partial S}$ are appreciable, for a non-optimal portfolio*

$$U(t', S(t')) - U(t, S(t)) \in \oslash\delta t \Leftrightarrow \begin{cases} t - t' = \oslash\delta t \\ S(t) - S(t') = \oslash\delta t. \end{cases}$$

Verify that the thickness of this external set is much thinner than the thickness at the optimum given by (10.20).

Exercise 10.3.4. *Consider the Geometric Brownian motion S_t with $S_0 = 1$ and $\mu = 0$. Then it is a martingale, i.e. the mean and all the conditional expectations are constant in time. Assume that σ is appreciable. Use* (10.18) *to show that up to some unlimited τ almost surely $S_t \simeq 0$ for all unlimited $t \leq \tau$; this in spite of the martingale property, i.e. $\mathrm{E}S_\tau$ being equal to $S_0 = 1$. Conclude that in the long run the almost sure losses must be compensated by rare big gains (N.B. With careful asymptotic analysis the property may be shown to hold for all unlimited t).*

10.4 Further applications of external numbers

This section considers applications of external numbers given by other authors. Subsection 10.4.1 considers optimization in the presence of uncertainties and Subsection 10.4.2 considers statistics with insecure hypotheses. Only a brief account is given, with references.

10.4.1 Near-optimization with uncertainties

In his thesis [146] Nam Van Tran obtains many results on linear and nonlinear near-optimization, i.e. optimization with uncertainties modelled by neutrices and optimal values up to a neutrix. The results concern both problems in one and in more variables, and include a Lagrange condition in terms of external numbers. We consider here only a simple example of linear optimization with constraints in terms of external numbers, related to the parameter method for solving flexible systems.

Let $\varepsilon > 0$ be infinitesimal. Consider first the linear programming problem

$$f(x, y) = -x + y \to \min, \tag{10.22a}$$

subject to the constraints

$$D := \begin{cases} (1 + \varepsilon\oslash)x + (1 + \varepsilon\pounds)y \leq 1 + \varepsilon\pounds \\ x \geq \oslash \\ y \geq \varepsilon\oslash . \end{cases} \tag{10.22b}$$

Let (ξ, η) be an external vector; we identify it with the set $\{(x, y) : x \in \xi \wedge y \in \eta\}$. We say that it is *feasible* if $(\xi, \eta) \subseteq D$. A value $f(\xi, \eta)$ is *nearly minimal* if (ξ, η) is feasible and $f(x, y) \geq f(\xi, \eta)$ for all $(x, y) \in D$, in the sense of inequalities of external numbers.

Going to representatives, we get the associated linear programming problem

$$f(x, y) = -x + y \to \min, \tag{10.23a}$$

subject to the constraints

$$D' := \begin{cases} x + y \leq 1 \\ x, y \geq 0. \end{cases} \tag{10.23b}$$

By an obvious geometrical method, we find that the point $A = (1, 0)$ is an optimal solution of the problem (10.23a)-(10.23b), with optimal value $f(1, 0) = -1 + 0 = -1$. Then an optimal solution of (10.22)-(10.22b) must be of the form $(\overline{\xi}, \overline{\eta}) = (1, 0) + N$, where N is a two-dimensional neutrix, which must be sufficiently small to be feasible. We note that the clause $x \geq \oslash$ is automatically satisfied close to A. Then the shape of N is obtained by writing the first and third constraints as inclusions for external numbers (ξ, η). Then we have

$$\begin{cases} (1+\varepsilon\oslash)\xi + (1+\varepsilon\pounds)\eta \subseteq 1+\varepsilon\pounds \\ \eta \subseteq \varepsilon\oslash \,. \end{cases} \tag{10.24}$$

The solution by the parameter method (see Exercise 10.4.1) yields

$$\begin{bmatrix} \overline{\xi} \\ \overline{\eta} \end{bmatrix} = \begin{bmatrix} 1 \\ 0 \end{bmatrix} + \varepsilon\pounds \begin{bmatrix} 1 \\ 0 \end{bmatrix} + \varepsilon\oslash \begin{bmatrix} 0 \\ 1 \end{bmatrix}. \tag{10.25}$$

It is easy to see that $(\overline{\xi}, \overline{\eta}) \subseteq D$, i.e. $(\overline{\xi}, \overline{\eta})$ is feasible. Substitution in the objective function f gives the near-minimal value $f_{\text{opt}} := -1 + \varepsilon\pounds$. As a final remark, observe that $f(\tilde{\xi}, \tilde{\eta}) = f(\overline{\xi}, \overline{\eta})$ with $\tilde{\xi} = 1 + \varepsilon\pounds, \tilde{\eta} = \varepsilon\pounds$, giving an external point with larger width such that also $f(x, y) \geq f(\tilde{\xi}, \tilde{\eta})$ on D; however $(\tilde{\xi}, \tilde{\eta})$ is not feasible.

10.4.2 On statistical estimation of uncertainties

In a series of articles [14, 15, 16, 17, 18, 20], Jacques Bosgiraud addressed the problem of statistical analysis starting from hypotheses with uncertainties modelled by external intervals. One may think about a random variable X with positive outcome of which one suspects that values close to zero will be rare. Then one could test the null hypothesis H_0: X is infinitesimal, against the alternative hypothesis H_1: $X \gtrapprox 0$.

The use of external numbers leads to various interesting results. Next to the facilitation of asymptotic calculations with many parameters, non-trivial properties of external numbers were applied in various situations. The H_0 hypothesis of being infinitesimal of the example above corresponds to $X \in \oslash$ and with respect to tests with particular thresholds H_0: $X \leq \varepsilon$ we have $\oslash = \sup\{\varepsilon : \varepsilon \simeq 0 \wedge \varepsilon > 0\} = \inf\{\varepsilon : \text{st}(\varepsilon) \wedge \varepsilon > 0\}$. The exponential small neutrix $\pounds e^{-@\omega}$ appeared in problems of large deviations, and the external vectors of Subsection 10.4.1 appeared in multivariate statistics.

Going into detail requires an extensive account of statistical tools; instead we consider a property treated by Bosgiraud in [19]: lognormal distributions with small standard deviation look very much like the usual normal distribution. *A priori* this is not obvious because the first concerns only positive numbers, and the second the whole of $\mathbb{R}$. His argument among others uses the notion of mass of Chapter 7.

A random variable X taking only positive values has a *lognormal* distribution, if $\log(X)$ has a normal distribution. Examples with an (almost) lognormal distribution are given by the random variables S_t of the discrete Geometric Brownian Motion of Section 10.3 in case the drift μ is limited and the volatility σ is appreciable (see Exercise 10.4.3).

In general the density f of the lognormal distribution is given by

$$f(x) = \frac{1}{bx\sqrt{2\pi}} e^{-\frac{(\log x - a)^2}{2b^2}} \qquad (x > 0). \tag{10.26}$$

Here $a \in \mathbb{R}$ and $b > 0$. In the just mentioned case of the discrete Geometric Brownian Motion we may take $a = {}^\circ\mu$ and $b = {}^\circ\sigma$. The expectation $\overline{\mu}$ of X is given by

$$\overline{\mu} := E(X) = \int_{-\infty}^{\infty} x f(x) dx = e^{a+\frac{b^2}{2}}, \tag{10.27}$$

and its variance $\overline{\sigma}^2$ by

$$\overline{\sigma}^2 := E(X^2) - (E(X))^2 = \int_{-\infty}^{\infty} x^2 f(x) dx - e^{2a+b^2} = e^{2a+b^2}(e^{b^2} - 1). \tag{10.28}$$

Let g be the normal density with mean $\overline{\mu}$ and standard deviation $\overline{\sigma}$, i.e.

$$g(x) = \frac{1}{\sqrt{2\pi}\overline{\sigma}} \cdot e^{-\frac{(x-\overline{\mu})^2}{2\overline{\sigma}}}. \tag{10.29}$$

To model that the lognormal distribution is very similar to the normal distribution for small variance, we let a be limited and b be infinitesimal. Then the similarity is formalized in the following theorem.

Theorem 10.4.1. *Let a be limited and b be infinitesimal, and f be given by* (10.26). *Let $\overline{\mu}$ as in* (10.27), *$\overline{\sigma}$ as in* (10.28), *and let g be given by* (10.29). *Then*

1. *$\overline{\mu} \simeq e^a$ and $\overline{\sigma} \sim e^a b$.*
2. *The densities f and g have equal mass $M := \overline{\mu} + \pounds\overline{\sigma}$, and $f(x) \sim g(x)$ for all $x \in M$.*

Proof. 1. It follows from (10.27) that $\overline{\mu} \simeq e^a$ and

$$\overline{\sigma} = e^{a+\frac{b^2}{2}} \sqrt{e^{b^2} - 1} \in (1 + \oslash)e^a \sqrt{(1+\oslash)b^2} \subseteq (1+\oslash)e^a b.$$

2. Put $x = \overline{\mu} + y\overline{\sigma}$ with limited y. Then

$$\overline{\mu} + y\overline{\sigma} = e^{a+\frac{b^2}{2}} + y e^{a+\frac{b^2}{2}} \sqrt{e^{b^2} - 1} \in e^{a+\frac{b^2}{2}}(1 + (1+\oslash)yb),$$

and $x \sim e^a$ by Part 1. Again applying Part 1 we find, by first-order approximation of the logarithm

$$\begin{aligned} f(x) &= \frac{1}{bx\sqrt{2\pi}} \cdot e^{-\frac{(\log(\overline{\mu}+y\overline{\sigma})-a)^2}{2b^2}} \\ &\in \frac{1}{(1+\oslash)be^a\sqrt{2\pi}} \cdot e^{-\frac{\left(a+\frac{b^2}{2}+(1+\oslash)yb-a\right)^2}{2b^2}} \\ &= \frac{1+\oslash}{\sqrt{2\pi}\overline{\sigma}} \cdot e^{-\frac{(yb+\oslash b)^2}{2b^2}} = \frac{1+\oslash}{\sqrt{2\pi}\overline{\sigma}} \cdot e^{-\frac{y^2}{2}} = \frac{1+\oslash}{\sqrt{2\pi}\overline{\sigma}} \cdot e^{-\frac{(x-\overline{\mu})^2}{2\overline{\sigma}^2}}. \end{aligned}$$

Hence $f(x) \sim g(x)$ for all $x \in M$, which is consequently also the mass of f. □

Exercises

Exercise 10.4.1. *Consider the optimization problem* (10.22).

1. *Verify that internal and external points outside the domain D are not feasible.*

2. *Solve the system* (10.24) *by the parameter method.*

3. *Let* $(\overline{\xi}, \eta)$ *be given by* (10.25). *Show that* $f(x,y) \geq f(\overline{\xi}, \overline{\eta})$ *for all* $(x,y) \in D$.

Exercise 10.4.2. ([50, p. 313]). *Consider two agents a and b. Both are interested in maximizing the utility with respect to two goods x and y, and to this end they are willing to exchange. The utility function of both is given by $u(x,y) = xy$. The goods are present in limited quantities. Agent a is only loosely interested, and is indifferent to infinitesimal changes. Agent b is an expert, and study of the market leads to a fixed infinitesimal $\varepsilon > 0$ such that indifference exists only for utilities which differ within $\pounds\varepsilon$.*

1. *Show that both indifference relations are equivalence relations.*

2. *Sketch the indifference domains in the plain, in the form of hyperboles with unequal "thicknesses".*

3. *Assume the initial allocation for a is $I_a = (2,6)$ and the initial allocation for b is $I_b = (5,1)$. We represent this in the "Edgeworth-box" [148], $(0,0), (7,0), (7,7), (0,7)$. Here the indifference curves of a enter from the perspective $(0,0)$, giving the thickened hyperboles $xy \simeq 6 \cdot 2 = 12$, and the indifference curves of a enter from the opposite perspective $(7,7)$, giving the thickened hyperboles $(x-7)(y-7) = 5 + \pounds\varepsilon$. Near-optimizing for both, show that, exchanging 2 unities of each goods leads to the point $(4,4)$ in the Edgeworth-box, corresponding to allocation $F_a = (4,4)$ for a and $F_b = (3,3)$ for b, increasing the utilities for both appreciably, in an almost optimal way; in a sense it is almost a "Pareto optimum"* [148].

4. *Still, during the negotiation process, agent b noticed the reduced alertness of agent a. Indicate an allocation, which improves the utility for b, such that the corresponding deterioration is hidden for a.*

Exercise 10.4.3. *Let S_t be the discrete Geometric Brownian Motion given by* (10.18) *with limited μ and appreciable volatility σ. Let y be appreciable, and possible to attain at appreciable time t by a trajectory λ of S. By* (10.17) *such a trajectory corresponds to a trajectory of the Wiener walk W. As said in 10.3, the Wiener walk is almost normally distributed with density $\frac{e^{-\frac{x^2}{2t}}}{\sqrt{2\pi t}}$.*

1. *Show that if $\lambda(t) = y$, the corresponding trajectory of the Wiener walk ends up at a value x satisfying $x \simeq \frac{\log(y) - \left(\mu - \frac{\sigma^2}{2}\right)}{\sigma}$, which is limited.*

2. *Deduce that the probability distribution of the discrete Geometric Brownian Motion is nearly lognormal, a least for appreciable y.*

Exercise 10.4.4. *Consider a random variable on the natural numbers X, having the Poisson distribution with unlimited parameter $\lambda \in \mathbb{N}$, i.e.*

$$\Pr\{X = n\} = \frac{e^{-\lambda}\lambda^n}{n!}.$$

It is well-known that $\mathrm{E}(\mathrm{X}) = \mathrm{Var}(X) = \lambda$, *so its mass is* $\lambda + \pounds\sqrt{\lambda}$. *Show, for example using Stirling's formula, that its probability distribution becomes nearly normal after the rescaling* $n \mapsto \frac{n-\lambda}{\sqrt{\lambda}}, x \mapsto \sqrt{\lambda}x$.

Chapter 11

External numbers as a complete arithmetical solid

11.1 The axioms 277
  11.1.1 Algebraic axioms 277
  11.1.2 Generalized Completeness axiom 280
  11.1.3 Arithmetical axioms 281
11.2 A formal construction of the external numbers 282
11.3 The solid $\mathcal{E}$ as a model for the axioms 284
11.4 On the axioms for the external numbers 291

In this chapter we present an axiomatic system for the external numbers. The axioms are divided into three groups. The first group deals with algebraic properties and consists of axioms for addition, axioms for multiplication, axioms for the order relation, axioms relating addition and multiplication, axioms of existence guaranteeing among other things that models are richer than fields and axioms on the value of products of magnitudes. The second group consists of an axiom scheme on a generalized Dedekind completion property. A third and final group of axioms consists of arithmetical axioms, including a scheme, on natural numbers and an Archimedean property.

A structure satisfying the algebraic axioms is called a *solid*, when the generalized completeness holds it is called a *complete solid*, and when all axioms hold it is called a *complete arithmetical solid*.

The algebraic axioms extend the semigroup axioms of Chapter 3. They point out some algebraic similarities between the external numbers and other well-known structures such as (totally ordered) groups, rings and fields. For example, the axioms for addition resemble the group axioms except that now each element has an individualized neutral element in the form of a magnitude, corresponding in the intended interpretation of external numbers to the neutrix part of that external number. In analogy with what happens within rings and fields, addition and multiplication have almost the same basic structure. Moreover, the order relation is total and compatible with the operations of addition and multiplication. A notable difference however is that the connection between addition and multiplication via the distributive law now contains a correcting factor as in Theorem 3.3.1. Solids also postulate the existence of (minimal) neutral elements for addition and multiplication (0 and 1), of a maximal individualized neutral element (denoted M), and of non-trivial magnitudes besides 0 and M. Numbers are called *precise* if their magnitude part is 0. A decomposition axiom allows us to write any element as the sum of a precise element and a magnitude.

The remaining axioms are related to the more advanced properties of the external numbers given in Chapter 4. These axioms include multiplicative properties of neutrices (which are still algebraic properties), a generalized Dedekind completeness property, and an Archimedean property. The multiplicative axioms are inspired by the results of Koudjeti and Van den Berg on idempotent neutrices. These axioms determine the product of idempotent neutrices, in fact of all neutrices, because, in analogy with Theorem 4.5.1, it is postulated that every neutrix is a multiple of an idempotent neutrix (see [42, Section 5]).

The generalized Dedekind completeness axiom is stated in the form of a scheme in order to remain in a first-order language. The structure of real numbers $\mathbb{R}$ is usually characterized up to isomorphism in a second-order language, as the unique Dedekind complete ordered field, or equivalently as the unique Archimedean complete ordered field in which Cauchy sequences converge. However second-order properties of nonstandard models of the reals are less obvious.

In order to deal with the Archimedean property some Peano-like axioms, including a scheme on induction are assumed. As shown in [42, Section 7], this Archimedean property implies that models must include a copy of the nonstandard integers, hence of the nonstandard rationals, and due to the generalized Dedekind completeness it must be possible to embed models in the nonstandard reals.

In [42] this axiomatic system was proved to be equiconsistent with ZFC via the construction of a nonstandard model $^*\mathbb{R}$, which has the form of an adequate ultralimit for a bounded version of Nelson's syntactical Reduction Algorithm [92] to hold. A sketch of this construction is given in Section 11.3. For an overview of ZFC and for a definition of adequate ultralimit we refer to Appendix A.

Up to isomorphism, once the magnitudes are specified the set of non-precise numbers of a complete arithmetical solid is completely determined as sums of a nonstandard rational and a magnitude. For the set of precise elements of a model there are upper and lower bounds; in fact the precise elements are situated between the nonstandard rationals and the nonstandard reals. As a result, complete arithmetical solids come rather close to a syntactical characterization of the external numbers.

Our approach bears similarity with the usual axiomatic presentation of the real numbers by field axioms, an axiom on completeness, and possibly an axiom on the Archimedean property.

This chapter resumes the work developed in [41, 40, 44, 42]. We refer the reader to these publications for the details.

11.1 The axioms

The axioms are presented in a first-order language. Addition, multiplication and order will be presented in the language $\{+, \cdot, \leq\}$; later on we add a unary predicate N to deal with the natural numbers. Magnitudes are individualized neutral elements. As such the individualized neutral elements are unique. The proof is similar to the proof of the uniqueness of neutral elements in groups (see [41]). It is often convenient to use the functional notation $e(x)$ to indicate the individualized neutral element of the element x. The individualized neutral elements for multiplication (unities) are also unique and we may use the functional notation $u(x)$ (see the commentary after Axiom 11.10). With respect to the individualized neutral element the symmetrical element is also unique. We may denote it by $s(x)$ or $-x$ in the case of addition and $d(x)$ or $\frac{1}{x}$ in the case of multiplication.

11.1.1 Algebraic axioms

Axioms for addition and multiplication are similar and gave rise to the notion of *assembly* in [41]. An assembly is a completely regular semigroup (union of groups), in which the magnitude operation is linear. A structure satisfying Axioms 11.1 - 11.29 is called a *solid* in [43] (see also [44]). The algebraic axioms are related to the properties of the external numbers given in Chapters 3 and 4.

1. **Axioms for addition.**

Axiom 11.1. $\forall x \forall y \forall z (x + (y + z) = (x + y) + z)$.

Axiom 11.2. $\forall x \forall y (x + y = y + x)$.

Axiom 11.3. $\forall x \exists e (x + e = x \wedge \forall f (x + f = x \Rightarrow e + f = e))$.

Axiom 11.4. $\forall x \exists s (x + s = e(x) \wedge e(s) = e(x))$.

Axiom 11.5. $\forall x \forall y (e(x + y) = e(x) \vee e(x + y) = e(y))$.

Axiom 11.3 expresses the idea that every element has its own (additive) imprecision, which can also be seen as a "magnitude" or as a "generalized zero"; non-trivial existence is postulated in Axiom 11.27 below. The element e of Axiom 11.3 is unique. Indeed, if e' satisfies Axiom 11.3, one has $e' = e' + e = e + e' = e$. This justifies the functional notation used in Axiom 11.4 and beyond. The element s is also unique and may be considered functional as well. Indeed, if s' satisfies Axiom 11.4 one has $s' = s' + e(s') = s' + e(x) = s' + x + s = x + s' + s = e(x) + s = e(s) + s = s$.

We may interpret $e(x)$ as the "imprecision" or "error" of x. Similarly we interpret $u(x)$, given by Axiom 11.8 below as a generalized unity element and as a multiplicative or relative imprecision. The functional notation used in Axioms 11.9 and 11.10 is justified in an analogous way (see Propositions 3.2.5 and 3.2.6).[1]

It follows from Axioms 11.3 and 11.5 that $e(x+y)=e(x)+e(y)$; i.e. the magnitude operation is linear.

2. **Axioms for multiplication**

Axiom 11.6. $\forall x \forall y \forall z (x(yz)=(xy)z)$.

Axiom 11.7. $\forall x \forall y (xy=yx)$.

Axiom 11.8. $\forall x \neq e(x) \exists u (xu = x \wedge \forall v (xv = x \Rightarrow uv = u))$.

Axiom 11.9. $\forall x \neq e(x) \exists d (xd = u(x) \wedge u(d) = u(x))$.

Axiom 11.10. $\forall x \neq e(x) \forall y \neq e(y) (u(xy) = u(x) \vee u(xy) = u(y))$.

Axioms 11.8 and 11.10 imply that $u(xy)=u(x)u(y)$.

3. **Order axioms**

Axioms 11.11-11.14 state that "$\leq$" is a total order relation. Axiom 11.15 states that the order relation is compatible with addition. The last two axioms state that the order relation is compatible with the multiplication by positive elements. With respect to classical order axioms, essentially the only new axiom is Axiom 11.16. This axiom states that if an element is "small", in the sense that it gets absorbed when added to a certain magnitude, then it is also smaller than that magnitude in terms of the order.

Axiom 11.11. $\forall x (x \leq x)$.

Axiom 11.12. $\forall x \forall y (x \leq y \wedge y \leq x \Rightarrow x = y)$.

Axiom 11.13. $\forall x \forall y \forall z (x \leq y \wedge y \leq z \Rightarrow x \leq z)$.

Axiom 11.14. $\forall x \forall y (x \leq y \vee y \leq x)$.

Axiom 11.15. $\forall x \forall y \forall z (x \leq y \Rightarrow x + z \leq y + z)$.

Axiom 11.16. $\forall x \forall y (y + e(x) = e(x) \Rightarrow (y \leq e(x) \wedge -y \leq e(x)))$.

Axiom 11.17. $\forall x \forall y \forall z ((e(x) < x \wedge y \leq z) \Rightarrow xy \leq xz)$.

Axiom 11.18. $\forall x \forall y \forall z ((e(y) \leq y \leq z) \Rightarrow e(x)y \leq e(x)z)$.

It follows from Axiom 11.16 that for magnitudes the order can be characterized in terms of addition. Indeed, as shown in [44, Proposition 2.6] we have that

$$\forall x, y\, (e(x) \leq e(y) \Leftrightarrow e(x) + e(y) = e(y))\,. \tag{11.1}$$

[1] In the intended model, where elements are external numbers, $e(x)$ corresponds to the neutrix part of x and $u(x)$ to its relative uncertainty (see Propositions 3.2.2 and 3.2.3).

4. **Axioms concerning addition and multiplication**

The first three axioms express properties of magnitudes. Axiom 11.19 states that the product of an element and a magnitude is a magnitude. Axiom 11.20 gives the magnitude of the product and Axiom 11.21 the magnitude of the individualized unity. Axiom 11.22 states that the distributive law holds up to a magnitude. Due to this restriction one needs to specify the symmetrical element of the product of two elements as done in Axiom 11.23.

Axiom 11.19. $\forall x \forall y \exists z (e(x)y = e(z))$.

Axiom 11.20. $\forall x \forall y (e(xy) = e(x)y + e(y)x)$.

Axiom 11.21. $\forall x \neq e(x)(e(u(x)) = e(x)d(x))$.

Axiom 11.22. $\forall x \forall y \forall z (xy + xz = x(y + z) + e(x)y + e(x)z)$.

Axiom 11.23. $\forall x \forall y (-(xy) = (-x)y)$.

5. **Axioms of existence**

Axiom 11.24 gives the existence of a (unique) minimal magnitude denoted by 0. Elements p such that $e(p) = 0$ are called *precise*. Axiom 11.25 gives the existence of a (unique) minimal unity denoted by 1. Axiom 11.26 states that there exists a maximal magnitude M; in fact, when constructing a model it corresponds to its domain. Axiom 11.27 states that there exist magnitudes other than 0 and M, implying that the domain of the model can no longer be a field. Axiom 11.28 states that every element is the sum of a precise element and a magnitude. Axiom 11.29 states that two magnitudes are separated by an element which is not a magnitude. Such an element is called *zeroless*. It follows from the existence of zeroless elements that $1 \neq 0$, hence also that a solid must contain a copy of $\mathbb{Q}$.

Axiom 11.24. $\exists m \forall x (m + x = x)$.

Axiom 11.25. $\exists u \forall x (ux = x)$.

Axiom 11.26. $\exists M \forall x (e(x) + M = M)$.

Axiom 11.27. $\exists x (e(x) \neq 0 \wedge e(x) \neq M)$.

Axiom 11.28. $\forall x \exists a (x = a + e(x) \wedge e(a) = 0)$.

Axiom 11.29. $\forall x \forall y (x = e(x) \wedge y = e(y) \wedge x < y \Rightarrow \exists z (z \neq e(z) \wedge x < z < y))$.

6. **Axioms on the product of magnitudes**

The next axiom needs some preparatory notions. A magnitude e is *idempotent* if $ee = e$. Let y be an idempotent magnitude such that $1 < y$. An *ideal* z of y is a magnitude such that $z \leq y$ and $\forall p (e(p) = 0 \wedge 0 \leq p < y \Rightarrow pz \leq z)$.

An ideal x of y is *maximal* if $x < y$ and all ideals z such that $x \leq z \leq y$ satisfy $z = x$ or $z = y$.

In a semantic setting, the ideals defined above are elements and not sets, such as in the usual algebraic interpretation of ideals of a ring. The two notions of ideal are indeed closely related. Maximal ideals happen to be idempotent. The existence of maximal ideals in the setting of rings is equivalent to the Axiom of Choice. The existence of maximal ideals in terms of magnitudes is a consequence of Axiom 11.32.

By Axiom 11.19 the product of magnitudes is a magnitude. The value of the product is obtained by relating them to idempotent magnitudes. Axiom 11.31 states that a magnitude is the product of a precise element and an idempotent magnitude. The value of all products of idempotent magnitudes is determined by Axiom 11.30 (see [42, Section 5]).

Axiom 11.30. *Let y be an idempotent magnitude such that $1 < y$ and x be the maximal ideal of y. Then $xy = x$.*

Axiom 11.31. $\forall x(x = e(x) \Rightarrow \exists p \exists y(e(p) = 0 \wedge y = e(y) \wedge yy = y \wedge x = py))$.

11.1.2 Generalized Completeness axiom

We give an axiom on generalized Dedekind completeness in order to capture the properties of Dedekind completeness of the external numbers expressed in Chapter 4. It comes in the form of a scheme, for a definite class of formulas.

Let k be a natural number. Let $\Phi(x_1, ..., x_k)$ be a formula of the language $\{+, \cdot, \leq\}$ with free variables $x_1, ..., x_k$. The formula Φ is called *restricted* if each quantifier ranges over precise elements.

Axiom 11.32 states that a lower halfline defined by a restricted formula $A(x)$ of a free precise variable x has a lowest upper bound which is the sum of a precise element and a magnitude e. This magnitude corresponds to the set of precise elements which leave the halfline invariant under addition. The completeness property can be extended to halflines of non-precise elements, generating three types of halflines instead of two as shown in [42, Theorem 3.2].

Axiom 11.32 (Generalized Dedekind completeness). *Let A be a restricted formula (possibly with non-precise parameters) allowing for a free precise variable x, and such that*

$$\exists x A(x) \wedge \forall x \forall y (A(x) \wedge y < x \Rightarrow A(y)). \tag{11.2}$$

Then one of the following holds:

1. $\exists \sigma \forall x (A(x) \Leftrightarrow x \leq \sigma)$.
2. $\exists \tau \forall x (A(x) \Leftrightarrow \forall t (t + e(\tau) = \tau \Rightarrow x < t)$.

It can be shown that 1 and 2 are mutually exclusive, and that σ and τ are unique. They are called the *weak least upper bound* of A and are denoted, with some abuse of notation, by zup A. The first condition expresses the lower-halfline property. If A is an arbitrary non-empty property, one may define A' by

$$A'(x) :\equiv \exists y(e(y) = 0 \wedge x \leq y \wedge A(y)).$$

Then A' satisfies the condition of being an halfline. The notion of weak supremum can be extended by defining $\operatorname{zup} A = \operatorname{zup} A'$. Working with upper halflines one may define in a similar way a *weak greatest lower bound*, denoted winf. Both notions can be appropriately extended to restricted formulas of a non-precise variable (see [42, Theorem 3.2]). We use this possibility in the following.

We define $\Phi(e)$ respectively $\Psi(f)$ by

$$e + e = e \wedge e < 1,$$
$$f + f = f \wedge 1 < f.$$

Then we define, again with abuse of language

$$\begin{aligned} \oslash &= \operatorname{zup} \Phi, \\ \pounds &= \operatorname{winf} \Psi. \end{aligned} \tag{11.3}$$

The element $\pounds$ is the minimal magnitude greater than 1 and the element $\oslash$ is the maximal magnitude less than 1. One can show that $\pounds$ and $\oslash$ are idempotent and that $\oslash$ is the maximal ideal of $\pounds$ in the sense of Axiom 11.30. It results from the latter axiom that $\oslash\pounds = \oslash$.

11.1.3 Arithmetical axioms

The last group of axioms allows us to distinguish between non-Archimedean ordered structures and structures with a (nonstandard) Archimedean property. The language is extended with a unary predicate, denoted by N, allowing for a free precise variable x. The idea is to interpret $N(x)$ as "x is a natural number". In this sense Axiom 11.33 states that there are no negative natural numbers, 0 is a natural number, the successor of a natural number is a natural number, and that there is no natural number between a natural number and its successor.

Axiom 11.33 (Natural numbers).

$$\begin{aligned} &\forall x(x < 0 \Rightarrow \neg N(x)) \wedge \\ &\qquad N(0) \wedge \forall x(N(x) \Rightarrow \forall y(x < y < x + 1 \Rightarrow \neg N(y)) \wedge N(x + 1)). \end{aligned}$$

The fact that within nonstandard analysis one can only apply induction to internal formulas means, in our context, that all parameters must be natural numbers and also that all references to non-precise elements such as $\pounds$ or $\oslash$

must be banned. To do so we allow induction in one precise variable, only for properties with quantifications over precise variables and with natural numbers as possible parameters.

Axiom 11.34 (Induction). *Let A be a property expressed with the symbols 0, 1, $+$, $\cdot$ and $\leq$, allowing for a free precise variable x and quantifications only over precise variables with all its parameters y satisfying $N(y)$. Then*

$$(A(0) \wedge \forall x(N(x) \Rightarrow (A(x) \Rightarrow A(x+1))) \Rightarrow \forall x(N(x) \Rightarrow A(x)).$$

The last axiom postulates the Archimedean property for the natural numbers given by Axiom 11.33.

Axiom 11.35 (Archimedean property).

$$\forall x \forall y(0 < x < y < M \Rightarrow \exists z(N(z) \wedge zx > y)).$$

Due to the existence of non-trivial magnitudes, within ordinary set theory ZFC any solid must be non-Archimedean. Indeed, let x be such that $0 < e(x) < M$. By Axiom 11.29 there exists y such that $e(x) < y < M$. Then $e'(x) := \frac{e(x)}{y} < 1$. Now $e'(x) + e'(x) = e'(x)$, and because the induction scheme holds, one obtains that $ne'(x) = e'(x)$ for all $n \in \mathbb{N}$. As a consequence $ne'(x) < 1$ for all $n \in \mathbb{N}$. However, there exists also an Archimedean field that gives rise to a solid: the external numbers.

11.2 A formal construction of the external numbers

In [42] it was shown that Axioms 11.1-11.35 are consistent by constructing a model in ZFC extending a particular nonstandard model of the real numbers; in fact, by taking a sufficiently saturated nonstandard model ${}^*\mathbb{R}$ of the real numbers, elementary equivalent to $\mathbb{R}$. Within this model one considers cosets with respect to convex subgroups which are definable by Σ_1 or Π_1 formulas. The resulting structure, whose construction we sketch below, is denoted by $\mathcal{E}$. We refer to Appendix A for the relevant background.

Let ${}^*\mathbb{R}$ be a sufficiently saturated nonstandard model of the real numbers which is elementary equivalent to $\mathbb{R}$. Let $Z_n = \bigcup_{k\leq n} \mathcal{P}^k(\mathbb{R})$ and $Z = \bigcup_{n\in\mathbb{N}} Z_n$ be the Zakon-Robinson superstructure [113, 124, 57]. Let ${}^*Z$ be an adequate ultralimit of Z [92]. Interpreting the elements of Z as standard, bounded versions of the axioms (I), (S), and (T) of Nelson's IST, as well as his Reduction Algorithm hold in this structure. In particular, the *Saturation Principle* given by Theorem 4.1.1 holds. In the context of the superstructure this implies that if $X \in Z$ and $s : X \to {}^*\mathbb{R}$, then s has always an internal extension $\tilde{s} : {}^*X \to {}^*\mathbb{R}$.

Definition 11.2.1. *We denote by $\mathcal{N}$ the set consisting of* $^*\mathbb{R}$ *and of all convex subgroups of* $^*\mathbb{R}$ *of the form* $\bigcup_{x\in X}[-s_x, s_x]$ *or* $\bigcap_{x\in X}[-s_x, s_x]$*, where* $X \in Z$ *and* $s : X \to {}^*\mathbb{R}$*. We call an element of* $\mathcal{N}$ *a* neutrix.

Without restriction of generality we may suppose that X is ordered and s is increasing in the case of unions and decreasing in the case of intersections.

Definition 11.2.2. *Let* A, B *be neutrices. With some abuse of language we call the set* $\{a \ {}^*\!+ b : a \in A \wedge b \in B\}$ *the* Minkowski sum *of* A *and* B *and the set* $\{a \ {}^*\!\cdot b : a \in A \wedge b \in B\}$ *the* Minkowski product *of* A *and* B*. Usually we simply write* $A + B$ *instead of* $A \ {}^*\!+ B$ *and* $A \cdot B$ *instead of* $A \ {}^*\!\cdot B$*.*

Definition 11.2.3. *The set* $\mathcal{E} := \{a + A : a \in {}^*\mathbb{R} \wedge A \in \mathcal{N}\}$ *is called the* set of external numbers.

If $\alpha = a + A$ is an external number, it is tacitly understood that $a \in {}^*\mathbb{R}$ and $A \in \mathcal{N}$. We may write $N(\alpha)$ instead of A and call it the *neutrix part* of α. It is not difficult to see that the neutrix part is a well-defined function from $\mathcal{E}$ to $\mathcal{N}$ (see Exercise 11.2.1).

Obviously $\alpha = y + A$ for any element $y \in \alpha$. The Minkowski sum and product of Definition 11.2.2 can be extended to external numbers in the following way.

Definition 11.2.4. *Let* $\alpha = a + A$ *and* $\beta = b + B$ *be two external numbers, the* sum *and* product *of* α *and* β *are defined as follows:*

$$\alpha + \beta = a + b + A + B,$$
$$\alpha \cdot \beta = ab + aB + bA + AB.$$

It is easy to see that the definition does not depend on the choice of representatives (see Exercise 11.2.2).

Let A, B be neutrices and $(s_x)_{x\in X}, (t_y)_{y\in Y}$ be families of elements of $^*\mathbb{R}$, with $X, Y \in Z$. Because we are only considering convex subgroups of $^*\mathbb{R}$ of the form $\bigcup_{x\in X}[-s_x, s_x]$ or $\bigcap_{x\in X}[-s_x, s_x]$, we need to show that the sum and product operations do not increase the complexity. Clearly, with addition complexity does not increase because one always has $A+B = A$ or $A+B = B$. We now consider multiplication. There is clearly no increase in complexity if both A and B are unions, or are intersections. Indeed

$$\bigcup_{x\in X}[-s_x, s_x] \cdot \bigcup_{y\in Y}[-t_y, t_y] = \bigcup_{(x,y)\in X\times Y}[-s_x, s_x]\cdot[-t_y, t_y],$$

and

$$\bigcap_{x\in X}[-s_x, s_x] \cdot \bigcap_{y\in Y}[-t_y, t_y] = \bigcap_{(x,y)\in X\times Y}[-s_x, s_x]\cdot[-t_y, t_y],$$

By the following proposition there is also no increase in complexity in the case where A is of the form $\bigcup_{x\in X}[-s_x, s_x]$ and B is of the form $\bigcap_{y\in Y}[-t_y, t_y]$.

Proposition 11.2.1. *Let $X, Y \in Z$. Let $A = \bigcup_{x\in X}[-s_x, s_x]$ and $B = \bigcap_{y\in Y}[-t_y, t_y]$ be neutrices where $(s_x)_{x\in X}, (t_y)_{y\in Y}$ are families of elements of ${}^*\mathbb{R}$. Then $AB = \bigcup_{w\in W}[-u_w, u_w]$ or $AB = \bigcap_{w\in W}[-u_w, u_w]$, where $(u_w)_{w\in W}$ is a family of elements of ${}^*\mathbb{R}$ with $W = X$ or $W = Y$.*

Proof. See Exercise 11.2.3. □

Definition 11.2.5. *Let ${}^*\leq$ be the order relation on ${}^*\mathbb{R}$. Given $\alpha, \beta \in \mathcal{E}$, we write with some abuse of language $\alpha \leq \beta$, if and only if*

$$\forall x \in \alpha \exists y \in \beta (x \ {}^*\!\leq y). \tag{11.4}$$

Let $\alpha \in \mathcal{E}$. Let $Q_\alpha = \{x \in \mathcal{E} : x \leq \alpha\}$. Then $\alpha \leq \beta$ if and only if $Q_\alpha \subseteq Q_\beta$. Note that if $\alpha \cap \beta = \emptyset$, formula (11.4) is equivalent to

$$\forall x \in \alpha \forall y \in \beta (x \ {}^*\!< y).$$

With some abuse of language, we will call $(\mathcal{E}, +, \cdot, \leq)$ the *structure of external numbers*.

Exercises

Exercise 11.2.1. *Show that the neutrix part function is a well-defined function from $\mathcal{E}$ to $\mathcal{N}$.*

Exercise 11.2.2. *Show that the addition and multiplication of external numbers are well-defined.*

Exercise 11.2.3. *Prove Proposition 11.2.1.* [Hint: Use halflines].

11.3 The solid $\mathcal{E}$ as a model for the axioms

In this section we show that the external numbers, as defined in the previous section, are a model for the axioms of Section 11.1. We work progressively, starting with the algebraic axioms of a solid.

Theorem 11.3.1. *The structure $(\mathcal{E}, +, \cdot, \leq)$ satisfies Axioms 11.1-11.29.*

In order to prove the theorem, we verify first that the axioms for addition and the axioms for multiplication are satisfied. Then we show that the order axioms and the axioms relating addition and multiplication are satisfied and finally we show that the existence axioms are verified.

Proposition 11.3.2. *The structure $(\mathcal{E}, +, \cdot, \leq)$ satisfies Axioms 11.1-11.10.*

Proof. The proof is essentially given in the same way as the results in Section 3.2 (see also [41, Section 4]). □

Proposition 11.3.3. *The structure $(\mathcal{E}, +, \cdot, \leq)$ satisfies Axioms 11.11-11.18.*

Proof. Working with halflines, it is immediate to see that the order relation is reflexive, transitive, antisymmetric and total. Then Axioms 11.11-11.14 are satisfied.

Axiom 11.15 follows from Theorem 3.1.18. As regards Axiom 11.16, assume that $\alpha + N(\beta) = N(\beta)$, i.e. $a + A + B = B$. Then $a + A \subseteq B$. Hence $\alpha \leq N(\beta)$. Axiom 11.17 is a consequence of Part 4 of Theorem 3.1.18 and Axiom 11.18 follows from Part 5 of Theorem 3.1.18. □

We turn now to the axioms which relate addition and multiplication.

Proposition 11.3.4. *The structure $(\mathcal{E}, +, \cdot, \leq)$ satisfies Axioms 11.19-11.23.*

Proof. Axioms 11.20 and 11.23 follow from Proposition 3.2.8. Axiom 11.21 follows directly from the definitions and from Theorem 1.4.2. Axiom 11.22 follows from Theorem 3.3.1. We still must show that Axiom 11.19 is satisfied. Let $A \in \mathcal{N}$ and $\beta = b + B \in \mathcal{E}$. One has

$$A(b + B) = bA + AB = \max(bA, AB).$$

Clearly $bA \in \mathcal{N}$, and $AB \in \mathcal{N}$ follows from Proposition 11.2.1. □

We consider now the group of axioms on existence. We prove first the existence of representatives of the special elements m, u and M.

Proposition 11.3.5. *The structure $(\mathcal{E}, +, \cdot, \leq)$ satisfies Axioms 11.24-11.26.*

Proof. The proposition follows by putting $m = 0$, $M = {}^*\mathbb{R}$, and $u = 1$ respectively. □

Axiom 11.27 states the existence of magnitudes between the smallest element m and the largest element M. With generalized Dedekind completeness we defined a largest magnitude $\oslash$ such that $0 < \oslash < 1$ and a smallest magnitude $£$ such that $1 < £ < M$ (here and below the symbol $<$ simply indicates "less than and not equal", and not "less than and separated from", like in the previous chapters). We will interpret $\oslash$ and $£$ in the following way, where we identify $\mathbb{N}$ with the standard integers of ${}^*\mathbb{R}$.

Definition 11.3.1. *We define* $\Lambda = \bigcup_{n \in \mathbb{N}} {}^*[-n, n]$ *and* $\Theta = \bigcap_{n \in \mathbb{N}} {}^*\left[-\frac{1}{n}, \frac{1}{n}\right]$.

Theorem 11.3.6. *The external sets Θ and Λ are neutrices. One has $0 < \Theta < 1 < \Lambda < {}^*\mathbb{R}$. The interpretation of $\oslash$ is Θ and the interpretation of $£$ is Λ.*

Proof. Clearly Θ and Λ are neutrices. Because ${}^*\mathbb{R}$ is a superstructure of $\mathbb{R}$ there exists an infinitely large element ν in ${}^*\mathbb{R}$. Clearly $\nu \notin \Lambda$ hence $\Lambda \neq {}^*\mathbb{R}$. Also $0 \neq \frac{1}{\nu} \in \Theta$. Obviously $\Theta < 1 < \Lambda$, because $1 \in \Lambda$ and $1 \notin \Theta$. Let L, I be the interpretations of $\pounds$ and $\oslash$ respectively. Then L, I must be neutrices. There does not exist a proper convex subset of $\mathbb{N}$ closed under addition so there does not exist a neutrix A such that $1 < A < \Lambda$. Also, there does not exist a neutrix B such that $\Theta < B < 1$. Hence $\Lambda \subseteq L$ and $I \subseteq \Theta$. Because $\Lambda + \Lambda = \Lambda$ and $1 < \Lambda$, by the definition of $\pounds$ one has $L \subseteq \Lambda$. Also, because $\Theta + \Theta = \Theta$ and $\Theta < 1$, by the definition of $\oslash$ one has $\Theta \subseteq I$. Hence $\Lambda = L$ and $\Theta = I$. □

Proposition 11.3.7. *The structure $(\mathcal{E}, +, \cdot, \leq)$ satisfies Axioms 11.27-11.29.*

Proof. By Theorem 11.3.6, Axiom 11.27 holds. Axiom 11.28 is trivially satisfied. Finally we turn to Axiom 11.29. Let $A, B \in \mathcal{N}$ be such that $A \neq B$. We may assume without loss of generality that $A \subset B$. Then there is a nonstandard real number b such that $b \in B$ and $b \notin A$. Furthermore, b may be supposed positive. We show that $A < b < B$. Indeed, because B is a group and b is positive one has $b < 2b \in B$. Hence $b < B$. Suppose that $b \leq A$. Then there exists $a \in A$ such that $b \leq a$. Because $0 \in A$ and $a \in A$, by convexity $b \in A$, a contradiction. Hence $A < b$. □

Proof of Theorem 11.3.1. The theorem follows by combining Proposition 11.3.2, Proposition 11.3.3, Proposition 11.3.4, Proposition 11.3.5 and Proposition 11.3.7. □

We will now prove that the Generalized Completeness Axiom 11.32 holds in $\mathcal{E}$. We deal with this axiom before the axioms on multiplication of magnitudes, because Generalized Completeness is needed to prove the existence of the maximal ideals of Axiom 11.30.

In Corollaries 4.3.4 and 4.3.5 a normal form for convex subsets of real numbers is stated. In the case of a (external) lower halfline this normal form indicates that its upper boundary is well-defined, in the form of a unique external number.

Let Z be the superstructure defined in the previous section. In order to prove that Axiom 11.32 holds we interpret formulas from the language $\{+, \cdot, \leq\}$ in the adequate ultralimit ${}^*Z$ and show that a bounded version of the Reduction Algorithm as well as the Saturation Principle hold in this structure.

Definition 11.3.2. *Let k be a natural number. Let $\Phi(x_1, ..., x_k)$ be a formula of* ZFC *with free variables $x_1, ..., x_k$. The formula Φ is called* bounded *(relatively to $Z \cup {}^*Z$) if all quantifications take the form $\forall x \in X$ or $\exists x \in X$, with $X \in Z$ or $X \in {}^*Z$.*

Definition 11.3.3. *Let k be a natural number. A bounded formula $\Phi(x_1, ..., x_k)$ is called* internal *(with some abuse of language) if all its quantifications take the form $\forall x \in X$ or $\exists x \in X$, with $X \in {}^*Z$.*

Let $\Phi(x_1, ..., x_k)$ be a restricted formula of the language $\{+, \cdot, \leq\}$. We will interpret Φ by a formula $\overline{\Phi}$ in the structure $\mathcal{E}$ by induction on the complexity of the formula, and show that $\overline{\Phi}$ is bounded. Observe that a term $t(x_1, ..., x_k)$ is the result of a finite number of additions and multiplications of the variables $x_1, ..., x_k$. Each variable x_i with $1 \leq i \leq k$ is interpreted by an element, say, α_i of $\mathcal{E}$; in particular, if x_i is precise, then $\alpha_i \in {}^*\mathbb{R}$. Then the interpretation $\overline{t}(\alpha_1, ..., \alpha_k)$ of t is the result of a finite number of additions and multiplications of the elements $\alpha_1, ..., \alpha_k$. An atomic formula is of the form $t(x_1, ..., x_k) \leq s(y_1, ..., y_m)$, where m is a natural number and s is a term with variables $(y_1, ..., y_m)$. Then its interpretation is of the form

$$\overline{t}(\alpha_1, ..., \alpha_k) \leq \overline{s}(\beta_1, ..., \beta_m), \tag{11.5}$$

with $\beta_1, ..., \beta_m \in \mathcal{E}$. It follows from Definition 11.2.1 that the α_i and β_j are either unions or intersections of families of intervals in ${}^*\mathbb{R}$ indexed by elements of sets which are elements of Z. Hence the inequality (11.5) is expressed by a bounded formula.

Clearly the negation of a bounded formula is a bounded formula, and the conjunction of bounded formulas is a bounded formula. Since quantifiers in restricted formulas of the language $\{+, \cdot, \leq\}$ range over precise elements, quantifiers in their interpretations range over ${}^*\mathbb{R}$, hence yield bounded formulas.

We conclude that the interpretation $\overline{\Phi}$ of Φ is bounded.

Nelson's Reduction Algorithm, properly adapted, transforms a bounded formula into a bounded formula of the form $\forall x \in X \exists y \in Y\, I(x, y)$, with $X, Y \in Z$ and $I(x, y)$ internal. Nelson's Reduction Algorithm uses three principles, Transfer (T), Idealization (I) and modified Standardization (S′), corresponding to formula (4.4), which is a consequence of (S).

By [92] the Transfer Axiom and the Idealization Axiom of IST, when relativized to ${}^*Z$, hold in ${}^*Z$ indeed. In our context (S′) takes the following form. Let $\Phi(x, y)$ be a bounded formula; this means that all quantifiers and parameters range over some Z_n. Let $m, n \in \mathbb{N}$ and X, Y such that $X \subseteq Z_m$ and $Y \subseteq Z_n$. Assume that $\forall x \in Z_m \exists y \in Z_n\, \Phi(x, y)$. Then there must exist a function $\tilde{y} \in Z$ such that $\forall x \in X\, \Phi(x, \tilde{y}(x))$. This is true, because if Φ is a formula of ZFC, by the Axiom of Choice there exists $\tilde{y} : X \to Y$ such that $\forall x \in X\, \Phi(x, \tilde{y}(x))$. Clearly $\tilde{y} \in Z$. So (S′) also holds in Z.

The bounded versions of the principles (I), (S′), (T) stated above hold in Z^*. This means that applying any of these principles to a bounded formula results in a bounded formula. By the reasoning in the paragraph above this is clearly true for (S′). We verify the property also for (T) and (I). Let in the formulas below Φ always be a bounded formula. Then (T) becomes

$$\forall y \in Y (\forall x \in X\, \Phi(x, y) \Leftrightarrow \forall x \in {}^*X\, \Phi(x, y)),$$

where $X, Y \in Z$ and Φ internal. Also (I) becomes

$$\forall w \in {}^*W (\forall v \in \mathcal{P}_{\text{fin}}(X) \exists y \in {}^*Y \forall x \in v\, \Phi(x, y, w)$$
$$\Leftrightarrow \exists y \in {}^*Y \forall x \in {}^*X\, \Phi(x, y, w)),$$

where $X, Y, W \in Z$, Φ internal and $\mathcal{P}_{\text{fin}}(X)$ is the set of all finite subsets of X. Note that $\mathcal{P}_{\text{fin}}(X) \in Z$.

So we have the following theorem.

Theorem 11.3.8. *Every bounded formula Φ is equivalent to a bounded formula of the form $\forall x \in X \exists y \in Y\, I(x, y)$, with $X, Y \in Z$ and $I(x, y)$ internal.*

Because the Saturation Principle is true in bounded IST it also holds in $^*\mathbb{Z}$. We may now apply Theorem 4.3.4 and Corollary 4.3.5 to show that generalized Dedekind completeness holds.

Theorem 11.3.9. *Axiom 11.32 holds in $\mathcal{E}$.*

Proof. The interpretation $\overline{A}(a)$ of $A(X)$, with $a \in {}^*\mathbb{R}$, is a bounded formula. Hence one can apply the Reduction Algorithm to $\overline{A}(a)$ to obtain an equivalent formula of the form $\forall u \in U \exists v \in V\, B(u, v, a)$, with $U, V \in Z$ and B internal. Since $\overline{A}(a)$ defines a lower halfline, by Theorem 4.2.1 this formula can be reduced to a formula of the form $\exists y \in Y\, C(y, a)$ or $\forall y \in Y\, C(y, a)$, with $Y \in Z$ and C internal. Then the result follows by Theorem 4.3.4 and Corollary 4.3.5. □

We show now that the two axioms on multiplication of magnitudes hold in the model $(\mathcal{E}, +, \cdot, \leq)$. We recall that magnitudes are interpreted by convex groups. The next proposition states that the interpretation of an idempotent magnitude larger than 1 is a ring with unity, the interpretation of an ideal in a solid is an ideal in the algebraic sense and that under these interpretations the product of an idempotent magnitude and its maximal ideal is equal to this maximal ideal.

We will use the following theorem from [42].

Theorem 11.3.10. *Let J be an idempotent magnitude such that $1 < J < M$ and let $I := \sup\left\{\frac{1}{\omega} : \omega \text{ precise} \wedge J < |\omega|\right\}$. Then $0 < I \leq \oslash$ and I is the maximal ideal of J. Moreover, the maximal ideal of M is equal to 0.*

Proof. See Exercise 11.3.6. □

Proposition 11.3.11. *Let S be a complete solid. Let $J \in S$ be an idempotent magnitude such that $1 < J$. Let I be an ideal of J. In the model $(\mathcal{E}, +, \cdot, \leq)$, the interpretation $\overline{J}$ of J is a ring and the interpretation $\overline{I}$ of I is an ideal of the ring $\overline{J}$. Moreover, if I is maximal, then $\overline{I} := \left\{\frac{1}{x} : x \in {}^*\mathbb{R} \wedge J < |x|\right\} \cup \{0\}$ is maximal and $\overline{I} \cdot \overline{J} = \overline{I}$.*

Proof. The interpretation $\overline{J}$ of J in $\mathcal{E}$ is an idempotent neutrix, which is clearly a ring. An ideal in our syntactical sense is a magnitude, so $\overline{I}$ is a neutrix. Because for all $y < J$ one has $yI \leq I$, by the Minkowski definition of the product $xz \in \overline{I}$ for all $x \in \overline{I}$ and $z \in \overline{J}$. This means that $\overline{I}$ is an ideal of $\overline{J}$ in the sense of rings.

Assume now that I is maximal; then $I := \sup\left\{\frac{1}{\omega} : \omega \text{ precise } \wedge J < |\omega|\right\}$ by Theorem 11.3.10. Let $K := \left\{\frac{1}{x} : x \in {}^*\mathbb{R} \wedge \overline{J} < |x|\right\} \cup \{0\}$. We show that $\overline{I} = K$. Suppose that there exists $y \in \overline{I}\backslash K$. Then $\left|\frac{1}{y}\right| < \overline{J}$. Hence there exists $u < I$ such that $\frac{1}{u} < J$, in contradiction with the definition of I. Hence $\overline{I} \subseteq K$. Suppose that there exists $z \in K\backslash\overline{I}$. Then $\frac{1}{z} < \overline{J}$, i.e. $\frac{1}{z} \in \overline{J}$, in contradiction with the definition of K. Hence $K \subseteq \overline{I}$ and we conclude that $\overline{I} = K$. Suppose the ring $\overline{J}$ has an ideal L with $\overline{I} \subset L \subset \overline{J}$. Let $x \in L\backslash\overline{I}$, $x < 1$ be positive. Because $\overline{I} = K$ we may find $y \in \overline{J}\backslash L$ such that $\frac{1}{x} < y$. Then $y^2 \in \overline{J}$. But $xy^2 \notin L$, since $y < xy^2$. So we have a contradiction. As a consequence $\overline{I}$ is the maximal ideal of the ring $\overline{J}$.

As observed above, $yz \in \overline{I}$ for all $y \in \overline{J}$ and $z \in \overline{I}$. Again by the Minkowski definition of the product, it holds that $\overline{I} \cdot \overline{J} \subseteq \overline{I}$. Clearly $\overline{I} \subseteq \overline{I} \cdot 1 \subseteq \overline{I} \cdot \overline{J}$. Hence $\overline{I} \cdot \overline{J} = \overline{I}$. □

Corollary 11.3.12. *The structure $(\mathcal{E}, +, \cdot, \leq)$ satisfies Axiom 11.30.*

Axiom 11.31 is verified by easily adapting the proof of Theorem 4.5.1 to the present setting.

Theorem 11.3.13. *The structure $(\mathcal{E}, +, \cdot, \leq)$ satisfies Axiom 11.31.*

Finally we prove that the axioms on the existence and behaviour of natural numbers hold in $\mathcal{E}$. We will require some preparatory results.

Lemma 11.3.14. *Let A be a structure satisfying Axioms 11.1-11.5 and Axioms 11.11-11.16. Let $x, y \in A$. If $x < e(x)$ then $x < e(y)$.*

Proof. See Exercise 11.3.1. □

Lemma 11.3.15. *Let S be a solid and let $x, z \in S$. If $e(x) < z$ and z is zeroless, there is a precise element t such that $e(x) < t \leq t + e(z) < z$.*

Proof. Let $z = p + e(z)$ with p precise. If $z < e(z)$ then $z < e(x)$ by Lemma 11.3.14, a contradiction. Hence $e(z) < z$, meaning that p is positive, so $0 < \frac{p}{2} < p$. One has $e(z) < \frac{p}{2}$. Indeed, if $\frac{p}{2} \leq e(z)$ one would have $p \leq 2e(z) = e(z)$ which is a contradiction. It follows that $\frac{p}{2} + e(z) < p \leq p + e(z) = z$. If $e(x) \leq e(z)$, we are done. If $e(z) < e(x)$, suppose that $\frac{p}{2} \leq e(x)$. Then $p \leq 2e(x) = e(x)$ and $z = p + e(z) \leq e(x) + e(x) = e(x)$, a contradiction. Hence $e(x) < \frac{p}{2} \leq \frac{p}{2} + e(z)$. □

Lemma 11.3.16. *The element z in Axiom 11.29 may be supposed precise.*

Proof. Let z be zeroless and such that $e(x) < z < e(y)$. By Lemma 11.3.15 there exists a precise t such $e(x) < t < z < e(y)$. □

Lemma 11.3.17. *Let S be a solid and $x, y \in S$ be such that $x < y$. Then there exists a precise element p such that $x < p < y$.*

Proof. Without restriction of generality we may assume that $x = e(x)$. The case $y = e(y)$ follows from Lemma 11.3.16 and the case where y is zeroless follows from Lemma 11.3.15. □

Theorem 11.3.18. *Let N be interpreted by $^*\mathbb{N}$, the set of non-negative non-standard integers of $^*\mathbb{R}$. Then Axioms 11.33-11.35 hold in $\mathcal{E}$.*

Proof. We interpret the symbols $+$, $\cdot$ and $\leq$ respectively by the addition, multiplication and order relation in $\mathcal{E}$. This corresponds with the addition $^*+$, the multiplication $^*\cdot$ and the order relation $^*\leq$ in $^*\mathbb{N}$. Then Axiom 11.33 holds because $^*\mathbb{N}$ does not contain negative numbers, $^*0 \in {}^*\mathbb{N}$ and whenever $n \in {}^*\mathbb{N}$, $n+1 \in {}^*\mathbb{N}$, but $y \notin {}^*\mathbb{N}$ for any $y \in {}^*\mathbb{R}$ with $n < y < n+1$. Axiom 11.34 states that induction is valid for each formula A with the symbols 0, 1, $+$ and $\cdot$, and precise variables which have the property N. Then its interpretation $^*A$ is a formula with the symbols ^{*0}, ^{*1}, $^*+$ and $^*\cdot$, with parameters interpreted by elements of $^*\mathbb{N}$, and quantifications ranging over $^*\mathbb{N}$. Because $^*\mathbb{N}$ is a model of Peano Arithmetic, Axiom 11.34 holds in $\mathcal{E}$ indeed. As regards Axiom 11.35, it follows from Lemma 11.3.17 that it is enough to show that the axiom holds for precise elements. Let $x, y \in {}^*\mathbb{R}$ be such that $0 < x < y < M$. By construction, for all $a \in {}^*\mathbb{R}$ there exists $n \in {}^*\mathbb{N}$ such that $a < n$. In particular there exists $m \in {}^*\mathbb{N}$ such that $\frac{y}{x} < m$. Hence $y < mx$, so Axiom 11.35 holds in $\mathcal{E}$. □

Theorem 11.3.19. *The structure $(\mathcal{E}, +, \cdot, \leq)$ satisfies Axioms 11.1-11.35.*

Proof. Directly from Theorem 11.3.1, Theorem 11.3.9, Corollary 11.3.12, Theorem 11.3.13 and Theorem 11.3.18. □

Corollary 11.3.20. *Axioms 11.1-11.35 are consistent with* ZFC.

Exercises

Exercise 11.3.1. *Prove Lemma 11.3.14.* [Hint: Use (11.1)].

Exercise 11.3.2. *Let S be a solid. Show that for every $x, y, z \in S$ it holds that*

1. $xy + xz = x(y + z) + e(xy + xz)$.
2. If $e(x)(y + z) = e(x)y + e(x)z$ then $xy + xz = x(y + z)$.

Exercise 11.3.3. *Let S be a solid and let $x, y \in S$ be zeroless. We define $x^0 = u(x)$ and $x^{n+1} = x^n x$. Show that*

1. $x^n x^m = x^{n+m}$.
2. $(x^n)^m = x^{nm}$.

3. $(xy)^n = x^n y^n$.

4. $\left(\frac{x}{y}\right)^n = \frac{x^n}{y^n}$.

Exercise 11.3.4. *Let S be a solid. Show that $xy + e(x)y = xy$.*

Exercise 11.3.5. *Let S be a solid. Show that for all $x, y, z \in S$ it holds that $e(x)(y+z) \leq e(xy+xz)$. Show, by giving a suitable example, that there exist external numbers x, y, z such that $e(x)(y+z) \neq e(xy+xz)$.*

Exercise 11.3.6. *Prove Theorem 11.3.10 using the following steps.*

1. *Let $0 < p$ be precise. Show that*

   (a) *If $p < I$ then $2p < I$.*

   (b) *If $p < I$ then $\sqrt{p} < I$.*

   (c) *If $1 \leq p < J$ then $I < 1/p$.*

2. *Show that $I < 1$ is an idempotent magnitude.*

3. *Let r, p be precise and such that $1 < p < J$ and $r < I$. Show that $rp < I$.*

4. *Prove that I is an ideal of J.*

5. *Show that for any ideal K of J it holds that either $K \leq I$ or $K = J$.*

11.4 On the axioms for the external numbers

As shown in [42] using a structure based on the Lightstone-Robinson field ${}^{\rho}\mathbb{R}$ (see [79]), Axioms 11.1-11.31 are not sufficient for a characterization of the external numbers. In an attempt to get as close as possible to a characterization of the external numbers using only first order axioms, the Axioms 11.32-11.35 were introduced.

It is to be expected that the set of precise numbers is not uniquely determined. Indeed, we would then have a first-order characterization of a (nonstandard) set of real numbers, for the axioms of Section 11.1 are stated within first-order logic. A result from [42] states that if a set of magnitudes is specified in a complete arithmetical solid, the set of non-precise numbers is completely determined as sums of nonstandard rationals and a nonzero magnitude.

For the set of precise numbers lower and upper bounds can be given. Every complete arithmetical solid contains a copy of a nonstandard model of Peano Arithmetic. As a consequence, in our framework we have a copy of the nonstandard rationals. By analogy to the construction of the reals via

Dedekind cuts it is possible to show that, up to isomorphism, the set of precise numbers is necessarily a nonstandard ordered field situated between the nonstandard rationals and the nonstandard reals [42, Theorem 7.17]; the field is Archimedean for the corresponding set of nonstandard natural numbers.

It is also possible to identify a notion of standard part, called *shadow* in [42], and use it to prove that a complete arithmetical solid has two built-in models of the rational numbers. This notion of shadow is based on the well-known construction of the standard reals as the quotient of the rationals by the infinitesimals. The set of shadows of the precise elements is situated between the set of shadows of the nonstandard rationals and the set of shadows of the nonstandard reals. This is to be compared with the well-known theorem expressing that an Archimedean ordered field necessarily lies between the rationals and the reals. The "standard" structure related to this field is also situated between the rationals and the reals. As a consequence, a complete arithmetical solid E can only be constructed in a nonstandard setting.

The fact that the algebraic axioms alone are not sufficient for a characterization of the external numbers is shown by exhibiting a proper substructure ${}^{\rho}\mathcal{E}$ of the structure $\mathcal{E}$ of Section 11.2 which also satisfies all the algebraic axioms. In this way is also possible to obtain an interpretation of the symbols $\oslash$, respectively $\pounds$ as defined in (11.3) different from the infinitesimals, respectively the limited numbers.

Indeed, let $\rho \in {}^*\mathbb{R}, \rho > 0$ be infinitely large. We define $G := \bigcup_{n\in\mathbb{N}} [-\rho^n, \rho^n]$ and $H := \bigcap_{n\in\mathbb{N}} \left[-\frac{1}{\rho^n}, \frac{1}{\rho^n}\right]$. Clearly G and H are idempotent. The field G/H is in fact the field ${}^{\rho}\mathbb{R}$ studied by Lightstone and Robinson in [79].

Definition 11.4.1. *We define ${}^{\rho}\mathcal{I}$ as the set of all convex sets $I \subseteq {}^*\mathbb{R}$ of the form $\bigcup_{n\in\mathbb{N}} [-p_n, p_n]$ or $\bigcap_{n\in\mathbb{N}} \left[-\frac{1}{p_n}, \frac{1}{p_n}\right]$, with $p_n > 0$, and $\frac{p_{n+1}}{p_n}$ increasing and such that $p_0 = 1$, $p_1 \geq \rho$ and $\frac{p_{n+1}}{p_n} \geq p_n^{\frac{1}{n}}$ for all $n \in \mathbb{N}\setminus\{0\}$. We let ${}^{\rho}\mathcal{N}$ be the set of all neutrices of the form qI where $q \in {}^*\mathbb{R}$ and $I \in {}^{\rho}\mathcal{I}$, and ${}^{\rho}\mathcal{E}$ as the set of elements of $\mathcal{E}$ of the form $r + L$ where $r \in {}^*\mathbb{R}$ and $L \in {}^{\rho}\mathcal{N} \cup \{0\} \cup \{{}^*\mathbb{R}\}$.*

Proposition 11.4.1. *The set ${}^{\rho}\mathcal{I}$ consists of idempotent neutrices, with minimal element greater than 1 equal to G and maximal element less than 1 equal to H. Moreover, ${}^{\rho}\mathcal{I}$ is closed under addition and multiplication and satisfies Axiom 11.30.*

Proposition 11.4.2. *The set ${}^{\rho}\mathcal{E}$ satisfies Axioms 11.1-11.31.*

The verification of these two propositions is straightforward [42]. By Proposition 11.4.1, the symbol $\oslash$ can be interpreted by H, and the symbol $\pounds$ can be interpreted by G. Indeed, in ${}^{\rho}\mathcal{E}$ the set of neutrices less than 1 has a weak supremum, in fact a maximum, in the form of H, while G is the weak infimum (minimum) of the set of neutrices larger than 1.

To show whether the generalized Dedekind completeness axiom holds in ${}^{\rho}\mathcal{E}$ is still an open problem. In order to do so, one should establish that definable lower halflines in ${}^{\rho}\mathcal{E}$ have a weak supremum. This requires a deeper study of polynomials of external numbers. The difficulty arises from the fact that since distributivity does not hold in full generality, the product of polynomials does not need to be a polynomial.

Nevertheless, the introduction of natural numbers via the arithmetical axioms permits us to distinguish between $\mathcal{E}$ and ${}^{\rho}\mathcal{E}$. Indeed, the set $\mathcal{E}$ contains a copy of ${}^*\mathbb{N}$, and induction holds in ${}^*\mathbb{N} \cap \pounds$, but not in ${}^*\mathbb{N} \cap G$. For example, in ${}^*\mathbb{N} \cap G$ the domain of function $x \mapsto 2^x$ is closed under the successor function, but this function is not total.

The axiomatic approach described above gives rise to an alternative way to build nonstandard real numbers in which, unlike Nelson's approaches [92, 94] the infinitesimals are not postulated through a new undefined symbol, but by the existence of magnitudes. The precise elements of a complete arithmetical solid satisfy the axioms of ZFL given in Chapter 1, moreover the axioms of ENA$^-$ when restricted to the language $\{+, \cdot\}$ [42]. It is even closely related to the axiomatic system ENA, for in a complete arithmetical solid external numbers are elements and permit us to define in a simple way external sets.

Appendix A

Background on Nonstandard Analysis

A.1	On the foundations of external sets		295
A.2	Set theoretical Nonstandard Analysis		296
	A.2.1	ZFC	296
	A.2.2	Theories for internal sets: IST and BST	299
	A.2.3	Theories for external sets: HST	301
		A.2.3.1 HST axioms	303
A.3	Model theoretical nonstandard analysis		305
	A.3.1	The superstructure approach	306

A.1 On the foundations of external sets

Infinitesimals have been present in mathematics, in one way or the other, at least since Archimedes' *The Method of Mechanical Theorems*, and without having a proper and rigorous formulation until the work of Abraham Robinson [111, 112]. Robinson and Zakon [113] later gave a simpler, purely set-theoretical approach using Model Theory and superstructures [113] (published in [86]), which was further developed in [155]. Also in [86], Kreisel [77] raised the following questions[1]:

> 1. Is there a simple formal system (in the usual sense, that is, with a recursive, preferably finite, list of rules and axiom schemata) in which existing practice of nonstandard analysis can be codified? And if the answer is positive:
>
> 2. Is this formal system a conservative extension of current systems of analysis (in which existing practice of standard analysis has been codified)?

As it turns out, the answer to both questions is positive and several set theories that axiomatize the nonstandard methods have been made since (see [71] for a book on axiomatic nonstandard theories and also [33] for a survey paper of nonstandard set theories). Nonstandard set theories can be classified

[1] Kreisel says in [77] that the first question posed itself as soon as Robinson published [111]. As for the second question he says that Robinson posed a related question, concerning nonstandard arithmetic, at a meeting in London in 1965.

into two groups: *internal theories*, which axiomatize standard and internal sets only and *external theories*, which axiomatize external sets as well. As such, external theories tend to be more intricate and complicated.

In Section A.2 we recall the axioms and some basic set-theoretical notions of ZFC, the usual framework for (standard) mathematics. Actually, almost all nonstandard theories are conservative extensions of ZFC. This means that every theorem of the nonstandard theory expressible in the language of ZFC is also a theorem of ZFC. However, due to the so-called *Hrbáček's paradoxes*, in order to deal with external sets some axioms which hold for common set theory need to be adapted. Nonstandard theories also typically include some sort of transfer in order to connect the standard and the (internal) nonstandard worlds. In IST, which was described in Chapter 4, transfer also implies that all uniquely defined objects are standard [92, p. 1166]. We review the nonstandard internal set theory BST which may be seen as a bounded version of IST. The theory BST has an extension to the external theory HST [71] which allows us to construct neutrices and external numbers that are in general external sets.

In Section A.3 we recall the model-theoretical construction of nonstandard analysis via superstructures by Robinson and Zakon [113, 155] and discuss how the external numbers may be developed in an appropriate superstructure.

A.2 Set theoretical Nonstandard Analysis

A.2.1 ZFC

The most common axiomatic set theory and thus the most common foundational background for mathematics is the theory Zermelo-Fraenkel-Choice which is usually abbreviated ZFC. For a thorough and modern treatment of ZFC we refer to [66]. In ZFC every object is a set. This theory has a very simple language, containing (apart from the logical symbols) only the binary membership predicate "$\in$". One should read $x \in y$ as 'x is a member of y' meaning that x is an element appearing in the set y. Bounded formulas give rise to sets while unbounded formulas give rise to classes. Classes in ZFC are informally defined as extensions of formulas in the following way. If $\varphi(x, p_1, ..., p_n)$ is a formula with parameters $p_1, ..., p_n$ then $\mathcal{C} = \{x : \varphi(x, p_1, ..., p_n)\}$ is a *class*. The members of a class $\mathcal{C}$ are the sets x that satisfy $\varphi(x, p_1, ..., p_n)$. We say that $\mathcal{C}$ is definable from $p_1, ..., p_n$. If $\varphi(x)$ has no parameters p_i then we say that the class $\mathcal{C}$ is *definable*. The reason to consider classes is because classes are simpler to deal with than formulas. In the following we recall the axioms of ZFC with some comments.

Extensionality

The first axiom postulates the intuitive notion that sets with the same elements are equal. This axiom is also useful to prove that a given set is unique.

$$\forall X \forall Y (\forall x (x \in X \Leftrightarrow x \in Y) \Leftrightarrow X = Y).$$

Pair

Given two sets a and b there is a set (unique by extensionality) containing exactly a and b.

$$\forall a, b \exists A \forall x (x \in A \Leftrightarrow x = a \vee x = b).$$

One calls *ordered pair* to the (unique) set $(x, y) := \{\{a\}, \{a, b\}\}$[2].

Separation (Scheme)

The elements x of a set X that verify the property $\varphi(x)$ form a new set Y (unique by extensionality). For each formula the following is an axiom.

$$\forall X \exists Y \forall x (x \in Y \Leftrightarrow (x \in X \wedge \varphi(x))).$$

By separation a subclass of a set is also a set. It follows from separation that there is a set with no elements (unique by extensionality) called the *empty set* and denoted $\emptyset$. Moreover, if X and Y are sets the *intersection*

$$X \cap Y := \{x \in X : x \in Y\}$$

is also a set.

Union

The next axiom states that for every set A there is a (unique) set $B := \bigcup A$.

$$\forall A \exists B \forall x (x \in B \Leftrightarrow \exists X \in A (x \in X)).$$

One usually writes $X \cup Y$ instead of $\bigcup \{X, Y\}$.

Power Set

A set B is a *subset* of a set A if for every $x \in B$ one has that $x \in A$ and writes $B \subseteq A$. If $B \subseteq A$ and $A \neq B$ one says that B is a *proper subset* of A and writes simply $B \subset A$. For any set X there is a (unique) set $Y := \mathcal{P}(X)$ that contains all subsets of X. This set is called the *power set* of X.

$$\forall X \exists Y \forall x (x \in Y \Leftrightarrow x \subseteq X).$$

The Power set axiom allows us to define the following notions. One defines the (Cartesian) *product* of X and Y in the following way.

$$X \times Y := \{(x, y) \in \mathcal{P}(\mathcal{P}(X \cup Y)) : x \in X \wedge y \in Y\}.$$

A *binary relation* R between two sets X and Y is a subset of the Cartesian product $X \times Y$; i.e. it is a set of ordered pairs. A binary relation f between two sets X and Y is a *function* if $(x, y) \in f$ and $(x, z) \in f$ implies $y = z$. A *partial order relation* "$\leq$" over a set X is a binary relation satisfying the following properties:

---

[2]This definition was given by Kazimierz Kuratowski [78] in 1921.

(O1) $\forall a \in X(a \leq a)$ (*reflexivity*).

(O2) $\forall a, b \in X(a \leq b \wedge b \leq a \Rightarrow a = b)$ (*antisymmetry*).

(O3) $\forall a, b, c \in X(a \leq b \wedge b \leq c \Rightarrow a \leq c)$ (*transitivity*).

A partial order relation is called a *total order relation* if it also satisfies

(O4) $\forall a, b \in X(a \leq b \vee b \leq a)$ (*totality*).

A *well-ordering* of a set is a total ordering of it according to which every non-empty subset has a least element.

Infinity
The next axiom states that there is an infinite set.

$$\exists X(\emptyset \in X \wedge \forall x(x \in X \Rightarrow x \cup \{x\} \in X)).$$

Replacement (Scheme)
The following axiom states that if a class $F = \{(x, y) : \varphi(x, y, p)\}$, where p is a parameter, is functional, then for every set X, $F(X)$ is a set.

$$\forall x \forall y \forall z(\varphi(x, y, p) \wedge \varphi(x, z, p) \Rightarrow y = z)$$
$$\Rightarrow \forall X \exists Y \forall y(y \in Y \Leftrightarrow (\exists x \in X)\varphi(x, y, p)).$$

A set A is called *transitive* if whenever $x \in A$, and $y \in x$, then $y \in A$. A set S which is strictly well-ordered with respect to $\in$ is an *ordinal* if and only if every element of S is also a subset of S. The class of all ordinals (which is not a set) is denoted *Ord*. An ordinal $\alpha = \beta + 1 = \beta \cup \{\beta\}$ is called a *successor ordinal.* If α is not a successor ordinal, then $\alpha = \sup\{\beta : \beta < \alpha\}$ is called a *limit ordinal.* In the presence of the following axiom it is possible to prove that $\alpha \in Ord$ if and only if α is a transitive set of transitive sets.

Regularity

$$\forall S(S \neq \emptyset \Rightarrow (\exists x \in S)(S \cap x = \emptyset)).$$

This means that every non-empty set has an $\in$-minimal element and, as a consequence, there can be no infinite sequences

$$... \in x_2 \in x_1 \in x_0.$$

Regularity is very useful in the construction of models but less relevant for the development of ordinary mathematics. Indeed, Regularity's main feature is to give a "nice picture" of the universe of sets [33]. The universe of sets V is given by the *Von Neumann cumulative hierarchy* described below. The cumulative hierarchy is a collection of sets V_α indexed by *Ord* in the following way:

- $V_0 := \emptyset$;
- $V_{\alpha+1} := \mathcal{P}(V_\alpha)$;
- For any limit ordinal α, $V_\alpha := \bigcup_{\beta<\alpha} V_\alpha$;
- $V := \bigcup_{\alpha \in Ord} V_\alpha$.

The axiom of Regularity implies that every set is obtained at some level of the cumulative hierarchy over the empty set; i.e. every set is in some V_α [66]. One defines the *rank* of x as the least α such that $x \in V_{\alpha+1}$.

Choice

Let S be a non-empty family of disjoint sets. A *choice function* for S is a function $f : \mathcal{P}(S) \setminus \{\emptyset\} \to S$ such that $f(X) \in X$, for every $X \in S$. The following axiom states that every family of non-empty sets has a choice function.

$$\forall S \exists Y \forall X \in S \setminus \{\emptyset\} \, (\exists z (Y \cap X = \{z\})).$$

Given X, Y sets and $R \subseteq X \times Y$ a relation, one can state the Axiom of Choice in the following equivalent way.

$$\forall x \in X \exists y \in Y \, [(x, y) \in R] \Rightarrow \exists f \in Y^X \forall x \in X \, [(x, f(x)) \in R] \, ,$$

where Y^X denotes the set of all functions f from X to Y.

A.2.2 Theories for internal sets: IST and BST

The philosophical position of the internal approach to nonstandard analysis (both IST, as presented in Chapter 4 and BST [69, 70, 71] described below) is that

> we do not enlarge the world of mathematical objects in any way, we merely construct a richer language to discuss the same objects as before. [91]

Having a richer language enables one to express new properties about the mathematical objects that are not possible to express in the language of ZFC.

Both IST and BST are conservative extensions of ZFC and therefore all ZFC axioms and (as long as one considers only formulas in the language of ZFC) therefore all ZFC theorems, definitions and constructions remain valid. Also, being conservative extensions, every statement in the language of ZFC that is provable in IST or BST can be proved in ZFC. Working in IST is very much like working in ZFC and perhaps for this reason it was so far, to our knowledge, the only axiomatic nonstandard theory to be used in practice. See for example [50, 34, 136, 130, 35].

In [95] Nelson showed that in IST is valid a Reduction Algorithm, that converts external formulas into internal equivalent formulas, for standard values of the parameters.

The consistency of IST was shown in [92] by using *adequate ultrafilters.* We explain this notion below.

An *ultrafilter* over a set X is a collection $\mathcal{U}$ of subsets of X such that

1. $\emptyset \notin \mathcal{U}$;

2. If $A, B \in \mathcal{U}$ then $A \cap B \in \mathcal{U}$;

3. If $A \in \mathcal{U}$ and $A \subset B$ then $B \in \mathcal{U}$;

4. For all $A \subset X$, either $A \in \mathcal{U}$ or $X \setminus A \in \mathcal{U}$.

We denote the set of all ultrafilters over X by $U(X)$.

Let $(A_i)_{i \in X}$ be a family of sets. The *generalized Cartesian product* $\prod_{i \in X} A_i$ is defined in the following way

$$\prod_{i \in X} A_i := \left\{ f : X \to \bigcup_{i \in X} A_i : \forall i (f(i) \in A_i) \right\}.$$

Let $\mathcal{U}$ be an ultrafilter over a set X. Two elements a, b of the generalized Cartesian product $\prod_{i \in X} A_i$ are said to be $\sim_{\mathcal{U}}$-equivalent if the set of indices for which they are equal is in the ultrafilter, in symbols

$$a \sim_{\mathcal{U}} b \Leftrightarrow \{i : a_i = b_i\} \in \mathcal{U}.$$

The $\sim_{\mathcal{U}}$-equivalence class of a is the set $[a] = \{b : a \sim_{\mathcal{U}} b\}$. The *ultraproduct* is defined as the set of $\sim_{\mathcal{U}}$-equivalence classes, i.e.

$$\prod_{\mathcal{U}} A_i := \left\{ [a] : a \in \prod_{i \in X} A_i \right\}.$$

An *ultrapower* is an ultraproduct for which all the factors A_i are equal. Let V be a set. Let V_* be an ultrapower of V. For $\xi \in V_*$, let

$$t(\xi) := \{E \subset V : \xi \in E_*\}.$$

As it turns out $t(\xi)$ is an ultrafilter on V, so we have a natural mapping $t : V_* \to U(V)$ and we call V_* an *adequate ultrapower* of V if t is surjective. Let $V = V_0$ be a set and for each $n \in \mathbb{N}$ let V_{n+1} be an ultrapower of V_n. Let $j_n : V_n \to V_{n+1}$ be the natural injection and let V_* be the direct limit of the sequence

$$V_0 \xrightarrow{j_0} V_1 \xrightarrow{j_1} V_2 \xrightarrow{j_2} \cdots$$

Identifying each V_n as a subset of V_{n+1} via the injection j_n we have $V_* = \bigcup_{n \in \mathbb{N}} V_n$. We call V_* an *ultralimit* of V and if for every $n \in \mathbb{N}$ it holds that

V_{n+1} is an adequate ultrapower of V_n we say that V_* is an *adequate ultralimit* of V. These notions of adequate ultrapower and adequate ultralimit were introduced by Nelson in [92] where it was also shown that every set V has an adequate ultralimit.

Some care is needed when working in IST, because external predicates are not allowed to define subsets. The violation of this rule is called *illegal set formation*. This is somewhat inconvenient for our purpose because it implies that there are no external sets in IST. As argued in [70, 71], the axiomatic system IST must be modified in order to admit an extension to external sets in a reasonable way. This is done by postulating a boundedness axiom that expresses that every set belongs to a standard set.

Boundedness

$$\forall x \exists^{\mathrm{st}} y \, (x \in y),$$

and since this contradicts idealization the following (bounded) form is taken instead:

Bounded Idealization

$$\forall^{\mathrm{st}} Y \left(\forall^{\mathrm{stfin}} F \exists y \in Y (\forall x \in F \, R(x,y) \Leftrightarrow \exists b \in Y \forall^{\mathrm{st}} x \, R(x,b)) \right),$$

for every $\in$-formula R. This yields the subsystem BST, which corresponds to the bounded sets of IST.

Notice that in usual analysis all sets are bounded. This means that BST is equivalent to IST in the known applications of IST. Nevertheless BST is somewhat more suitable for foundational purposes, as argued in [69, 71]. For example, in BST is valid a theorem [69] (valid in IST only for bounded formulas [95]) that allows us to convert every formula φ in an equivalent formula φ' that is of the form $\exists^{\mathrm{st}} \forall^{\mathrm{st}} \psi$, where ψ is an $\in$-formula (formulas of this type are called $\sum_2^{\mathrm{st}}$ formulas). This theorem permits to apply Nelson's *Reduction Algorithm* [95] without any restrictions.

A.2.3 Theories for external sets: HST

A "perfect" external set theory (a nonstandard set theory that includes external sets) should satisfy some requirements. It should be a conservative extension of classical mathematics (usually ZFC) so that all classical mathematical theorems and constructions remain valid. The theory should also allow us to perform nonstandard constructions in its full generality and therefore include a strong version of Saturation (called Idealization in IST and Bounded Idealization in BST) and transfer principles. Finally it should allow us to build, for any given set, the standard set of all its standard elements, which amounts to Standardization. This means that ideally it should come close to an extension of IST allowing external sets and quantification over external formulas. However, as pointed out by Karel Hrbáček [64] such a theory cannot exist because of the so-called *Hrbáček's paradoxes*. In fact, the axiom of Regularity cannot be extended to the external universe. To see that let $\not\infty$ denote the external set of infinitely large real numbers. Observe that for all ω in the (non-empty) external set $\not\infty \cap \mathbb{N}$, one has $\not\infty \cap \mathbb{N} \cap \omega \neq \emptyset$. Additionally, if one wishes to formulate a nonstandard set theory with IST-style Saturation, the Replacement axiom in the external universe contradicts both Power set and Choice (see [71, Theorem 1.3.9, Remark 1.3.10]). This is to be expected since the set $\not\infty \cap \mathbb{N}$ has no minimum but could be well-ordered. A set is of *standard size* if it is the image of the set of all standard elements of a standard set[3]. Now, whenever two sets are well-ordered there is an order preserving embedding of one into the other (see [66, Theorem 2.8]). Clearly the class of all standard ordinals ${}^\sigma Ord$ cannot be embedded into a set x; otherwise ${}^\sigma Ord$ would be a set by Replacement. Then there is an embedding of x into ${}^\sigma Ord$. In fact, to an initial segment of ${}^\sigma Ord$. This means that x is of standard size. As a consequence, sets which are not of standard size cannot be well-ordered (see [71, Theorem 1.3.1]).

The absence of the Regularity axiom is maybe not so important. Regularity is usually given so that every set is obtained at some level of the cumulative hierarchy over $\emptyset$ as mentioned in Section A.2 and has no great impact on which theorems are true. This "nice picture" of the universe is contested by some mathematicians and a suitable anti-foundation axiom can be taken instead (see for example [1]).

In [64] Hrbáček considered already two possibilities to avoid the problems with Power set and Choice. The first one was to omit both Power set and Choice for external sets, leading to the system $\mathfrak{NS}_1$. The second one was to omit the Replacement axiom for external sets, which lead to his theory $\mathfrak{NS}_2$[4]. A third possibility was developed by Kanovei and Reeken (see [69, Part 3] and [70, Chapter 6]). The idea is to restrict saturation by a standard infinite cardinal κ in order to reinstate the Power set axiom. This is a system of partially saturated external sets which modifies the system HST (described below), called HST_κ. This may be a solution for many practical purposes but not a solution as a foundational system for nonstandard mathematics.

The theory BST possesses an extension to HST [70, 71], which formulates within first-order language essential aspects of the behaviour of standard, internal and external sets within a nonstandard model, much as in Hrbáček's system $\mathfrak{NS}_1$. The system HST is conservative over ZFC [64, 69] and equiconsistent with both BST and ZFC (see [71, Chapter 5]).

A set in HST is called *internal* if it is an element of a standard set (see also the Boundedness axiom).

Below we use (definable) classes; they only should be interpreted as abbreviations of formulas with sets. Two important definable classes in HST are

[3]In HST, a set X is of standard size if and only if X is well-ordered (see Section 1.3a in [71])

[4]Hrbáček proposed still a third theory called $\mathfrak{NS}_3$ that we do not consider here because it is not a conservative extension of ZFC.

the class of standard sets

$$\mathbb{S} = \{x : \mathrm{st}(x)\}$$

and the class of internal sets

$$\mathbb{I} = \{x : \exists y(\mathrm{st}(y) \wedge x \in y)\}.$$

Internal sets of HST do not correspond entirely to the internal sets of IST. In fact, $(\mathbb{I}, \in, \mathrm{st})$ is a model of BST but not a model of IST (see [71, Chapter 3]).

The axioms of HST are divided into three categories. The first deals with axioms which are valid for all sets, the second deals with axioms which are valid for standard or internal sets and the third deals with axioms which are valid for standard size sets. As aforementioned, to be of standard size means to be an image of the set of all standard elements of a standard set. Let X be a set. So a set is of standard size if it is of the form $\{f(x) : x \in X \cap \mathbb{S}\}$.

As mentioned above in HST the Power set axiom does not hold for external sets. However, if X is an internal set then $\mathcal{P}_{\mathrm{int}}(X) = \mathcal{P}(X) \cap \mathbb{I}$ exists. So the Power set axiom holds for internal sets. Moreover if a set is of standard size then $\mathcal{P}(X)$ is also of standard size.

A.2.3.1 HST axioms

Axioms for all sets

The axioms of this group are valid for all sets. These axioms are similar to the respective ones of ZFC with the difference that in HST they are presented in the full language $\{\in, \mathrm{st}\}$. This implies in particular, by the axiom of Separation, that the theory HST deals with external sets; for example if X is standard and infinite $\{x \in X : \mathrm{st}(x)\}$ is an external set.

1. *Extensionality*

$$\forall X \forall Y (\forall x (x \in X \Leftrightarrow x \in Y) \Leftrightarrow X = Y).$$

2. *Pair*

$$\forall a, b \exists A \forall x (x \in A \Leftrightarrow x = a \vee x = b).$$

3. *Union*

$$\forall A \exists B \forall x (x \in B \Leftrightarrow \exists X \in A (x \in X)).$$

4. *Infinity*

$$\exists X (\emptyset \in X \wedge \forall x (x \in X \Rightarrow x \cup \{x\} \in X)).$$

5. *Separation*

$$\forall X \exists Y \forall x (x \in Y \Leftrightarrow (x \in X \wedge \varphi(x))).$$

6. *Collection*

$$\forall X \exists Y \forall x \in X (\exists y \varphi(x, y) \Rightarrow \exists y \in Y \varphi(x, y)).$$

The Power set, Regularity and Choice axioms of ZFC are not valid in general. This is because, as mentioned above, adding any of these axioms (if considered in the full language of HST) leads to a contradiction.

Axioms for standard and internal sets

In this group as well as in the next there are axioms which are not valid for all sets. The first axiom scheme states that all ZFC axioms, when restricted to standard parameters are valid in HST.

1.
$$\mathrm{ZFC}^{\mathrm{st}}.$$

This means, in particular, that the following are axioms of HST:

(a) *Regularity*st

$$\forall^{\mathrm{st}} S(S \neq \emptyset \Rightarrow (\exists^{\mathrm{st}} x \in S)(x \cap S \neq \emptyset)).$$

(b) *Power Set*st

$$\forall^{\mathrm{st}} X \exists^{\mathrm{st}} Y \forall^{\mathrm{st}} x(x \in Y \iff x \subseteq X).$$

(c) *Choice*st

$$\forall^{\mathrm{st}} S \exists^{\mathrm{st}} Y \forall^{\mathrm{st}} x \in S \setminus \{\emptyset\}\, (\exists^{\mathrm{st}} z(Y \cap x = \{z\})).$$

The fact that every axiom of ZFC restricted to standard sets is an axiom of HST means that the class $\mathbb{S}$ models ZFC.

2. *Transfer*

$$\forall^{\mathrm{st}} x_1 ... \forall^{\mathrm{st}} x_n (\varphi^{\mathrm{st}}(x_1, ..., x_n) \Leftrightarrow \varphi^{\mathrm{int}}(x_1, ..., x_n)),$$

where φ is an arbitrary closed $\in$-formula containing only standard parameters. This means that the universe $\mathbb{I}$ is an elementary extension of $\mathbb{S}$ in the ZFC language.

3. *Transitivity of* $\mathbb{I}$

$$\forall^{\mathrm{int}} x \forall y (y \in x \Rightarrow \mathrm{int}(y)).$$

The next axiom states that the class $\mathbb{I}$ is regular. This means that sets in HST are built over $\mathbb{I}$ in a way similar to the Von Neumann hierarchy of sets in ZFC over the empty set.

4. *Regularity over* $\mathbb{I}$

$$\forall X \neq \emptyset \exists x \in X (x \cap X \subset \mathbb{I}).$$

5. *Standardization*

$$\forall X \exists^{\mathrm{st}} Y (X \cap \mathbb{S} = Y \cap \mathbb{S}).$$

This axiom implies that the only sets consisting entirely of standard sets are of the form $Y \cap \mathbb{S}$, where $Y \in \mathbb{S}$.

Axioms for sets of standard size

1. *Saturation*

   If $\mathcal{A} \subset \mathbb{I}$ is a standard size set then

$$((\forall X, Y \in \mathcal{A} \Rightarrow X \cap Y \in \mathcal{A}) \wedge (X \in \mathcal{A} \Rightarrow X \neq \emptyset)) \Rightarrow \bigcap \mathcal{A} \neq \emptyset.$$ [5]

2. *Standard Size Choice*

   Choice is available in the case where the domain of the choice function f is of standard size.

   Let X be a set of standard size and F a function on X. Then

$$\forall x \in X(F(x) \neq \emptyset) \Rightarrow \exists f \forall x(f(x) \in F(x)).$$

3. *Dependent Choice*

   Any non-empty partially ordered set without maximal elements includes a non-empty linearly ordered subset (sequence) where every element has its immediate successor[6].

It is possible to maintain the notation of IST in the way that the traditional symbols for the uniquely defined objects of a ZFC universe are attributed to the objects of the internal sub-universe.

In HST neutrices and external numbers are incorporated in the following way. Each neutrix, being a convex (bounded) subgroup of $\mathbb{R}$, is a genuine set within HST. Neutrices are defined by Π^{st} or Σ^{st} formulas, with reference to standard sets of all possible cardinals [130]. As shown in [70] this implies that we may not speak of the set of all neutrices. In fact, the neutrices form a definable class because quantification within HST may range over the whole universe. Then the algebraic operations on neutrices and external numbers are also definable classes, as well as the function which associates to each external number its neutrix.

A.3 Model theoretical nonstandard analysis

Model theory is a branch of mathematical logic that deals with models for axiomatic systems. Good introductory references on the subject are for example [26, 88, 62]. A *model* for a theory can be seen as the assignment of meaning to the symbols of the language of the theory. So, for a given set of axioms a

[5]It is possible to show the equivalence of Saturation in terms of non-empty intersection and a form using functions as used in Chapter 4.

[6]See [71, p. 19], for an alternative formulation of this axiom.

model is a mathematical set that satisfies the axioms. A *nonstandard model* is a model of a theory which is not isomorphic to the intended model (standard model). If the standard model is infinite and the language is first-order then by the Löwenheim–Skolem theorem [119] it has nonstandard models. Moreover, the nonstandard models can be chosen as *elementary extensions* (or elementary substructures) of the intended model. In particular, $\mathbb{R}$, the set of real numbers, has nonstandard models $^*\mathbb{R}$ in which all first-order properties are the same. This fact was deeply explored by Robinson [111, 112], leading to the foundation of nonstandard analysis. Nonstandard analysis was to a great extent inspired by Leibniz's ideas and intuitions towards the use of infinitesimal and infinitely large quantities. These "ghosts of departed quantities" [9] had been used informally throughout the early stages of the development of calculus. They were posteriorly abandoned when the $\varepsilon - \delta$ based notions of analytic concepts such as limit, continuity, derivative and integral were introduced by the works of Cauchy, Bolzano and Weierstrass. In Robinson's words [112]:

> It is shown in this book that Leibniz's ideas can be fully vindicated and that they lead to a novel and fruitful approach to classical Analysis and to many other branches of mathematics. The key to our method is provided by the detailed analysis of the relation between mathematical languages and mathematical structures which lies at the bottom of contemporary model theory.

A.3.1 The superstructure approach

Robinson and Zakon [113, 155] (see also [124] or [57] for a more recent treatment of these constructions) gave a simpler approach to nonstandard analysis using set-theoretical objects called superstructures. A *superstructure* $V(S)$ over a set S is defined in the following way:

$$\begin{aligned} V_0(S) &= S \\ V_{n+1}(S) &= V_n(S) \cup \mathcal{P}(V_n(S)) \\ V(S) &= \bigcup_{n\in\mathbb{N}} V_n(S). \end{aligned}$$

Superstructures of the empty set consist of sets of finite rank in the cumulative hierarchy and therefore do not satisfy the Infinity axiom. Making $S = \mathbb{R}$ will suffice for virtually any construction necessary in analysis.

Bounded formulas are formulas where the range of the quantified variables is restricted to a given set y; i.e. quantifiers occur in the form $\forall x(x \in y \Rightarrow \dots)$ or $\exists x(x \in y \wedge \dots)$. A *nonstandard embedding* is a mapping $* : V(X) \to V(Y)$ from a superstructure $V(X)$ called the *standard model*, into another superstructure $V(Y)$, called *nonstandard model*, satisfying the following:

- ${}^*X = Y$
- *Transfer Principle*

  For every bounded formula $\varphi(x_1, ..., x_n)$ and elements $a_1, ..., a_n \in V(X)$, the property φ is true for $a_1, ..., a_n$ in the standard model if and only if it is true for ${}^*a_1, ..., {}^*a_n$ in the nonstandard model:

$$\langle V(X), \in \rangle \models \varphi(a_1, ..., a_n) \Leftrightarrow \langle V(Y), \in \rangle \models \varphi({}^*a_1, ..., {}^*a_n).$$

- *Non-triviality*

  For every infinite set A in the standard model, the set $\{{}^*a : a \in A\}$ is a proper subset of ${}^*A$.

A set X is *internal* if and only if X is an element of ${}^*A$ for some element A of $V(\mathbb{R})$. Let X be a set with $A = \{A_i\}_{i \in I}$ a family of subsets of X. Then the collection A has the *finite intersection property*, if any finite subcollection $\{A_i\}_{i \in J}$, with $J \subseteq I$ has a non-empty intersection. A model is *κ-saturated* if whenever $\{A_i\}_{i \in I}$ is a collection of internal sets with the finite intersection property and the cardinal of I is less than or equal to κ it holds that $\bigcap_{i \in I} A_i \neq \emptyset$.

In the model-theoretical context, a nonstandard model of the theory of real numbers ${}^*\mathbb{R}$ generates external neutrices in an obvious way, since ${}^*\mathbb{R}$ contains such external convex subgroups. It is convenient to choose a superstructure large enough to be able to apply the usual operations of analysis on neutrices and external numbers. It is also convenient to suppose the superstructure to be κ-saturated, for some infinite limit cardinal κ. Neutrices defined over a (standard) set X of cardinality less than or equal to κ satisfy strong properties. Indeed, it is not difficult to adapt the proofs of axiomatic nonstandard analysis in [130, 76] to derive firstly that these neutrices are of the lowest complexity, i.e. are of the form $\bigcup_{x \in X}[-a_x, a_x]$ and $\bigcap_{x \in X}[-a_x, a_x]$, secondly that every neutrix N of this form is the multiple $N = \lambda I$ by a hyperreal number λ of an *idempotent* neutrix I (i.e. with $I.I = I$) and thirdly that every definable external lower halfline of ${}^*\mathbb{R}$ is bounded from above by an external number, which is cofinal with it, or just beyond [130, 76].

The superstructure approach allows one to use the methods of nonstandard analysis without paying much attention to the foundational details. However, it has some limitations if one wishes to use the nonstandard methods in their full generality. Superstructures can only model a fragment of ZFC. Take the following example from [33]. Let $N \subseteq X$ be a copy of the natural numbers. Consider the function F with domain N such that $n \mapsto \{...\{\emptyset\}...\}$ (n brackets). Then F is definable in $V(X)$ but $\text{rng}(F) \notin V(X)$. Another objection to the superstructure approach's non-unicity is the fact that different problems will sometimes require different superstructures and different embeddings. The problem of non-unicity also occurs in the approach using adequate ultralimits. Also, the model-theoretical concepts do not seem to be very appealing to most mathematicians. Finally one can argue that nonstandard analysis, in principle, is not concerned with superstructures.

Appendix B

Solutions to selected exercises

Chapter 1 309
Chapter 2 315
Chapter 3 315
Chapter 4 316
Chapter 5 318
Chapter 6 319
Chapter 7 319
Chapter 8 320
Chapter 9 321
Chapter 10 322
Chapter 11 323

Chapter 1

1.1.1. Suppose that $\varepsilon^{\frac{1}{m}}$ is not infinitesimal, i.e. $\exists^{\mathrm{st}} n \in \mathbb{N}\left(\frac{1}{n} < \varepsilon^{\frac{1}{m}}\right)$. Then $\frac{1}{n^m} < \varepsilon$, a contradiction. Also $\varepsilon \leq \frac{1}{m}$ for all standard $m \in \mathbb{N} \setminus \{0\}$. Hence $e^{-\frac{1}{\varepsilon}} = \frac{1}{e^{\frac{1}{\varepsilon}}} < \frac{1}{e^m} \leq \frac{1}{m}$ is also infinitesimal. Finally, put $\log(\varepsilon) = x$; then $\varepsilon = e^{-|x|}$, with $x < 0$. If x is limited, then $e^{|x|}$ is limited, hence ε cannot be infinitesimal. Hence x is unlimited, and $\frac{1}{\log(\varepsilon)}$ is infinitesimal.

1.1.2. Assume first that $x \simeq y$. Then there exists $\varepsilon \simeq 0$ such that $y = x+\varepsilon$. Then $\frac{y}{x} = 1+\frac{\varepsilon}{x} \simeq 1$. Hence $x \sim y$. Assume now that $x \sim y$. Then there exists $\delta \simeq 0$ such that $\frac{x}{y} = 1 + \delta$. Hence $x = y + y\delta$. We have that $y\delta \simeq 0$ by Part 2 of Theorem 1.1.5. Hence $x \simeq y$.

1.1.3. We have $e^{-\omega} < \frac{1}{\omega} < 1000000 < \log(\omega) < \omega^{\frac{1}{2}} < \omega^3 < e^{\omega} < \omega! < \omega^{\omega}$. Observe that $\sin(\omega)$ may be appreciable or infinitesimal, or even zero if $\frac{\omega}{2\pi} \in \mathbb{N}$. However, $|\sin(\omega)| \leq 1$.

1.1.4. Let $\omega > 0$ be unlimited. Then $\frac{\omega^2}{\omega} = \omega$ is unlimited, $\frac{\omega}{\omega} = 1$ is appreciable and $\frac{\omega}{\omega^2} = \frac{1}{\omega}$ is infinitesimal.

1.1.5. Let $\omega > 0$ be unlimited. An example where the sum of a family of infinitesimals is appreciable is given by $\sum_{n=1}^{\omega} \frac{1}{\omega}(= 1)$. An example where the sum of a family of infinitesimals is unlimited is $\sum_{n=1}^{\omega^2} \frac{1}{\omega}(= \omega)$. Also, $\sum_{n=1}^{\omega} \frac{1}{\omega^2}\left(= \frac{1}{\omega}\right)$ is infinitesimal; note that the sum of an unlimited family of infinitesimal numbers is clearly infinitesimal if those infinitesimals are all equal to 0.

1.1.7. Since x, y are limited, $\exists^{st} n \in \mathbb{N}(|x| \leq n)$ and $\exists^{st} m \in \mathbb{N}(|y| \leq m)$. By Theorem 1.1.10 if $x \simeq 0$, then $e^x \simeq 1$. Hence $e^{x-y} \simeq 1$, since $x - y \simeq 0$. Now e^x is appreciable by Part 5 of Theorem 1.1.2 and Theorem 1.1.3. Because $e^x - e^y = e^x(1 - e^{y-x})$, $e^{y-x} \simeq 1$, $1 - e^{y-x} \simeq 0$ and $e^x(1 - e^{y-x}) \simeq 0$. Then $e^x \simeq e^y$. If $\omega \simeq +\infty$, then $e^{\omega + \frac{1}{\omega}} - e^{\omega} = e^{\omega}\left(e^{\frac{1}{\omega}} - 1\right) \sim \frac{e^{\omega}}{\omega}$, which is unlimited.

1.1.8. We show the result for $x > 0$, the case where $x < 0$ is analogous. Observe that, since $\sin(x)$ is clearly appreciable, $\sin(x) \sim x \Leftrightarrow \frac{\sin(x)}{x} \simeq 1$. We have that

$$|\sin(x) - x| \leq \frac{x^2}{2}\left(\max_{0 \leq y \leq x} |-\sin(y)|\right) \leq \frac{x^2}{2} \simeq 0.$$

Hence $\left|\frac{\sin(x)}{x} - 1\right| \leq \frac{x}{2} \simeq 0$ and we conclude that $\frac{\sin(x)}{x} \simeq 1$.

1.1.9. We have that $\omega \log\left(1 + \frac{1}{\omega}\right) \sim 1$. Then $\omega \log\left(1 + \frac{1}{\omega}\right) \simeq 1$ and

$$\left(1 + \frac{1}{\omega}\right)^{\omega} \simeq e^{\omega \log\left(1 + \frac{1}{\omega}\right)} \simeq e.$$

The case where x is limited is similar.

1.2.1. The set $\{0, ..., \omega\}$ is finite, being a bounded internal subset of $\mathbb{N}$. Hence there does not exist an internal $1-1$-mapping onto $\{1, ..., \omega\}$. The mapping defined by $f(n) = n + 1$, for standard n and $f(n) = n$, for nonstandard n is $1 - 1$, but external. The set ${}^{\sigma}\mathbb{N}$ is not finite, for finiteness is only defined for internal sets.

1.2.3. The function f is continuous, but must be external, for it does not satisfy the Intermediate Value Theorem.

1.2.4. One has $\oslash = \bigcap_{st(n)} \left[-\frac{1}{n}, \frac{1}{n}\right]$, $\pounds = \bigcup_{st(n)}[-n, n]$, $@ = \bigcup_{st(n)} \left[\frac{1}{n}, n\right]$, and $\not\infty = \bigcap_{st(n)}[n, +\infty)$.

1.2.5. All are external. With the exception of 8. and 9., all are groups, being symmetrical and stable by multiplication by 2. The intersection of 10. with the horizontal axis is a halo, and its intersection with the vertical axis is a galaxy, so it is neither a galaxy nor a halo.

1.2.6. The intersection with lines of limited slope are halos, and with lines of unlimited slope are galaxies.

1.2.8. See also Exercise 1.1.10.

1. One has $e^{-x} \sim e^{-\frac{n}{\omega}} \sim e^{-\frac{n+1}{\omega}}$, because $e^{-x} \simeq e^{-\frac{n}{\omega}} \simeq e^{-\frac{n+1}{\omega}}$, and they are appreciable. Moreover

$$e^{-\frac{n}{\omega}} \frac{1}{\omega} \leq \int_{\frac{n}{\omega}}^{\frac{n+1}{\omega}} e^{-x} dx \leq e^{-\frac{n+1}{\omega}} \frac{1}{\omega},$$

so $\int_{\frac{n}{\omega}}^{\frac{n+1}{\omega}} e^{-x} dx \sim e^{-\frac{n}{\omega}} \frac{1}{\omega}$ because $e^{-\frac{n}{\omega}} \frac{1}{\omega} \sim e^{-\frac{n+1}{\omega}} \frac{1}{\omega}$.

2. By Cauchy's sum formula $\sum_{0 \leq n \leq \omega} e^{-\frac{n}{\omega}} \frac{1}{\omega} \sim \int_0^1 e^{-x} dx$. We have near-equality, because the quantities are appreciable.

4.

$$\sum_{0\leq n\leq\omega}\left(1+\frac{-\frac{n}{\omega}}{\omega}\right)^{\omega}\sim\sum_{0\leq n\leq\omega}e^{-\frac{n}{\omega}}\sim\omega\sum_{0\leq n\leq\omega}e^{-\frac{n}{\omega}}\frac{1}{\omega}\sim\omega\int_0^1 e^{-x}dx=\omega(e^{-1}-1).$$

1.2.9. Because $f(x,0)=0$ and $\frac{\partial f}{\partial y}(x,y)>0$ is appreciable, by integration for appreciable y

$$f(t,y)=\begin{cases}\gtrapprox 0, & y\gtrapprox 0\\ \simeq 0, & y\simeq 0\\ \lessapprox 0, & y\lessapprox 0.\end{cases}$$

By continuity, f must be onto $\oslash$, so $\{(x,y):f(x,y)\simeq 0\}$ is a halo for all $x\in[0,1]$, and then also $\{(x,y):0\leq x\leq 1\wedge f(x,y)\simeq 0\}$ must be a halo. Then its complement in $[0,1]\times[-1,1]$ must be a galaxy, hence also the set $\{(x,y):0\leq x\leq 1\wedge f(x,y)$ appreciable$\}$.

1.3.1. One has st(0) by Axiom 1.8. Assume that st(n) and st(m) for all $m\leq n$. Then st($n+1$) by Axiom 1.9. Hence st(m) for all $m\leq n+1$. By External Induction one concludes Theorem 1.3.2.

1.3.2 Let $(B_n)_{n\in\mathbb{N}}$ be a strictly decreasing internal sequence such that $H=\bigcap_{\mathrm{st}(n)\in\mathbb{N}}B_n$ and $\nu\in\mathbb{N}$ be unlimited. Define $f:X\to\mathbb{R}$ by

$$f(x)=\begin{cases}\max\left\{\frac{1}{n}:x\notin B_n\right\}, & \exists n\in\mathbb{N}(x\notin B_n)\\ \frac{1}{\nu}, & \text{otherwise.}\end{cases}$$

Let $x\in H$. Then $x\in B_n$ for all st(n). Hence, if $\min\{n:x\notin B_n\}$ is well-defined, it is unlimited, and $\max\left\{\frac{1}{n}:x\notin B_n\right\}\simeq 0$. Else $f(x)=\frac{1}{\nu}$, hence always $x\in f^{-1}(\oslash)$. Also, $x\in f^{-1}(\oslash)$ implies that $x\in B_n$ for all st(n), i.e. $x\in\bigcap_{\mathrm{st}(n)\in\mathbb{N}}B_n$. Combining, $H=f^{-1}(\oslash)$.

Conversely, $f^{-1}(\oslash)=f^{-1}\left(\bigcap_{\mathrm{st}(n)}\left[-\frac{1}{n},\frac{1}{n}\right]\right)=\bigcap_{\mathrm{st}(n)}f^{-1}\left[-\frac{1}{n},\frac{1}{n}\right]$. Because f is onto $\oslash$, the sequence $\left(f^{-1}\left[-\frac{1}{n},\frac{1}{n}\right]\right)_{n\in\mathbb{N}}$ is certainly strictly decreasing for unlimited n, and by Cauchy's Principle down to some st(m). As a consequence $\bigcap_{\mathrm{st}(n)}f^{-1}\left[-\frac{1}{n},\frac{1}{n}\right]$ must be external, hence a halo. This proves Part 2 of Theorem 1.3.3.

1.3.3. By injectivity, $(f[-n,n])_{n\in\mathbb{N}}$ is strictly increasing at least on ${}^{\sigma}\mathbb{N}$, and $\left(f\left[-\frac{1}{n},\frac{1}{n}\right]\right)_{n\in\mathbb{N}}$ is strictly decreasing at least on ${}^{\sigma}\mathbb{N}$. We concude by Theorem 1.3.3.

1.3.4.

1. By continuity, f is onto $\oslash$, so $f^{-1}(\oslash)$ is a halo. Analogous arguments hold for Parts 2 and 5.

3. This set is equal to the internal set $\tan\left[\frac{\pi}{2}-\varepsilon,+\infty\right)$.

4. $f(\mathbb{R}^+)$ is equal to the internal set $\left[-\frac{\pi}{2},\frac{\pi}{2}\right]$.

1.3.5. For $x<0$ unlimited we have $R_\omega(x)=e^x-\left(1+\frac{x}{\omega}\right)^\omega\simeq\left(1+\frac{x}{\omega}\right)^\omega$. Let $\omega>0$ be unlimited. Then $R_\omega(2\omega)\simeq e^{2\omega}-2^\omega$ and $R_\omega(-2\omega)\simeq-2^\omega$

are certainly unlimited, so $R_\omega(x)$ is onto $\oslash^+$ and $\pounds^+$, or $\oslash$ and $\pounds$ depending on whether ω is odd or even. As a consequence $\{x : R_\omega(x) \simeq 0\}$ is a halo and $\{x : R_\omega(x) \text{ limited}\}$ is a galaxy, and then $\{x : R_\omega(x) \text{ is appreciable}\}$ must also be a galaxy. If $x = -2\omega + u$ for limited u we have $R_\omega(x) \simeq \pm\left(1 - \frac{u}{\omega}\right)^\omega \simeq \pm e^u$. Together with a monotonicity argument one concludes that $\{x \leq 0 : R_\omega(x) \text{ appreciable}\} = -2\omega + \pounds$.

1.4.1. $e^\oslash = 1 + \oslash$, $e^{-\not\infty} = \oslash^+$, $\log(\oslash^+\backslash\{0\}) = -\not\infty$, $\log(\pounds^+\backslash\{0\}) = (-\infty, \pounds]$, $\frac{1}{2+\varepsilon\pounds} = \frac{1}{2} + \varepsilon\pounds$, $e^{\pounds\varepsilon} = 1 + \varepsilon\pounds$, $@^\varepsilon = e^{\varepsilon\log(@)} = e^{\pounds\varepsilon} = 1 + \varepsilon\pounds$, $\sqrt{1+\varepsilon\pounds} = 1 + \varepsilon\pounds$.

1.4.2. We verify some entries which are less direct. We let $\varepsilon \simeq 0$, $\varepsilon > 0$ and $\omega \simeq +\infty$.

1. $(\oslash^+)^{\oslash^+}$ contains $0 = 0^\varepsilon$ and $1 = \varepsilon^0$. Hence $(\oslash^+)^{\oslash^+} = [0, 1]$. Similarly for $(\pounds^+)^{\oslash^+}$.

2. $\left(\frac{1}{\log(\omega)}\right)^{\frac{\log(\omega)}{\log(\log(\omega))}} = \frac{1}{\omega}$. This enables us to prove that $(\oslash^+)^{\not\infty} = \oslash^+$.

3. $@^{\oslash^+} = e^{\oslash^+ \log(@)} = e^{\oslash^+ \pounds} = e^\oslash = 1 + \oslash$.

4. Using $e^\varepsilon > 1 + \varepsilon$ one proves that $(\pounds^+)^{\oslash^+} = [0, 1 + \oslash]$.

5. $\omega^0 = 1$, which permits us to prove that $\not\infty^{\oslash^+} = \not\infty^{\pounds^+} = [1, +\infty)$.

1.4.3. One could take, for $\omega \simeq +\infty$, the neutrix $\bigcup_{\mathrm{st}(n)}[-\omega^n, \omega^n]$, or $\bigcup_{\mathrm{st}(n)}[-e^{\omega^n}, e^{\omega^n}]$.

1.4.5. We have

$$\begin{aligned}
(\omega+1)(\omega+2)\cdots(\omega+u\sqrt{\omega}) &= \omega^{u\sqrt{\omega}}\left(1+\frac{1}{\omega}\right)\left(1+\frac{2}{\omega}\right)\cdots\left(1+\frac{u}{\sqrt{\omega}}\right)\\
&= \omega^{u\sqrt{\omega}} e^{\sum_{k=1}^{u\sqrt{\omega}} \log\left(1+\frac{k}{\omega}\right)}\\
&= u\sqrt{\omega} e^{\sum_{k=1}^{u\sqrt{\omega}} (1+\oslash)\frac{k}{\omega}}\\
&\subseteq u\sqrt{\omega} e^{(1+\oslash)\sum_{k=1}^{u\sqrt{\omega}} \frac{k}{\sqrt{\omega}}\frac{1}{\sqrt{\omega}}}\\
&= u\sqrt{\omega} e^{(1+\oslash)\int_0^u v\,dv}\\
&= u\sqrt{\omega} e^{(1+\oslash)u^2}\\
&\subseteq (1+\oslash)\omega^{u\sqrt{\omega}} e^{u^2}.
\end{aligned}$$

1.4.6. Put $v = \frac{u}{\sqrt{\omega}}$. Then

$$\begin{aligned}
\int_{-\frac{1}{\sqrt{\omega}}}^{\frac{1}{\sqrt{\omega}}} e^{-\omega v}(1+v)^{\omega} dv &= \frac{1}{\sqrt{\omega}} \int_{-1}^{1} e^{-\sqrt{\omega}u + \omega \log\left(1+\frac{u}{\sqrt{\omega}}\right)} du \\
&\in \frac{1}{\sqrt{\omega}} \int_{-1}^{1} e^{-\sqrt{\omega}u + \sqrt{\omega}u - (1+\oslash)\frac{u^2}{2}} du \\
&\subseteq \frac{1}{\sqrt{\omega}} \int_{-1}^{1} (1+\oslash) e^{-\frac{u^2}{2}} du \\
&\subseteq (1+\oslash) \int_{-1}^{1} e^{-\frac{u^2}{2}} du \\
&= (1+\oslash)\sqrt{\frac{2\pi}{\omega}} \cdot (\mathcal{N}(1) - \mathcal{N}(-1)).
\end{aligned}$$

1.4.7. As for the differential equation $\frac{dy}{dt} = y$,

$$y(t) = \varepsilon e^t \in @ \Leftrightarrow e^t \in \frac{@}{\varepsilon} \Leftrightarrow t \in -\log(\varepsilon) + £,$$

so $\varepsilon e^t \simeq 0 \Leftrightarrow t < -\log(\varepsilon) + £$.

For the differential equation $\frac{dy}{dt} = y^2$ we have

$$y(t) = \frac{\varepsilon}{1 - \varepsilon t} \in \oslash \Leftrightarrow t \in \frac{1}{\varepsilon} + \oslash.$$

1.5.1. Obviously the linear function $x \mapsto x$ is of class S^0. Applying Theorem 1.5.1 and External Induction it is seen that all monomials $x \mapsto x^n$ of standard degree n are of class S^0. If we multiply these x^n with a standard number a, again by Theorem 1.5.1 we see that $x \mapsto ax^n$ is of class S^0. With External Induction it is proved that all polynomials of standard degree n are of class S^0.

1.5.2. For $x \not\simeq \pm 1$.

1.5.3.

1. We have $€(x) = (1+\delta x)^{\frac{x}{\delta x}} = e^{\frac{x}{\delta x}\log(1+\delta x)} \in e^{\frac{x}{\delta x}\delta x(1+\oslash)} \subseteq e^{x+\oslash}$. Then $€(x) \simeq e^x$. Now the exponential function is of class S^0, so by Part 3 of Theorem 1.5.1 also the function €.

2. We have

$$F(x) = \prod_{0 \le \xi < x} (1+\xi\delta x) = e^{\sum_{0\le\xi<x} \log(1+\xi\delta x)} = e^{\sum_{0\le\xi<x} (1+\oslash)\xi\delta x}.$$

By Cauchy's multiplicative approximations lemma we have

$$\sum_{0\le\xi<x} (1+\oslash)\xi\delta x \subseteq (1+\oslash) \sum_{0\le\xi<x} \xi\delta x.$$

Now $\sum_{0\leq\xi<x}\xi\delta x$ is the sum of an arithmetic sequence, with sum $\frac{x(x-\delta x)}{2}\simeq\frac{x^2}{2}$. Hence $F(x)\simeq e^{\frac{x^2}{2}}$. Being nearly equal to a function of class S^0 for limited x, by Part 3 of Theorem 1.5.1 also the function F is of class S^0.

1.5.4. Let $h\simeq 0, h>0$. The assertions follow readily from the expressions and approximations

$$\frac{a(x+h)-ax}{h}=a,\qquad \frac{(x+h)^2-x^2}{h}=2x+h\simeq 2x,$$
$$\frac{\log(x+h)-\log(x)}{h}=\frac{\log\left(1+\frac{h}{x}\right)}{h}\simeq\frac{1}{x}.$$

1.5.5. For limited x, always $\frac{1}{\omega}\sin(\omega x)\simeq 0$, but for instance

$$\frac{\frac{1}{\omega}\sin(\omega\frac{\pi}{2\omega})-\sin(0)}{\frac{\pi}{2\omega}}=\frac{2}{\pi}\not\simeq 0=\frac{\frac{1}{\omega}\sin(\omega\frac{\pi}{\omega})-\sin(0)}{\frac{\pi}{\omega}}.$$

Hence $x\mapsto\frac{1}{\omega}\sin(\omega x)$ cannot be of class S^1.

1.5.6.

1. From the identity $\frac{\delta\,€(x)}{\delta x}=€(x)$ it follows that the function €, being of class S^0 by Exercise 1.5.3, is also of class S^1, and with this mechanism we show by External Induction that it is of class S^n for all standard n.

2. From the identity $\frac{\delta F(x)}{\delta x}=xF(x)$ it follows that the function F, being of class S^0 by Part 2 of Exercise 1.5.3, is also of class S^1, because $x\mapsto xF(x)$ is of class S^0 as a product of two functions of class S^0.

1.5.7.

3. $\sum_{0\leq x<1}x^2(1+\varepsilon e^{\sin(x)})\delta x\simeq\int_0^1 x^2dx=\frac{1}{3}$.

4. $\sum_{0\leq x<1}(1+x\delta x)^{\frac{1}{\delta x}}\delta x\simeq\int_0^1 e^x dx=e-1$.

1.5.8.

1. For all limited $b>1$, $\sum_{1\leq x<b}\frac{1}{x^2}\delta x\simeq\int_1^b\frac{1}{x^2}dx=\frac{b-1}{b}$. By the Fehrele Principle, the near-equality still holds for some unlimited $\beta>1$. Now

$$\sum_{b\leq x<\infty}\frac{1}{x^2}\delta x<\frac{1}{\beta^2}+\int_b^\infty\frac{1}{x^2}dx\simeq\frac{1}{\beta}\simeq 0.$$

Hence $\sum_{1\leq x<\infty}\frac{1}{x^2}\delta x=\sum_{1\leq x<b}\frac{1}{x^2}\delta x+\sum_{b\leq x<\infty}\frac{1}{x^2}\delta x\simeq\frac{\beta-1}{\beta}\simeq 1$.

2. $\sum_{-\infty<x<+\infty}e^{-\frac{x^2}{2}}\delta x\simeq\sqrt{2\pi}$.

3. $\sum_{0\leq x<\frac{1}{\delta x}}(1-x\delta x)^{\frac{1}{\delta x}}\delta x\simeq\int_0^\infty e^{-x}dx=1$.

1.5.9.

1. For $x\simeq y, x<y$, we have $F(y)-F(x)\leq(y-x)\sqrt{\delta x}\cdot\frac{1}{\sqrt{\delta x}}=y-x\simeq 0$. Hence F is S-continuous. Because $0=f(\sqrt{\delta x}-\delta x)\not\simeq f(\sqrt{\delta x})=\frac{1}{\sqrt{\delta x}}$, the function f is not S-continuous.

2. The measure of the support of f is $\frac{1}{\sqrt{\delta x}}\cdot\delta x=\sqrt{\delta x}\simeq 0$.

3. $F(1)=1$.

1.5.10. Interpreting $S_\omega(x)$ as a Riemann-sum with increment $\frac{2}{\omega}$, we have

$$S_\omega(x) = \frac{2}{\pi}\sum_{n=0}^{\frac{\omega-1}{2}} \frac{\sin\left(\frac{(2n+1)u}{\omega}\right)}{\frac{2n+1}{\omega}}\frac{2}{\omega} \simeq \frac{2}{\pi}\int_0^1 \frac{\sin(uy)}{y}dy = \frac{2}{\pi}\int_0^u \frac{\sin(t)}{t}dt.$$

Chapter 2

2.1.1. For limited t we have $\left(1-\frac{t}{x}\right)^m e^{-t} \simeq e^{-t}$. In addition e^{-t} majorizes $\left(1-\frac{t}{x}\right)^m e^{-t}$. Then apply the Lemma of Dominated Approximation to prove (2.7). To prove (2.8), for some (possibly negative) $\varepsilon \simeq 0$ we have $x \geq \omega(1+\varepsilon)$. Then $\left(1-\frac{t}{\omega(1+\varepsilon)}\right)^\omega \simeq e^{-t}$ for limited t and $\left(1-\frac{t}{\omega(1+\varepsilon)}\right)^\omega \leq e^{-\frac{t}{2}}$ for all $t \leq x$. Then apply again the Lemma of Dominated Approximation.

2.1.2. The absolute value of the terms of the Taylor series of the Bessel function is the square of the absolute value of the terms of the Taylor series of the exponential function. Because a value is appreciable if and only if its square is appreciable, for unlimited ω we find the same value for ω' $\left(=\frac{\omega}{e}+\frac{\log(\omega)}{2e}\right)$. The nearly geometric series for the approximation factor $A_\omega(x)$ at $x \in \omega + \pounds$ has the approximate reason $\frac{1}{e^2}$, so (2.3) becomes

$$R_{\omega-1}(\omega'+u) \simeq \frac{e^2}{e^2-1}e^{2eu},$$
$$R_{\omega-1}(-\omega'+u) \simeq \frac{e^2}{1+e^2}e^{-2eu}.$$

Also (2.4) remains the same.

2.2.1.

1. $\pounds$, 2. $@$, 3. $(\oslash, 1+\oslash)$, 4. $(\frac{\omega}{2}+\pounds\sqrt{\omega}) \cap \mathbb{N}$, 5. $(\omega+\pounds\sqrt{\omega}) \cap \mathbb{N}$.

2.2.2. Identical to the proof of Stirling's formula, noting that $\omega! = \Gamma(\omega+1)$.

2.3.1. $\pounds\varepsilon$, $\pounds\sqrt[3]{\varepsilon}$.

2.3.2. λ has a jump of thickness $\varepsilon @$ at -1, and by symmetry a jump (from $\oslash$ to $\not\oslash$) of thickness $\varepsilon\pounds$ at 1.

2.3.4. Bounds within $ae^{-@t}+\oslash$ respectively $ae^{-\frac{@t}{\varepsilon}}+\oslash$ are found when solving for a linear differential equation with constant coefficients.

2.4.2. 1. No solution, 2. $\pounds$, 3. $-1+\oslash$, 4. No solution, 5. No solution, 6. No solution, 7. $\omega+\sqrt{\omega}\oslash$, only for $\varepsilon \in \frac{\oslash}{\omega}$, 8. $-1+\pounds e^{-\frac{@}{\delta}}$, only for $\varepsilon \in e^{\frac{\oslash}{\delta}}$.

2.4.3. 1. No solution, 2. $\pounds$, 3. $\frac{\pounds}{\omega}$, 4. $(-1+\frac{\pounds}{\omega}) \cup (1+\frac{\pounds}{\omega})$.

2.4.4.

3. If $d \in I$, one has $I(I-d) = II = I$, and for $d > I$ it holds both that $\frac{I}{d}(\frac{I}{d}-d) = \frac{I}{d^2}+I = I$ and $(d+\frac{I}{d})(d+\frac{I}{d}-d) = (d+\frac{I}{d})\frac{I}{d} = I+\frac{I}{d^2} = I$.

4. $\oslash \cup (1+\oslash)$, $\frac{\pounds}{\omega} \cup (\omega+\frac{\pounds}{\omega})$ and $\pounds \cup (\omega+\pounds)$.

Chapter 3

3.1.1. We use Proposition 3.1.1. The only non-obvious verifications are whether $2M_\varepsilon = M_\varepsilon$ and $2m_\varepsilon = m_\varepsilon$. Let $x \in M_\varepsilon$. Then for all standard $n \in \mathbb{N}$

it holds that $|x| < \varepsilon^n$. By permanence the property must also hold for some unlimited number ω, i.e. $|x| < \varepsilon^\omega$. Hence $|2x| < \varepsilon^{\omega-1}$ and we conclude that $|2x| < \varepsilon^n$, for all standard $n \in \mathbb{N}$. As for m_ε, let $y \in m_\varepsilon$. Then there exists some standard $n \in \mathbb{N}$ such that $|y| < e^{-\frac{1}{n\varepsilon}}$. Consequently $|2y| < e^{-\frac{1}{(n+1)\varepsilon}}$. Since $n+1$ is standard we are done.

3.1.3. Suppose that $\alpha \cap \beta \neq \emptyset$. Then there is a real number x such that $x \in \alpha$ and $x \in \beta$. By Proposition 3.1.8 we may write $\alpha = x + A$ and $\beta = x + B$. Hence, if $\max(A, B) = A$, $\beta \subseteq \alpha$ and if $\max(A, B) = A$, $\alpha \subseteq \beta$.

3.1.4. Only 1., 4., and 6. are true.

3.1.5.

1. Assume $\alpha - \beta < \gamma$. If $N(\beta) \subseteq N(\gamma)$ we have $\beta + \gamma = c + b + N(\gamma) = b + \gamma$. Then $\alpha + N(\beta) < \gamma + b = \gamma + \beta$. If $N(\gamma) \subset N(\beta)$, then $N(\gamma+\beta) = N(\beta) \subseteq N(\alpha) + N(\beta) = N(\alpha - \beta)$. We prove first that $(\alpha + N(\beta)) \cap (\gamma + \beta) = \emptyset$. Suppose that there exists $x \in (\alpha + N(\beta)) \cap (\gamma + \beta)$. Then $\alpha + N(\beta) = x + N(\alpha) + N(\beta)$ and $\beta + \gamma = x + N(\beta)$. It follows that $\beta + \gamma \subseteq \alpha + N(\beta)$. In particular, $b + c \in \alpha + N(\beta)$ which implies that $c \in \alpha - \beta < \gamma$, a contradiction. Also, we have $a \in \alpha + N(\beta), b + c \in \beta + \gamma$ and $a < b + c$ by the assumption. Hence $\alpha + N(\beta) < \beta + \gamma$.

2. The proof is analogous to the proof of Part 1.

3. It follows from Part 1 that $\alpha \leq \alpha + N(\beta) < \beta + \gamma$.

3.2.1. Let $\alpha = a + A$ be an external number and assume that $\beta = b + B$ is zeroless. Then by (3.2) we have $e(\alpha)u(\beta) = A\left(1 + \frac{B}{b}\right) = A + \frac{AB}{b} = A = e(\alpha)$.

3.2.2. Assume that $\alpha = a + A$ is zeroless. Then by Proposition 3.2.6 and (3.2)

$$e(\alpha)d(\alpha) = A\left(\frac{1}{a} + \frac{A}{a^2}\right) = \frac{A}{a} = e(u(\alpha)).$$

3.3.2. By formula (3.8)

$$(\alpha + \beta)\gamma M = \gamma(\alpha + \beta)M = \gamma(\alpha M + \beta M) = \alpha\gamma M + \beta\gamma M.$$

3.3.3.

1. $\mathcal{R}(\alpha) = \varepsilon \oslash$.
2. $\mathcal{R}(\alpha) = \frac{\oslash}{-5} = \oslash$.
3. $\mathcal{R}(\alpha) = \frac{\varepsilon^2 \pounds}{-\frac{2}{\varepsilon}} = \varepsilon^3 \pounds$.
4. $\mathcal{R}(\alpha) = \frac{\pounds}{\omega}$.
5. $\mathcal{R}(\alpha) = 0$.
6. $\mathcal{R}(\alpha) = \mathbb{R}$.

3.3.4. $\mathcal{R}\left(\frac{\alpha}{\beta}\right) = \varepsilon \oslash$.

3.3.7-3.3.9. Use External Induction.

Chapter 4

4.1.1.

1., 2. The limit and the set of fixed points are well-defined from standard objects, hence standard.

3. $\arctan[0,\omega)$ is internal nonstandard, as the image by a bijection of the internal nonstandard set $[0,\omega)$; $\sin[0,\omega) = [-1,1]$ is standard.

4. $\{\mathrm{st}(n) \in \mathbb{N} : n \text{ even}\}$ is external.

4.1.3. Assume the real function f is continuous at the standard point x. Let $\varepsilon > 0$ be standard and $y \simeq x$. By Transfer, there exists standard $\delta > 0$ such that $|f(z) - f(x)| < \varepsilon$ for all z with $|z - x| < \delta$. Hence $|f(y) - f(x)| < \varepsilon$. Because ε is arbitrary, one has $f(y) \simeq f(x)$. Hence f is S-continuous at x.

Conversely, assume the f is S-continuous at x. Let $\varepsilon > 0$ be standard. Then there exists $\delta > 0$ such that $|f(z) - f(x)| < \varepsilon$ for all z with $|z - x| < \delta$; any infinitesimal will do. By Transfer the latter property is true for every $\varepsilon > 0$. Hence f is continuous at x.

4.1.6. λ may be nonstandard if f is nonstandard, but by uniqueness it must be standard if f is standard. The suggestion to add "ex-standard" solutions of differential equations to the remarkable solutions like equilibium points or cycles, stems from Georges Reeb; it was also inspired by Nelson's theorem (see Exercise 4.1.9), which implies that there are at most a finite number of them.

4.1.7. With some geometry one shows that there is a $1-1$-continuous curve passing by every standard finite sets of points of the plane $\mathbb{R}^2$. By Idealization there is a $1-1$-continuous curve passing by every standard point of the plane $\mathbb{R}^2$.

4.1.8. Let S be an infinite set. Put $B(x,y) :\equiv (x \in S \wedge y \in S \wedge x \neq y)$. Then for all standard finite subset z of S there exist some $y \in S \setminus \{z\}$, for S is infinite. By Idealization there exists a nonstandard element in S, the "wolf".

4.1.9. We have $\forall^{\mathrm{stfin}} z \exists y \forall x \in z(x \in y \wedge y \text{ finite})$. Then by Idealization $\exists F \forall^{\mathrm{st}} x(x \in F \wedge F \text{ finite})$. Such a set F cannot be standard, for then $F \in F$, satisfying Russell's Paradox.

4.1.10. Goethe seems to have grasped Idealization and Nelson's Theorem, the set F of Exercise 4.1.9 is not finite in the naive way.

4.1.11.

1. $(0,1)$.
2. $(0,+\infty)$.
3. $\emptyset$.
4. $\{-\sqrt{2}, \sqrt{2}\}$.
5. $\emptyset$.
6. $\{(x,y) \in \mathbb{R}^2 : x \in [0,1] \wedge y = e^x\}$.

4.1.12.

1. Let $B(u,p)$ be the internal relation defined by

$$B(u,p) :\equiv u \in P \wedge P \subseteq R \wedge P \text{ is a Cartesian product.}$$

Let $z \subseteq X \times Y$ be a standard finite set and let p be the projection of $X \times Y$ onto X and q be the projection of $X \times Y$ onto Y. Put $P_z = p(z) \times q(z)$. Then $P_z \subseteq R$. Hence $B(u, P_z)$ is verified by all $u \in z$. By Idealization we conclude that there exists an internal set P such that $B(u,P)$ for all standard $u \in X \times Y$. Then P is a rectangle $I \times J$ such that every standard $x \in X$ is an element of I, every standard $y \in Y$ is an element of J and $I \times J \subseteq R$.

2. Let X, Y be standard sets and $(A_x)_{x\in X}$ and $(B_y)_{y\in Y}$ be two internal families such that $G = \bigcup_{\mathrm{st}(x)\in X} A_x$ and $H = \bigcap_{\mathrm{st}(y)\in Y} B_y$. Let $R \subseteq X \times Y$ be the internal relation defined by $R(x,y) \Leftrightarrow A_x \subseteq B_y$. By Part 1, there exist internal sets I, J such that $\forall^{\mathrm{st}} x \in X(x \in I)$ and $\forall^{\mathrm{st}} y \in Y(y \in J)$. So

$$G = \bigcup_{\mathrm{st}(x)\in X} A_x \subseteq \bigcup_{x\in I} A_x \subseteq \bigcap_{y\in J} B_y \subseteq \bigcap_{\mathrm{st}(y)\in Y} B_y = J.$$

Let $S := \bigcup_{x\in I} A_x$. Then clearly S is an internal set such that $G \subseteq S \subseteq H$. By Cauchy's Principle we have that $G \subset S \subset H$. Hence $G \neq H$.

4.2.1. For example $[\oslash, \pounds]$.

4.2.2. $H = \bigcup\{\frac{1}{n} + \oslash : \mathrm{st}(n) \in \mathbb{N}\} \cup \oslash$ is a halo.

4.3.1. $(\oslash, y + \oslash]$.

4.3.2.

1. $@ = (\oslash, \pounds]$, 3. $-@ = [\pounds, \oslash)$.

4.5.1. Since J is idempotent we have $J : J = \{c \in \mathbb{R} : cJ \subseteq J\} = J$. Hence $II = (J:J)(J:J) = JJ = J = J:J = I$. Now, $IJ = (J:J)J = JJ = J$, so J is an ideal of I. The maximality of J is obvious.

Chapter 5

5.1.1. We only show the first claim. Clearly (5.1)⇒(5.2). To see that also (5.2)⇒(5.1) we let $\varepsilon > N$ be non-precise and $a \in \varepsilon$. Then $N < a \leq \varepsilon$. By our hypothesis there exists $n_0 \in \mathbb{N}$ such that for $n \geq n_0$ one has $|u_n - \alpha| < a \leq \varepsilon$. Hence (5.1) holds and therefore (5.2)⇒(5.1). We conclude that the two conditions are equivalent.

5.1.2. Assume first that $(u_n + M) \underset{N}{\longrightarrow} \alpha$. Then there exists $n_0 \in \mathbb{N}$ such that for $n \geq n_0$ we have $|(u_n + M) - \alpha| < \varepsilon$. Then for $n \geq n_0$ it holds that $|u_n - \alpha| \leq |(u_n + M) - \alpha| < \varepsilon$. Hence $u_n \underset{N}{\longrightarrow} \alpha$. Assume now that $u_n \underset{N}{\longrightarrow} \alpha$. Then there exists $n_0 \in \mathbb{N}$ such that for $n \geq n_0$ we have $|u_n - \alpha| < \frac{\varepsilon}{2}$. Clearly $\frac{\varepsilon}{2} > N$. Then $|(u_n + M) - \alpha| \leq |(u_n + N) - \alpha| \leq |u_n - \alpha| + N < \frac{\varepsilon}{2} + \frac{\varepsilon}{2} = \varepsilon$. Therefore $(u_n + M) \underset{N}{\longrightarrow} \alpha$. Hence we have the desired equivalence.

5.1.3. The result is an immediate consequence of the fact that $||u_n| - |\alpha|| \leq |u_n - \alpha|$.

5.2.1. Let $\varepsilon > N(\alpha)$. Then for $n \geq n_0$ we have $|u_n - \alpha| = |\alpha - \alpha| = N(\alpha) < \varepsilon$. Hence $u_n \longrightarrow \alpha$.

5.2.3. By Part 1 of Proposition 5.1.4 we have $N(\alpha), N(\beta) \subseteq N$. Suppose that $\alpha + N \nleq \beta + N$. Then $\beta + N < \alpha + N$, hence also $\beta < \alpha$. It follows that $\varepsilon := \frac{\alpha-\beta}{2} > N$. So there exist $n_1, n_2 \in \mathbb{N}$ such that $|u_n - \alpha| < \varepsilon$ for $n \geq n_1$ and $|v_n - \beta| < \varepsilon$ for $n \geq n_2$. Let $n_3 = \max\{n_0, n_1, n_2\}$. Then for $n \geq n_3$ we have $v_n + N < \beta + \varepsilon = \beta + \frac{\alpha-\beta}{2} = \alpha - \frac{\alpha-\beta}{2} = \alpha - \varepsilon < u_n + N$. So $v_n < u_n$ for all $n \geq n_3$, which is a contradiction. Hence $\alpha + N \leq \beta + N$.

Chapter 6

6.1.1. Let $\alpha = a + A$ be an external number and let F be a flexible function such that $F(x) = f(x) + N_F(x)$ for all x and for some x_0 it holds that ${}^{N}_{M}\lim_{x \to x_0} F(x) = \alpha$. Suppose that $N \subset A$. Let $\varepsilon \in A$ be such that $\varepsilon > N$. Then there exists $\delta > M$ such that $|F(x) - \alpha| = |f(x) + N_F(x) - a + A| < \varepsilon$ whenever $0 < |x - x_0| < \delta$. It follows that $|f(x) - a| < \varepsilon$ and hence $|f(x) - a| \in A$. We conclude that if $0 < |x - x_0| < \delta$ then $\varepsilon \leq |f(x) - a + N_F(x) + A| = N_F(x) + A < \varepsilon$, which is a contradiction. Hence $A \subseteq N$.

6.1.2. We only prove that (6.1) $\Leftrightarrow$ 2. The direct implication is obvious. As for the converse implication assume 2. and let $\varepsilon > N$. Then there exists $a \in \mathbb{R}$ such that $\varepsilon > a > N$. Then, using our hypothesis, $|f(x) - \alpha| < a < \varepsilon$. Hence also 2. $\Rightarrow$ 6.1.

6.1.3. By Proposition 6.1.3 we may assume that $N(\alpha) = N$. Suppose that $\alpha > \beta + N$. Then $\alpha - \beta > N$. Let $N < \varepsilon < \alpha - \beta$. Then there exists $\delta > M$ such that $|F(x) - \alpha| < \varepsilon$. This implies that $\alpha - \varepsilon < F(x) + N$. Then $\alpha - (\alpha - \beta) < \alpha - \varepsilon < F(x) + N(\alpha)$. So $N + \beta < F(x) + N(\alpha)$. Hence $F(x) > \beta + N$, a contradiction. We conclude that $\alpha \leq \beta + N$.

6.1.4.

1. Let $\varepsilon > N$. Then $\frac{\varepsilon}{2} > N$ and there exists $\delta > M$ such that for all $x \in X$, if $0 < |x - x_0| < \delta$ then $|F(x) - \alpha| < \frac{\varepsilon}{2}$. So, $|F(x) - (\alpha + N)| = |F(x) - \alpha| + N < \frac{\varepsilon}{2} + \frac{\varepsilon}{2} = \varepsilon$. Hence $\alpha + N$ is an $M \times N$-limit of F at x_0.

2. By Proposition 6.1.2, we have that $N(\alpha) \subseteq N$, so if $\beta - \alpha \subseteq N$ then $\beta \subseteq \alpha + N$. Reasoning as in Part 1 we conclude that β is an $M \times N$-limit of F at x_0.

6.3.3. This is a consequence of the fact that if α is an $M \times N$-limit of F at x_0, then it is also an $M \times N$-outer limit of F at x_0.

6.3.5. We have $N_F(x) = \oslash$, for all $x \in \mathbb{R}$. Let $\varepsilon > \oslash$. Let $\delta = \frac{\varepsilon}{2}$. Then $\delta > \oslash$ and for all $x \in \mathbb{R}$, if $\oslash < |x - x_0| < \delta$, then

$$\begin{aligned} \left| \frac{F(x) - F(x_0)}{x - x_0} - 2x_0 + \oslash \right| &= \left| \frac{x^2 + \oslash - x_0^2 + \oslash}{x - x_0} - 2x_0 + \oslash \right| \\ &= \left| x + x_0 + \frac{\oslash}{x - x_0} - 2x_0 + \oslash \right| \\ &< |x - x_0| + \oslash < \frac{\varepsilon}{2} + \frac{\varepsilon}{2} = \varepsilon. \end{aligned}$$

Chapter 7

7.1.1.

1. $2\arctan(1 + \oslash)$.
2. $1 + \oslash$.
3. Not integrable.
4. $-1 + \oslash$.
5. $\oslash\omega\log(\omega)$.

7.2.1. For $x \in (2,3]$ define, say, $\phi(x) = \oslash$. Then $\int_0^3 \phi(x)dx = \oslash$.

7.2.2. We have $\overline{\int_0^1} f(x)dx = 1 + \oslash$ and, given standard $\varepsilon > 0$, using appropriate small intervals of standard length, the fact that $\mathbb{Q}$ is countable and the Fehrele principle, one shows that $\underline{\int_0^1} f(x)dx > 1-\varepsilon$, hence also $\underline{\int_0^1} f(x)dx > 1 + \oslash$. We conclude that f is outer-integrable.

7.2.3.

1. $\oslash\omega^2$.
2. $\oslash\omega^3$.
3. $£e^{\omega}$.
4. Dividing into subintervals of length $\frac{\pi}{\omega}$, and summing up, we find that $\int_0^{\oslash} \oslash \sin(\omega x)dx = \oslash$. Now $\int_0^{\oslash} \sin(\omega x)dx$ is not integrable, but $\int_0^{y} \sin(\omega x)dx$ takes for positive infinitesimal y values between 0 and $\frac{2}{\omega}$, which are absorbed by $\oslash$. Combining, $\int_0^{\oslash} (1+\oslash) \sin(\omega x)dx = \oslash$.

7.2.4.

1. $\frac{\pi^2}{6} + \oslash$.
2. $£$.
3. $\gamma + \oslash$.

7.4.3. The Euler expansion $\sum_{n=0}(-1)^n n!\varepsilon^n$ of the exponential integral $\int_0^{\infty} \frac{e^{-t}}{1+\varepsilon t}dt$ is obtained as a Borel transform of the geometric series expansion of $\frac{1}{1+x}$ with remainder $r_{n-1}(x) := \frac{(-1)^n x^n}{1+x}$. Hence

$$R_{n-1} = (-1)^n \int_0^{\infty} \frac{e^{-t}t^n}{1+\varepsilon t}dt.$$

In approximation we have (7.30).

1. This follows from the Lemma of Dominated Approximation, noting that $n\varepsilon \simeq 0$.
2. The proof is entirely similar to the proof of Proposition 7.4.13.
3. This is proved by comparing the remainder $R_{n-1}(\varepsilon)$ with the remainder with an index well after the index of the smallest term, say $R_{\left[\frac{1}{3\varepsilon}\right]}(\varepsilon)$. Counting backwards one sees that the error has been highly increased, see [130, p. 6].
4. It is proved working with the first neglected term

$$T_{n+1}(\varepsilon) = (-1)^{n+1}(n+1)!\varepsilon^{n+1}.$$

One has $\frac{R_n(\varepsilon)}{R_{n-1}(\varepsilon)} \in (1+\oslash)\frac{T_{n+1}(\varepsilon)}{T_n(\varepsilon)} = (n+1)\varepsilon$.

Chapter 8

8.2.1.

1. $2 + \varepsilon £$.
2. $-2 - \varepsilon + \varepsilon\oslash$.
3. $5 + \oslash$.
4. $\varepsilon\oslash$.

5. $-3+\varepsilon^2\oslash$.

8.2.2.

1. Not solvable. 2. $x=\oslash$.

8.2.3. Let $\varepsilon \simeq 0$.

1. Take for example $\gamma=\oslash$ and $\mathcal{A}=\begin{bmatrix}1+\varepsilon\pounds & 1+\varepsilon\pounds\\ 1 & 1\end{bmatrix}$. Then $\gamma^2\det(\mathcal{A})=\varepsilon\oslash$ and $\det(\gamma\mathcal{A})=\oslash$.

2. Take $\mathcal{A}=\begin{bmatrix}1+\oslash & \oslash\\ \oslash & 1+\oslash\end{bmatrix}$ and $\mathcal{B}=\begin{bmatrix}1+\varepsilon\pounds & \varepsilon\pounds\\ \varepsilon\pounds & 0\end{bmatrix}$. Then $\det(\mathcal{A})\det(\mathcal{B})=\varepsilon^2\pounds$ and $\det(\mathcal{AB})=\varepsilon\oslash$.

3. Let $\omega=\frac{1}{\varepsilon}$. Let $\mathcal{A}=\begin{bmatrix}2 & \oslash\\ \oslash & \varepsilon^2\end{bmatrix}$ and $\mathcal{B}=\begin{bmatrix}\omega\oslash & \pounds\\ 1 & \varepsilon\oslash\end{bmatrix}$. Then $\det((\mathcal{A}+\mathcal{B})^2)=\omega^2\oslash$ and $\det(\mathcal{A}+\mathcal{B})^2=\pounds$.

8.3.1.

1. $\left(\frac{3}{\varepsilon}+\varepsilon\oslash,-\frac{4}{\varepsilon}+2+4\varepsilon+\varepsilon\oslash,-\frac{2}{\varepsilon}-2+4\varepsilon+\varepsilon\oslash\right)$.

8.4.1.

4. $$\begin{bmatrix}\xi\\ \eta\\ \zeta\end{bmatrix}=\begin{bmatrix}\frac{7}{\varepsilon}+4\\ -\frac{4}{\varepsilon}-2\\ -\frac{2}{\varepsilon}-2\end{bmatrix}+\pounds\begin{bmatrix}4\varepsilon\\ 2+2\varepsilon\\ 1+2\varepsilon\end{bmatrix}+\oslash\begin{bmatrix}3+\varepsilon\\ 2+\varepsilon\\ 1+\varepsilon\end{bmatrix}+\varepsilon\oslash\begin{bmatrix}-2\\ 1\\ 1\end{bmatrix}.$$

8.4.3. With Gauss-Jordan elimination we obtain

$$\begin{cases}(1+\varepsilon\oslash)\xi+\varepsilon\pounds\eta+(-2+\oslash)\zeta\subseteq -1+\oslash\\ \varepsilon\oslash\xi+(1+\varepsilon\pounds)\eta+(2+\oslash)\zeta\subseteq -3+\oslash\\ \varepsilon\oslash\xi+\varepsilon\pounds\eta+\oslash\zeta\subseteq\oslash.\end{cases}$$

Neglecting the last line we get

$$\begin{cases}(1+\varepsilon\oslash)\xi+\varepsilon\pounds\eta+(-2+\oslash)\zeta\subseteq -1+\oslash\\ \varepsilon\oslash\xi+(1+\varepsilon\pounds)\eta+(2+\oslash)\zeta\subseteq -3+\oslash.\end{cases}$$

Solving and taking into account the principal restriction $\zeta\in\pounds$ we obtain

$$\begin{bmatrix}\xi\\ \eta\\ \zeta\end{bmatrix}=\begin{bmatrix}-1\\ -3\\ 0\end{bmatrix}+\oslash\begin{bmatrix}1\\ 0\\ 0\end{bmatrix}+\oslash\begin{bmatrix}0\\ 1\\ 0\end{bmatrix}+\pounds\begin{bmatrix}2\\ -2\\ 1\end{bmatrix}.$$

Chapter 9

9.2.2.

1. Note that necessarily $x\in\oslash\omega$. Then

$$\begin{aligned}\sqrt{\omega+x}-\sqrt{\omega}&=\frac{x}{\sqrt{\omega+x}+\sqrt{\omega}}\subseteq\frac{x}{\sqrt{\omega+\oslash\omega}+\sqrt{\omega}}\\ &=\frac{x}{2\sqrt{\omega}(1+\oslash)}.\end{aligned}$$

Because $\frac{1}{2(1+\oslash)} \subseteq M(\oslash) = @$, the equation $\frac{x}{\sqrt{\omega}} \in \oslash$ and the original equation have the same set of solutions, i.e. $\oslash\sqrt{\omega}$.

2. Again necessarily $x \in \oslash\omega$. Then

$$\log(\omega + x) - \log(\omega) = \log\left(1 + \frac{x}{\omega}\right) \subseteq \frac{x}{\omega}(1+\oslash).$$

Because $M(1+\oslash) = (1+\oslash)$, the original equation and the equation $\frac{x}{\omega} \in 1+\oslash$ have the same set of solutions, i.e. $(1+\oslash)\omega$.

9.2.3. If $x \in (1+\oslash)\frac{\omega}{e}$, the approximation factor $A(x)$ of (9.5) satisfies

$$A(x) \in \frac{1+\oslash}{1+\frac{1+\oslash}{e}} = \frac{e+\oslash}{e+1}.$$

So we need to solve $\frac{x^\omega}{\omega!} \in 1+\oslash$, which reduces to $x \in \omega!^{\frac{1}{\omega}}(1+\oslash)^{\frac{1}{\omega}}$. By Stirling's formula

$$\begin{aligned}\omega!^{\frac{1}{\omega}}(1+\oslash)^{\frac{1}{\omega}} &= \left(\omega^\omega e^{-\omega}\omega^{\frac{1}{2}}(2\pi)^{\frac{1}{2}}(1+\oslash)\right)^{\frac{1}{\omega}} \\ &= \frac{\omega}{e} e^{\frac{\log(\omega)+\frac{1}{2}\log(2\pi)+\log(1+\oslash)}{\omega}} \\ &= \frac{\omega}{e} + \frac{\log(\omega)}{2e} + \frac{\log(2\pi)}{2e} + \oslash.\end{aligned}$$

Putting $x_\omega = \frac{\omega}{e} + \frac{\log(\omega)}{2e} + \frac{\log(2\pi)}{2e} + \oslash$ we see that

$$x_{\omega+1} - x_\omega = \frac{1}{e} + \frac{\log(\omega+1) - \log(\omega)}{2e} + \oslash = \frac{1}{e} + \oslash.$$

9.2.6.

1. Put $f(x) = \frac{1}{x}$. Then

$$f\left(\omega + \frac{£}{\omega^2}\right) = \frac{1}{\omega + \frac{£}{\omega^2}} = \frac{1}{\omega\left(1+\frac{£}{\omega^3}\right)} = \frac{1}{\omega} + \frac{£}{\omega^4} \sqsubset \frac{£}{\omega^2}.$$

Then $\omega - \frac{1}{\omega} + \frac{£}{\omega^2}$ solves the equation by Theorem 9.2.8 and Corolllary 9.2.6.

2. Put $f(x) = \sqrt[3]{x}$. Then $f(£e^{@\omega}) = £e^{@\omega}$. Since f is increasing, $x + \sqrt[3]{x} \in £e^{@\omega}$ and $x \in £e^{@\omega}$, have the same set of solutions, i.e. $£e^{@\omega}$.

Chapter 10

10.1.1. Take an unlimited $\omega \in \mathbb{N}$.

1. Declare **st** the elements of $\omega £ \cap \mathbb{N}$.
2. Declare **st** the elements of $£e^{@\omega} \cap \mathbb{N}$.

10.2.1. For example, $u_n = 1$ for limited $n \in \mathbb{N}$, and $u_n = 2$ for unlimited $n \in \mathbb{N}$.

10.2.2. $u_n = 4 + \varepsilon £ + c\left(\frac{1+\varepsilon £}{2}\right)^n$ is $\varepsilon £$-asymptotically stable for all $c \in \mathbb{R}$.

10.2.3. 1 is an $\oslash$-asymptotically stable solution, but not $\pounds$-asymptotically stable, because 0 is $\pounds$-asymptotically unstable.

10.2.4.

1. Working backwards in time.
2. By 1. $|\mu(t)|$ cannot be maximal.
3. By 1. and 2.
4. By Theorem 9.3.2 and 3
5. Working, as in 1 - 4, with infinitesimal deviations larger than $\varepsilon\pounds$.
6. By permanence.
7. Here $a = 0$.

10.3.3. Due to the first-order term of the Taylor-expansion being appreciable.

10.3.4. By permanence (10.18) is still valid up to some unlimited $\tau \in \mathbb{R}$. By the Mass Concentration Lemma almost surely $W(\tau) \in \pounds\sqrt{\tau}$. Then almost surely $S_\tau \in (1+\oslash)S_0 \cdot e^{-\frac{\sigma^2}{2}\tau+\sigma\pounds\sqrt{\tau}} \subset \pounds e^{-@\tau} \subset \oslash$. To obtain mean 1, this can only be compensated by gains at least of order $\pounds e^{@\tau}$, of infinitesimal probability.

10.4.2. Proposing $(4-\sqrt{\varepsilon}, 4-\sqrt{\varepsilon})$ instead of $(4,4)$ in the Edgeworth-box is considered advantageous by agent b, but the corresponding deterioration is not noticed by a.

10.4.4. Let $\omega \in \mathbb{N}$ be unlimited and y be limited. Then

$$\sqrt{\lambda}\,\Pr\{X = \lambda + y\sqrt{\lambda}\} \in \sqrt{\lambda}(1+\oslash)\frac{e^{-\lambda}(\lambda+y\sqrt{\lambda})^{\lambda+y\sqrt{\lambda}}}{e^{-\lambda}\lambda^{\lambda}\sqrt{\lambda}\sqrt{2\pi}} \subseteq (1+\oslash)\frac{e^{-\frac{y^2}{2}}}{\sqrt{2\pi}}.$$

It is well-known that $\mathrm{E}(X) = \mathrm{Var}(X) = \lambda$, so its mass is $\lambda + \pounds\sqrt{\lambda}$.

Chapter 11

11.3.3. In every case the proof is by External Induction.

11.3.4. We have that $xy + e(x)y = xy + e(xy) + e(x)y = xy + e(x)y + xe(y) + e(x)y = xy + e(x)y + xe(y) = xy + e(xy) = xy$.

11.3.5. Using the distributivity axiom on the first equality we have

$$\begin{aligned} x(y+z) - (xy+xz) &= x(y+z) - (x(y+z) + e(x)y + e(x)z) \\ &= e(x(y+z)) + e(x)y + e(x)z \\ &= e(x)(y+z) + xe(y+z) + e(x)y + e(x)z \\ &= e(x)(y+z) + xe(y) + xe(z) + e(x)y + e(x)z \\ &= e(x)(y+z) + e(xy) + e(xz). \end{aligned}$$

On the other hand, applying cancellation to the distributivity axiom and Exercise 11.3.4 we get

$$\begin{aligned} e(xy) + e(xz) &= x(y+z) - xy - xz + e(x)y + e(x)z \\ &= x(y+z) - xy - xz. \end{aligned}$$

Hence $e(xy+xz) = e(x)(y+z)+e(xy+xz)$ and we conclude that $e(x)(y+z) \leq e(xy+xz)$.

Let $x = \oslash$, $y = -1 + \varepsilon\oslash$ and $z = 1 + \varepsilon\oslash$, where ε is a fixed infinitesimal. Then clearly $e(x)(y+z) \neq e(xy+xz)$.

Bibliography

[1] P. Aczel. *Non-well-founded Sets.* Center for the Study of Language and Information Publication Lecture Notes. Cambridge University Press, 1988.

[2] R. M. Anderson. *Star-finite Probability Theory.* ProQuest LLC, Ann Arbor, MI, 1977. Thesis (Ph.D.)–Yale University.

[3] V. Anishchenko, T. Vadivasova, and G. Strelkova. *Deterministic Nonlinear Systems.* Springer Series in Synergetics. Springer, Cham, 2014. A short course.

[4] L. Bachelier. Theory of speculation [reprinted from *The random character of stock market prices* ([1900] 1964), MIT Press, Cambridge, MA, 17–78]. In *Financial Risk Measurement and Management*, volume 267 of *Internat. Lib. Crit. Writ. Econ.*, pages 135–196. Edward Elgar, Cheltenham, 2012.

[5] J. Barnes. Medicine, experience and logic. In J. Barnes, J. Brunschwig, M. Burnyeat, and M. Schofield, editors, *Science and Speculation.* Cambridge University Press, 1982.

[6] R. Bartle. *The Elements of Integration and Lebesgue Measure.* Wiley Classics Library. John Wiley & Sons, Inc., New York, 1995.

[7] E. Benoît. Random walks and stochastic differential equations. In *Nonstandard Analysis in Practice*, Universitext, pages 71–90. Springer, Berlin, 1995.

[8] E. Benoît, J.-L. Callot, F. Diener, and M. Diener. Chasse au canard. *Collectanea Mathematica*, 31, Fasc. 1–3:37–119, 1981.

[9] G. Berkeley. *The Analyst; Or a Discourse Addressed to an Infidel Mathematician. Wherein it is examined whether the Object, Principles, and Inferences of the modern Analysis are more distinctly conceived, or more evidently deduced, than Religious Mysteries and Points of Faith.* J. Tonson, 1734.

[10] Patrick Billingsley. *Convergence of Probability Measures.* Wiley Series in Probability and Statistics: Probability and Statistics. John Wiley

& Sons, Inc., New York, second edition, 1999. A Wiley-Interscience Publication.

[11] K. Bogenberger. *Adaptive fuzzy systems for traffic responsive and coordinated ramp metering.* Veröffentlichungen des Fachgebiets Verkehrstechnik und Verkehrsplanung der Technischen Universität München. Fachgebiet Verkehrstechnik und Verkehrsplanung der Technischen Universität München, 2001.

[12] A. Bohé. Free layers in a singularly perturbed boundary value problem. *SIAM Journal on Mathematical Analysis*, 21:1264–1280, 1990.

[13] A. Bohé. The shock location for a class of sensitive boundary value problems. *J. Math. Anal. Appl.*, 235(1):295–314, 1999.

[14] J. Bosgiraud. Tests statistiques non standard pour les modèles à rapport de vraisemblance monotone. I. *Ann. I.S.U.P.*, 44(2-3):87–101 (2001), 2000.

[15] J. Bosgiraud. Tests statistiques non standard sur des proportions. *Ann. I.S.U.P.*, 44(1):3–12, 2000.

[16] J. Bosgiraud. Tests statistiques non standard pour les modèles à rapport de vraisemblance monotone. II. *Ann. I.S.U.P.*, 45(2-3):61–78 (2002), 2001.

[17] J. Bosgiraud. Nonstandard chi-squared test. *J. Inf. Optim. Sci.*, 26(2):443–470, 2005.

[18] J. Bosgiraud. Nonstandard likelihood ratio test in exponential families. In *The Strength of Nonstandard Analysis*, pages 145–169. SpringerWienNewYork, Vienna, 2007.

[19] J. Bosgiraud. Des lois log-normales presque normales. In *Des nombres et des mondes*, pages 87–90. Hermann, Paris, 2012.

[20] J. Bosgiraud. Moderate deviations in $\mathbb{R}^k$. In *Des nombres et des mondes*, pages 91–98. Hermann, Paris, 2012.

[21] N. Bourbaki. Eléments de mathématique. XVII. Première partie: Les structures fondamentales de l'analyse. Livre I: Théorie des ensembles. Actualités Sci. Ind., no. 1212, Chapitre I: Description de la mathématique formelle. Chapitre II: Théorie des ensembles, Hermann & Cie, Paris, 1954.

[22] D. Brown and A. Robinson. Nonstandard exchange economies. *Econometrica*, 43:41–55, 1975.

[23] J.-L. Callot. *Bifurcations du portrait de phase des équations differentielles linéaires du second ordre ayant pour type l'équation d'Hermite.* PhD thesis, Publ. I.R.M.A., Strasbourg, 1981.

[24] J.-L. Callot. Chasse au canard. III. Les canards ont la vie brève. *Collect. Math.*, 32(2):99–114, 1981.

[25] T. Chambers. On vagueness, sorites, and Putnam's "Intuitionistic Strategy". *Monist*, 81:343–348, 1998.

[26] C. C. Chang and H. J. Keisler. *Model theory*, volume 73 of *Studies in Logic and the Foundations of Mathematics*. North-Holland Publishing Co., Amsterdam, third edition, 1990.

[27] U. Clotz. *Untersuchungen einiger topologischer Sachverhalte und Konstruktionen in HST*. PhD thesis, Bergische Universität Wuppertal, 2009.

[28] P. Cobreros. Supervaluationism and necessarily borderline sentences. *Disputatio*, 3(25):41–49, 2008.

[29] N. Cutland, editor. *Nonstandard analysis and its applications*, volume 10 of *London Mathematical Society Student Texts*. Cambridge University Press, Cambridge, 1988. Papers from a conference held at the University of Hull, Hull, 1986.

[30] N. Cutland, E. Kopp, and W. Willinger. From discrete to continuous stochastic calculus. *Stochastics Stochastics Rep.*, 52(3-4):173–192, 1995.

[31] N. de Bruijn. *Asymptotic Analysis*. North Holland, 1961.

[32] K. DeRose. Contextualism: An explanation and defense. In J. Greco and E. Sosa, editors, *The Blackwell Guide to Epistemology*, pages 187–205. Blackwell, 1999.

[33] M. di Nasso. On the foundations of nonstandard mathematics. *Math. Japon.*, 50(1):131–160, 1999.

[34] F. Diener and M. Diener, editors. *Nonstandard Analysis in Practice*. Universitext. Springer-Verlag, Berlin, 1995.

[35] F. Diener and G. Reeb. *Analyse non standard*, volume 40 of *Collection Enseignement des Sciences [Collection: The Teaching of Science]*. Hermann, Paris, 1989.

[36] M. Diener. The canard unchained or how fast/slow dynamical systems bifurcate. *Math. Intelligencer*, 6(3):38–49, 1984.

[37] M. Diener. Regularizing microscopes and rivers. *SIAM J. Math. Anal.*, 25(1):148–173, 1994.

[38] M. Diener and I. van den Berg. Halos et galaxies: une extension du lemme de Robinson. *C. R. Acad. Sci. Paris Sér. I Math.*, 293(8):385–388, 1981.

[39] B. Dinis. Old and new approaches to the sorites paradox. In Nguyễn Minh Xuân et al., editor, *Proceedings of the Annual Scientific Meeting of Thang Long University 2017–2018, Vietnam*, pages 5–21.

[40] B. Dinis. *Axiomatics for the External Numbers of Nonstandard Analysis and modelation of uncertainties.* PhD thesis, Universidade de Évora, 2013.

[41] B. Dinis and I. van den Berg. Algebraic properties of external numbers. *J. Logic & Analysis*, 3(9):1–30, 2011.

[42] B. Dinis and I. van den Berg. Axiomatics for the external numbers of nonstandard analysis. *J. Logic & Analysis*, 9(7):1–47, 2017.

[43] B. Dinis and I. van den Berg. On the quotient class of non-archimedean fields. *Indagationes Mathematicae*, 28(4):784–795, 2017.

[44] B. Dinis and I. van den Berg. Characterization of distributivity in a solid. *Indagationes Mathematicae*, 29(2):580 – 600, 2018.

[45] B. Dinis, N. Van Tran, and I. van den Berg. On flexible sequences. *Acta Mathematica Vietnamica*, January 2019. https://doi.org/10.1007/s40306-018-00303-4.

[46] W. Eckhaus. *Asymptotic analysis of singular perturbations*, volume 9 of *Studies in Mathematics and its Applications.* North-Holland Publishing Co., Amsterdam-New York, 1979.

[47] J. Fenstad, editor. *Selected Works in Logic.* Oslo: Universitetsforlaget, 1970.

[48] K. Fine. Vagueness, truth and logic. *Synthese*, 30(3-4):265–300, 1975.

[49] J. Fodor and E. Lepore. What cannot be evaluated cannot be evaluated and it cannot be supervalued either. *Journal of Philosophy*, 93(10):516–535, 1996.

[50] A. Franco de Oliveira and I. van den Berg. *Matemática não-standard, uma introdução com aplicações.* Fundação Calouste Gulbenkian, 2007.

[51] G. Frege. *Translations from the Philosophical Writings of Gottlob Frege.* Rowman and Littlefield, Totowa, N.J., third edition, 1980. Edited and with a preface by Peter T. Geach and Max Black, Index by E. D. Klemke.

[52] M. Gevrey. Sur la nature analytique des solutions des équations aux dérivées partielles. Premier mémoire. *Annales scientifiques de l'École Normale Supérieure*, 35:129–190, 1918.

[53] M. Gevrey. *Œuvres de Maurice Gevrey.* Éditions du Centre national de la recherche scientifique, 1970.

[54] K. Gödel. Über formal unentscheidbare Sätze der Principia Mathematica und verwandter Systeme I. *Monatshefte für Mathematik und Physik*, 38(1):173–198, Dec 1931.

[55] K. Gödel. Zum Intuitionistischen Aussagenkalkül. *Anzeiger der Akademie der Wissenschaften in Wien*, 69:65–66, 1932.

[56] J. Goethe. *Gedichte: Ausgabe letzter Hand 1827.* Hofenberg, 2016.

[57] R. Goldblatt. *Lectures on the hyperreals: An introduction to nonstandard analysis*, volume 188 of *Graduate Texts in Mathematics.* Springer-Verlag, New York, 1998.

[58] S. Gottwald. *A treatise on many-valued logics*, volume 9 of *Studies in Logic and Computation.* Research Studies Press, Ltd., Baldock, 2001.

[59] P. Hájek, L. Godo, and F. Esteva. A complete many-valued logic with product-conjunction. *Arch. Math. Logic*, 35(3):191–208, 1996.

[60] G. Hardy. *Orders of infinity. The Infinitärcalcül of Paul du Bois-Reymond.* Hafner Publishing Co., New York, 1971. Reprint of the 1910 edition, Cambridge Tracts in Mathematics and Mathematical Physics, No. 12.

[61] J. Harthong. Le moiré. *Adv. in Appl. Math.*, 2(1):24–75, 1981.

[62] W. Hodges. *Model theory*, volume 42 of *Encyclopedia of Mathematics and its Applications.* Cambridge University Press, Cambridge, 1993.

[63] R. V. Hogg, J. W. McKean, and A. T. Craig. *Introduction to Mathematical Statistics.* Pearson Education International. Pearson Education, 2005.

[64] K. Hrbáček. Axiomatic foundations for nonstandard analysis. *Fund. Math.*, 98(1):1–19, 1978.

[65] D. Hyde. The sorites paradox. In Giuseppina Ronzitti, editor, *Vagueness: A Guide*, pages 1–17. Springer Verlag, 2011.

[66] T. Jech. *Set Theory.* Springer Monographs in Mathematics. Springer-Verlag, Berlin, 2003. The third millennium edition, revised and expanded.

[67] J. Justino. *Nonstandard linear algebra with error analysis.* PhD thesis, Universidade de Évora, 2013.

[68] J. Justino and I. van den Berg. Cramer's rule applied to flexible systems of linear equations. *Electron. J. Linear Algebra*, 24(Special issue for the 2011 Directions in Matrix Theory Conference):126–152, 2012/13.

[69] V. Kanovei and M. Reeken. Internal approach to external sets and universes. I. Bounded set theory. *Studia Logica*, 55(2):229–257, 1995.

[70] V. Kanovei and M. Reeken. Mathematics in a nonstandard world I. *Mathematica Japonica*, 45:2:369–408, 1997.

[71] V. Kanovei and M. Reeken. *Nonstandard Analysis, Axiomatically.* Springer Monographs in Mathematics, 2014.

[72] R. Keefe. *Theories of Vagueness.* Cambridge University Press, 2000.

[73] H. Keisler. An infinitesimal approach to stochastic analysis. *Mem. Amer. Math. Soc.*, 48(297):x+184, 1984.

[74] S. Kleene. *Introduction to metamathematics.* D. Van Nostrand Co., Inc., New York, N. Y., 1952.

[75] F. Koudjeti. *Elements of External Calculus with an application to Mathematical Finance, Ph.D. thesis.* Labyrinth publications. Capelle a/d IJssel, The Netherlands, 1995.

[76] F. Koudjeti and I. van den Berg. Neutrices, external numbers, and external calculus. In F. and M. Diener, editors, *Nonstandard Analysis in Practice*, Universitext, pages 145–170. Springer, Berlin, 1995.

[77] G. Kreisel. Axiomatizations of nonstandard analysis that are conservative extensions of formal systems for classical standard analysis. In *Applications of Model Theory to Algebra, Analysis, and Probability (Internat. Sympos., Pasadena, Calif., 1967)*, pages 93–106. Holt, Rinehart and Winston, New York, 1969.

[78] C. Kuratowski. Fundamenta Mathematicae, 1, 161–171, *Sur la notion de l'ordre dans la Théorie des Ensembles*, http://eudml.org/doc/212966, 2, 1921.

[79] A. Lightstone and A. Robinson. *Nonarchimedean Fields and Asymptotic Expansions.* North-Holland Publishing Co., Amsterdam-Oxford; American Elsevier Publishing Co., Inc., New York, 1975. North-Holland Mathematical Library, Vol. 13.

[80] C. Lobry and P. Delfini. Formal controllability and physical controllability of linear systems. In *Analysis and optimization of systems (Antibes, 1990)*, volume 144 of *Lect. Notes Control Inf. Sci.*, pages 623–631. Springer, Berlin, 1990.

[81] C. Lobry, T. Sari, and S. Touhami. On Tykhonov's theorem for convergence of solutions of slow and fast systems. *Electron. J. Differential Equations*, pages No. 19, 22 pp. 1998.

[82] P. Loeb. Conversion from nonstandard to standard measure spaces and applications in probability theory. *Trans. Amer. Math. Soc.*, 211:113–122, 1975.

[83] J. Łukasiewicz. *Selected Works.* North-Holland Publishing Co., Amsterdam-London; PWN-Polish Scientific Publishers, Warsaw, 1970. Edited by L. Borkowski, Studies in Logic and the Foundations of Mathematics.

[84] R. Lutz. Rêveries infinitésimales. *Gazette des mathématiciens*, 34:79–87, 1987.

[85] R. Lutz and M. Goze. *Nonstandard analysis: A practical guide with applications*, volume 881 of *Lecture Notes in Mathematics.* Springer-Verlag, Berlin-New York, 1981. With a foreword by Georges Reeb.

[86] W. Luxemburg, editor. *Applications of Model Theory to Algebra, Analysis, and Probability.* Holt, Rinehart and Winston, New York-Montreal, Que.-London, 1969.

[87] K. Machina. Truth, belief, and vagueness. *Journal of Philosophical Logic*, 5(1):47–78, 1976.

[88] D. Marker. *Model Theory : An Introduction.* Graduate Texts in Mathematics. Springer New York, 2013.

[89] R. Moore. *Methods and applications of interval analysis*, volume 2 of *SIAM Studies in Applied Mathematics.* Society for Industrial and Applied Mathematics (SIAM), Philadelphia, Pa., 1979.

[90] R. Moore, R. Kearfott, and M. Cloud. *Introduction to Interval Analysis.* Society for Industrial and Applied Mathematics (SIAM), Philadelphia, PA, 2009.

[91] E. Nelson. Unfinished book on nonstandard analysis. Chapter 1, Internal Set Theory.

[92] E. Nelson. Internal set theory: a new approach to nonstandard analysis. *Bull. Amer. Math. Soc.*, 83:6:1165–1198, 1977.

[93] E. Nelson. *Predicative Arithmetic.* Mathematical Notes - Princeton University Press. Princeton University Press, 1986.

[94] E. Nelson. *Radically Elementary Probability Theory.* Annals of Mathematical Studies, vol. 117. Princeton University Press, Princeton, N. J., 1987.

[95] E. Nelson. The syntax of nonstandard analysis. *Ann. Pure Appl. Logic*, 38(2):123–134, 1988.

[96] A. Neumaier. *Interval Methods for Systems of Equations.* Encyclopedia of Mathematics and its Applications. Cambridge University Press, 1991.

[97] F. W. J. Olver and W. Rheinbolt. *Asymptotics and Special Functions.* Elsevier Science, 2014.

[98] R. O'Malley, Jr. *Introduction to singular perturbations.* Academic Press [A subsidiary of Harcourt Brace Jovanovich, Publishers], New York-London, 1974. Applied Mathematics and Mechanics, Vol. 14.

[99] G. Peano. The Principles of Arithmetic, presented by a new method. In J. van Heijenoort, editor, *From Frege to Gödel. A Source Book in Mathematical Logic, 1879–1931*, pages 83–97. Harvard University Press, Cambridge, Mass., 1967.

[100] H. Poincaré. Sur les intégrales irrégulières: Des équations linéaires. *Acta Math.*, 8(1):295–344, 1886.

[101] H. Poincaré. *Les Méthodes Nouvelles de la Mécanique Céleste.* Dover Publications, 1957.

[102] E. Post. Introduction to a general theory of elementary propositions. *Amer. J. Math.*, 43(3):163–185, 1921.

[103] D. Prawitz. *Natural Deduction. A Proof-theoretical Study.* Acta Universitatis Stockholmiensis. Stockholm Studies in Philosophy, No. 3. Almqvist & Wiksell, Stockholm, 1965.

[104] G. Priest. The logic of paradox. *J. Philos. Logic*, 8(2):219–241, 1979.

[105] H. Putnam. Vagueness and alternative logic. *Erkenntnis*, 19(1-3):297–314, 1983.

[106] W. Quine. Two dogmas of empiricism. *Philosophical Review*, 60(1):20–43, 1951.

[107] S. Rashid. *Economies with Many Agents: An Approach using Nonstandard Analysis.* Johns Hopkins University Press, Baltimore, MD, 1987.

[108] S. Read and C. Wright. Hairier than Putnam thought. *Analysis*, 45(1):56–58, 1985.

[109] C. Reder. Observation macroscopique de phénomènes microscopiques. In *Actes de l'École d'Été: Analyse Non Standard et Représentation du Réel (Oran, 1984)*, pages 195–244. OPU, Algiers, 1985.

[110] G. Reeb. La mathématique non standard vieille de soixante ans? *Cahiers Topologie Géom. Différentielle*, 22(2):149–154, 1981. Third Colloquium on Categories, Part III (Amiens, 1980).

[111] A. Robinson. Non-standard analysis. *Nederl. Akad. Wetensch. Proc. Ser. A 64*, 23:432–440, 1961.

[112] A. Robinson. *Non-standard Analysis.* North-Holland Publishing Co., Amsterdam, 1966.

[113] A. Robinson and E. Zakon. A set-theoretical characterization of enlargements. In W.A.J. Luxemburg, editor, *Applications of Model Theory to Algebra, Analysis and Probability*, pages 109–122. Holt, Rinehart and Winston, New York, 1969.

[114] B. Russell. Vagueness. *Australasian Journal of Philosophy*, 1(2):84–92, 1923.

[115] T. Sari. Stroboscopy and averaging. In *Colloque Trajectorien à la Mémoire de Georges Reeb et Jean-Louis Callot (Strasbourg-Obernai, 1995)*, volume 1995/13 of *Prépubl. Inst. Rech. Math. Av.*, pages 95–124. Univ. Louis Pasteur, Strasbourg, 1995.

[116] S. Schwartz. Intuitionism and Sorites. *Analysis*, 47(4):179–183, 1987.

[117] S. Shapiro. *Vagueness in Context.* Oxford University Press, 2006. Reviewed by Matti Eklund, available online at http://ndpr.nd.edu/news/vagueness-in-context/.

[118] S. Shapiro. Reasoning with slippery predicates. *Studia Logica*, 90(3):313–336, 2008.

[119] T. Skolem. Über die nicht-charakterisierbarkeit der zahlenreihe mittels endlich oder abzählbar unendlich vieler aussagen mit ausschliesslich zahlenvariablen. *Fundam. Math.*, 23:150–161, 1934. Reprinted in [47].

[120] N. Smith. *Vagueness and Degrees of Truth.* Oxford University Press UK, 2013.

[121] M. Sobel. Asymptotic confidence intervals for indirect effects in structural equation models. *Sociological Methodology*, 13:290–312, 1982.

[122] R. Sorensen. *Blindspots.* Oxford University Press, 1988.

[123] R. Sorensen. *Vagueness and Contradiction.* Oxford University Press, 2001.

[124] K. Stroyan and W. Luxemburg. *Introduction to the Theory of Infinitesimals.* Academic Press [Harcourt Brace Jovanovich, Publishers], New York-London, 1976. Pure and Applied Mathematics, No. 72.

[125] D. Teodorović. Fuzzy logic systems for transportation engineering: the state of the art. *Transportation Research Part A: Policy and Practice*, 33(5):337–364, 1999.

[126] A. Tikhonov. Systems of differential equations containing small parameters in the derivatives. *Mat. Sbornik N. S.*, 31(73):575–586, 1952.

[127] M. Tye. Sorites paradoxes and the semantics of vagueness. *Philosophical Perspectives*, 8:189–206, 1994.

[128] É. Urlacher. Un système rapidement oscillant. In *Mathématiques finitaires & analyse non standard, Tome 1, 2*, volume 31 of *Publ. Math. Univ. Paris VII*, pages 235–256. Univ. Paris VII, Paris, 1989.

[129] A. Urquhart. Many-valued logic. In D. Gabbay and F. Guenther, editors, *Handbook of Philosophical Logic, Vol. III.* D. Reidel Publishing Co., 1986.

[130] I. van den Berg. *Nonstandard Asymptotic Analysis.* Lecture Notes in Mathematics 1249. Springer-Verlag, 1987.

[131] I. van den Berg. An external probability order theorem with applications. In *Nonstandard Analysis in Practice*, Universitext, pages 171–183. Springer, Berlin, 1995.

[132] I. van den Berg. *Principles of Infinitesimal Stochastic and Financial Analysis.* World Scientific Publishing Co., Inc., River Edge, NJ, 2000.

[133] I. van den Berg. Asymptotics of families of solutions of nonlinear difference equations. *Log. Anal.*, 1(2):153–185, 2008.

[134] I. van den Berg. A decomposition theorem for neutrices. *Annals of Pure and Applied Logic*, 161(7):851 – 865, 2010.

[135] I. van den Berg. Discretisations of higher order and the theorems of Faà di Bruno and DeMoivre-Laplace. *J. Logic & Analysis*, 5(6):1–35, 2013.

[136] I. van den Berg and M. Diener. Diverses applications du lemme de Robinson en analyse non standard. *C. R. Acad. Sci. Paris Sér. I Math.*, 293(10):501–504, 1981.

[137] I. van den Berg and V. Neves. *The Strength of Nonstandard Analysis.* Springer, 2007.

[138] J. van der Corput. Introduction to the neutrix calculus. *J. Analyse Math.*, 7:281–399, 1959/1960.

[139] J. G. van der Corput. Neutrices. *J. Soc. Indust. Appl. Math.*, 7:253–279, 1959.

[140] J. G. van der Corput. Neutrix calculus. I. Neutrices and distributions. *Nederl. Akad. Wetensch. Proc. Ser. A 63 = Indag. Math.*, 22:115–123, 1960.

[141] J. G. van der Corput. Neutrix calculus. $\mathbf{II}_A$., II_B., II_C. *Nederl. Akad. Wetensch. Proc. Ser. A 64 = Indag. Math.*, 23:1–14, 15–24, 25–37, 1961.

[142] J. G. van der Corput. Neutralized values. I. Neutralized values with one neutrix. *Nederl. Akad. Wetensch. Proc. Ser. A 69=Indag. Math.*, 28:387–401, 1966.

[143] J. G. van der Corput. Neutralized values. II. Neutralized values with more than one neutrix. *Nederl. Akad. Wetensch. Proc. Ser. A 69=Indag. Math.*, 28:402–411, 1966.

[144] M. Van Dyke. *Perturbation Methods in Fluid Mechanics.* Parabolic Press, 1975.

[145] B. van Fraassen. Singular terms, truth-value gaps, and free logic. *Journal of Philosophy*, 63(17):481–495, 1966.

[146] N. Van Tran. *Optimization with flexible objectives and constraints.* PhD thesis, Universidade de Évora, 2017. http://hdl.handle.net/10174/23324.

[147] N. Van Tran and I. van den Berg. A parameter method for linear algebra and optimization with uncertainties. (submitted), 2018.

[148] H. Varian. *Microeconomic Analysis.* Norton, New York, 3rd edition, 1992.

[149] A. Vretblad. *Fourier analysis and its applications*, volume 223 of *Graduate Texts in Mathematics.* Springer-Verlag, New York, 2003.

[150] T. Williamson. *Vagueness.* Routledge, 1994.

[151] T. Williamson. *Knowledge and its Limits.* Oxford University Press, 2000.

[152] L. Wittgenstein. *Philosophical Investigations.* The Macmillan Co., New York, 1953. Translated by G. E. M. Anscombe.

[153] L. Wittgenstein. *The Blue and Brown Books.* Harper & Row, 1958.

[154] L. Zadeh. Fuzzy sets. *Information and Control*, 8:338–353, 1965.

[155] E. Zakon. A new variant of non-standard analysis. In *Victoria Symposium on Nonstandard Analysis (Univ. Victoria, Victoria, B.C., 1972)*, pages 313–339. Lecture Notes in Math., Vol. 369. Springer, Berlin, 1974.

[156] H.-J. Zimmermann. *Fuzzy Set Theory—and Its Applications.* Kluwer Academic Publishers, Boston, MA, second edition, 1992. With a foreword by L. A. Zadeh.

Index

absolute value, 66
absorber, 63
adequate
  ultralimit, 300
  ultrapower, 300
admissible solution, 190
almost density, 180
appreciable
  number, 5
  with respect to, 63
approximation factor, 41
Archimedean property, 282
assembly, 71, 277
asymptotic numbers, 6
asymptotically differentiable, 218
asymptotically more precise, 73
attractive slow curve, 227

bounded formula, 88, 286, 306
Bounded Idealization, 301
bounded inclusion, 217
Boundedness Axiom, 301

canard, 22, 53, 230, 231
Cauchy criterion, 153
Chasles Relation, 164, 173
class
  S^0, 27
  S^n, 31
conservative extension, 4
continuity
  inner, 153
  M×N, 150
convergence
  local, 146
  N-convergence, 117, 118
  strong, 117, 118
  strong local, 147
  with respect to, 126
convex component, 95
cut, 94

determinant, 192
domain of infinitesimal
    approximation, 41
drain, 262

ENA^-, 17, 87, 88
ENA, 16
equation
  homogeneous, 55
  non-singular, 55
  singular, 55
error after truncation, 41
exact solution, 190
expansion in ε-shadow, 212
exploder, 63
exponential order, 185
external
  formula, 9
  set, 10
External Induction, 16, 88
external number, 21, 62, 283
external number
  module, 213
  neutrix part, 63, 283
  representative, 63
  zeroless, 21, 62
external numbers structure, 284

feasibility space, 201
feasible vector, 269
finite intersection property, 307
flexible

function, 100
function
integrable, 169
recurrence relation, 260
sequence, 104
system, 189
formula
external, 9
internal, 9
function
averageable, 180
flexible, 100
internal integrable, 164
precise, 100
representative, 100
selfneglecting, 217

galaxy, 11, 89
galaxy
epsilon, 12, 60
micro, 21
principal, 12
Gauss-Jordan eliminable, 196
Gauss-Jordan elimination, 196
Geometrical Brownian Motion, 265
Gevrey-bound, 60
Gibbs–Wilbraham phenomenon, 36

halfline
lower, 94, 99
upper, 94, 99
halo, 11, 89
halo
epsilon, 12, 60
micro, 21
Hrbáček's paradoxes, 296, 302

ideal, 111, 279
ideal languages, 247
Idealization, 86
idempotent, 279, 307
illegal set formation, 301
Induction, 282
infinitely
close, 6
large, 5
small, 5
infinitesimal
discretization, 32
families, 113
number, 5
inner-integrable, 167
integrable, 164, 169, 170
integrable
inner, 167
outer, 167
internal
formula, 9
set, 9
internal bounded formula, 286
internal bridge, 104, 125
internally representable, 100
IST, 1, 85

jump, 51
jump
positive, 51
thickness, 51

Laplace formula, 193
Laplace method, 182–184
Leibniz Rules, 3, 6
Lemma
Cauchy multiplicative approximations, 25
Concentration, 183
Dominated Approximation, 35
Mass Concentration, 48
Mass Concentration for densities, 178
Robinson, 14
limit
M×N, 140
outer, 156
strong, 118
limited
number, 5
Local Laplace method, 184
Loeb measure, 237

M×N
differentiability, 157

limit, 140
outer continuity, 150
mass, 47, 178
Matching for functions, 226
Matching for sequences, 225
maximal
ideal, 111, 280
valuation domain, 113
maximal solution, 190
mean, 180
microgalaxy, 21
microgalaxy
epsilon, 60
microhalo, 21, 211
microhalo
epsilon, 60
functional, 91
Minkowski operations, 20, 61, 283
minor, 193
model, 305
module, 213

N-asymptotical stability, 261
N-close, 139
N-limit point, 139
N-minimum, 160
N-stability, 261
naive integers, 3
near minimum, 160
near-indifference, 266
nearly minimal, 186, 269
neighbourhood, 139
neutrix, 20, 59, 283
neutrix
feasibility space, 191
idempotent, 57, 110, 307
part, 63, 100, 283
scalar, 59
nonstandard embedding, 306
nonstandard model, 306
normal form, 98
number
appreciable, 5
asymptotic, 6
infinitesimal, 5
limited, 5
precise, 276
unlimited, 5

open ball, 139
opposite
with respect to, 73
order of magnitude, 19
order scale, 212
outer continuity, 150
outer-integrable, 167
overspill, 11, 14

paradox
Hrbáček, 296, 302
Sorites, 244
parameter method, 191
permanence, 11
precise, 119, 276, 279
precise
more, 73
pregalaxy, 11, 89
prehalo, 11, 89
Principle
Cauchy, 11
Fehrele, 13, 89, 94
Saturation, 88, 282

quasi-linear equations, 215

reduced matrix, 190, 196
relative uncertainty, 55, 73
representative, 63, 100, 122
repulsive slow curve, 227
restricted formula, 280
Riccati-Hermite equation, 231
ring property, 109
root, 111

S-continuity, 27
S-convergence, 33, 35
S-differentiability, 29
S-integral, 32
S-local extrema, 31
S-maximal, 113
S-primitive, 33

S-Riemann integrable, 32
semigroup, 68
semigroup
  regular, 68
sequence
  bounded, 129
  Cauchy, 135
  cofinal, 108
  coinitial, 108
  eventually bounded, 129
  flexible, 104
  local, 125
  N-Cauchy, 135
  of representatives, 122
  precise, 104
  strongly N-Cauchy, 135
  zeroless, 131
set
  cofinal, 92
  coinitial, 92
  external, 10, 17
  external numbers, 283
  internal, 9, 307
  of differences, 96
  pre-external, 10
  standardized, 88
shadow, 211, 292
shadow
  expansion, 212
singular matrix, 196
singular perturbation, 227, 230
singular point, 227
slow curve, 227
solid, 277
solution
  admissible, 190
  exact, 55, 190
  maximal, 55, 190
Sorites paradox, 244
square flexible system, 190
standard model, 306
standard normal density, 180
standard size, 302
Standardization, 86
Stirling's formula, 49
strict rank, 200
strong connectives, 252
subdistributivity, 72
subsequence, 108
sum
  lower, 32
  Riemann, 32
  upper, 32
superstructure, 306
symmetrical
  exponential, 259
  logarithm, 259

tail, 47, 178
tail
  lower, 178
  upper, 178
Taylor series, 41
temporal stability, 263
Theorem
  Expansion, 221, 222
  Generalized Dedekind
    completeness, 97
  Koudjeti, 112
  Localization, 218–220
  Multiplicative approximation of
    functions, 25
  Nonstandard Borel-Ritt, 212
  Representation, 13, 18, 107
  Squeeze, 131, 146
  Substitution, 213
  Weak Intermediate Value, 151
  Weak Maximum Value, 152
thickness of the border, 96
Transfer, 86

ultrafilter, 300
ultralimit, 300
ultrapower, 300
ultraproduct, 300
unlimited
  number, 5

valuation domain, 113
Von Neumann hierarchy, 298

weak infimum, 98
weak least upper bound, 281
weak supremum, 98
Wiener walk, 265
winf, 98, 281
zeroless, 21, 62, 131, 279
ZFC, 3, 296
ZFL, 3
zup, 98, 281